TRANSITION
TO PROOFS

TRANSITION TO PROOFS

Simon Rubinstein-Salzedo

Euler Circle, USA

World Scientific

EW JERSEY · LONDON · SINGAPORE · BEIJING · SHANGHAI · HONG KONG · TAIPEI · CHENNAI · TOKYO

Published by

World Scientific Publishing Co. Pte. Ltd.

5 Toh Tuck Link, Singapore 596224

USA office: 27 Warren Street, Suite 401-402, Hackensack, NJ 07601

UK office: 57 Shelton Street, Covent Garden, London WC2H 9HE

Library of Congress Cataloging-in-Publication Data
Names: Rubinstein-Salzedo, Simon, author.
Title: Transition to proofs / Simon Rubinstein-Salzedo, Euler Circle, USA.
Description: New Jersey : World Scientific, [2023] | Includes bibliographical references and index.
Identifiers: LCCN 2023014992 | ISBN 9789811272080 (hardcover) |
    ISBN 9789811272325 (paperback) | ISBN 9789811272097 (ebook)
Subjects: LCSH: Number theory--Textbooks. | Proof theory--Textbooks.
Classification: LCC QA241 .R83 2023 | DDC 511.3/6--dc23/eng/20230605
LC record available at https://lccn.loc.gov/2023014992

British Library Cataloguing-in-Publication Data
A catalogue record for this book is available from the British Library.

For any available supplementary material, please visit
https://www.worldscientific.com/worldscibooks/10.1142/13301#t=suppl

Introduction

This book is the textbook for a class I have taught every year since 2018 at Euler Circle to high-school (and younger) students. The target audience I have in mind for this class is students who have some problem-solving experience doing tricky numerical-answer problems, but without much exposure to proofs. For students who do math competitions, this translates roughly into being able to qualify for the American Invitational Mathematics Examination. While the main goal of this class is to help students become comfortable with reading and writing proofs, I don't just want to teach my students the mechanics of proofs: I also want to teach them lots of interesting mathematics simultaneously. It takes longer to teach proofs and also interesting mathematics than just to teach proofs, so this class is a yearlong sequence, where we focus on proofs in number theory, combinatorics, and analysis in the fall, winter, and spring quarters, respectively, and these are parts 1, 2, and 3 of the book, respectively. We do not cover part 0 in the class, but it is there for reference for students who have not seen that material previously.

While we cover a decent amount of each of the three main topics (number theory, combinatorics, and analysis), this book and the corresponding class are not intended to be comprehensive studies in any of them. There are many important topics in all of these areas that are notably absent from this book. I was especially sorry not to be able to include quadratic reciprocity in the number theory section, but it seemed just a little bit beyond the scope of the class. That is okay though, I think: it won't be the last time students see these topics.

Students who attend Euler Circle for several years would likely take the fundamentals of higher mathematics sequence after this class. That class focuses on linear algebra, basic abstract algebra, real analysis, and point-set topology. It is especially helpful for the students to see analysis again at a slightly deeper level after completing the transition to proofs class.

For the most part, each chapter corresponds to one week of class. The problems at the end of each chapter are the homework problems for the week. Some of them are quite challenging, and while I want students to be able to solve all the problems, not all of them are able to do so, and that is okay. In my classes, each student is assigned a teaching assistant to work with, and the students submit their problem sets to their teaching assistant each week and discuss the problems, including how to improve their proofwriting. Going through this book with a mentor is the ideal use case for this book. This book can also be used for self-study, but it would be easier to use it in conjunction with a class. But then, which textbook wouldn't?

I would like to thank everyone who pointed out mistakes and improvements in previous versions of this book, thus helping it to improve. These people include all the students at teaching assistants from the classes, as well as Porter Adams, Travis Chen, Eric Gottlieb, Yunjiang Jiang, Urban Larsson, Andrew Lin, Nitya Mani, Thomas Roybal, Lynn Sokei, Enrique Treviño, Jonathan Webster, Roy Whelden, Mike Wills, Adeline Wong, and the anonymous referees.

Thanks also to everyone at World Scientific for making the publication of this book possible.

Contents

Part 2. Combinatorics   **139**

PART 0
Preliminaries

Chapter −1

Sets and set operations

−1.1   Sets

For us, a *set* is a collection of objects, or *elements*. This is not the real definition, and it leads to serious problems such as the infamous *Russell's paradox*, but it will serve our purposes well enough; truly valid definitions require much more work to build up and feel less satisfactory initially.

The typical way of writing a set is by listing its elements between braces. For example, the set containing only the two elements 1 and 5 is denoted $\{1,5\}$. The elements of a set can be more complicated than just numbers. For instance, $\{\clubsuit, \pi, \sqrt{-2}, \{5, \{5\}\}\}$ is a set, albeit one that isn't likely to be useful in very many contexts. Note that this set contains four elements: these elements are named $\clubsuit$, π, $\sqrt{-2}$, and $\{5, \{5\}\}$. Sometimes sets may contain too many elements to list out conveniently, so we can use an ellipsis (i.e. the symbol "...") to represent "and so forth." For example, we may write $\{1, 2, 3, \ldots, 100\}$ to mean the set consisting of the numbers from 1 to 100. We also allow sets to contain infinitely many elements, in which case listing them all is out of the question. In this case, we are forced to use an ellipsis or some other method to describe all the elements in our set.

The elements in a set are not ordered and cannot appear more than once. Thus, for instance, $\{1,1\}$ and $\{1\}$ are the same set, because they contain exactly the same elements. Similarly, $\{1,2\}$ and $\{2,1\}$ are the same.

There is also one very special set. This is the *empty set*, which we denote by $\varnothing$. This is the set containing no elements at all.

Many of the sets we consider in this book are sets of numbers. There are various special sets that are of particular importance to us. Here are some key ones:

- $\mathbb{N}$ is the set of *natural numbers*. This is the set

$$\mathbb{N} = \{1, 2, 3, \ldots\}.$$

- $\mathbb{Z}$ is the set of *integers*. This is the set

$$\mathbb{Z} = \{\ldots, -3, -2, -1, 0, 1, 2, 3, \ldots\}.$$

  The reason for the letter "Z" is that it is the first letter of the German word *Zahlen*, which means "number."

- $\mathbb{Q}$ is the set of *rational numbers*. This is the set consisting of all fractions, such as $-\frac{7}{3}$. It includes all the integers, because an integer such as 7 can be written as $\frac{7}{1}$. The "Q" stands for "quotient."

- $\mathbb{R}$ is the set of *real numbers*. This includes numbers such as π, which are not in $\mathbb{Q}$. Giving a precise definition of the real numbers is an involved process and will be discussed in Part 3 of this book.

- $\mathbb{C}$ is the set of *complex numbers*. These are all numbers of the form $a + bi$, where a and b are real numbers, and i is a square root of -1.

If we have a set S with an element s, we write $s \in S$, and we read this as "s is an element of S," or more simply as "s is in S." For instance, $1 \in \{1, 5\}$, and $1 \in \mathbb{Z}$. When s is not an element of S, we write $s \notin S$. For instance, $\frac{1}{2} \notin \mathbb{Z}$.

Now that we have element notation, we can use it to describe more sets easily. To do this, we put a description of the set in the braces. To do this, we write the type of elements, followed by a colon, followed by the rules they must satisfy. The general setup of a set written in this notation is

$$S = \{\text{variable(s)} : \text{condition(s)}\},$$

and we write this as "the set of (variables) such that (conditions)." For example, if we wish to consider the set of even numbers, we can write

$$S = \{n \in \mathbb{Z} : n \text{ is even}\} = \{\ldots, -6, -4, -2, 0, 2, 4, 6, \ldots\},$$

and we read this as "S is the set of integers n such that n is even," where the colon gets translated into "such that." Similarly, we can write the set of perfect squares as

$$S = \{n^2 : n \in \mathbb{Z}\} = \{0, 1, 4, 9, 16, \ldots\}.$$

Suppose we have a set S, which contains several elements. We can form a new set out of some of the elements of S. We call such a set a *subset*.

In terms of element notation, saying that T is a subset of S is the same as saying that whenever $t \in T$, then $t \in S$. When T is a subset of S, we write $T \subseteq S$. (Another common notation is $T \subset S$.) Note that there are two extreme cases of subsets. First, for any set S, we have $\varnothing \subseteq S$. This is because every element of the empty set (all zero of them!) is an element of S. At the other extreme, for any set S, we have $S \subseteq S$. When T is not a subset of S, we write $T \nsubseteq S$. When T is a subset of S and we specifically want to highlight that T is not all of S, we write $T \subsetneq S$.

Among the special sets we have highlighted, we have:

$$\varnothing \subsetneq \mathbb{N} \subsetneq \mathbb{Z} \subsetneq \mathbb{Q} \subsetneq \mathbb{R} \subsetneq \mathbb{C}.$$

−1.2 Set operations

Given two or more sets, we can form several new sets out of them via various different operations. The first of these operations is the *union*. Given two sets S and T, we can form a new set $S \cup T$, which consists of all the elements of S and all the elements of T. A set never contains the same element twice, so if we have an element that is in both S and T, it still only appears once in $S \cup T$. For instance, if $S = \{1, 2\}$ and $T = \{1, 4\}$, then $S \cup T = \{1, 2, 4\}$. This is illustrated pictorially by a Venn diagram, as shown in Figure −1.1. We can also apply the union construction to more than two sets—even infinitely many of them. In that case, the union consists of all the elements that are in at least one of the sets. For example, we might have a set S_n for each $n \in \mathbb{N}$, perhaps $S_n = \{n^2, n^3\}$. The union of all of these sets is

$$S_1 \cup S_2 \cup S_3 \cup \cdots = \{1, 1, 4, 8, 9, 27, \ldots\} = \{1, 4, 8, 9, 16, 25, 27, \ldots\},$$

and it consists of all the positive integers that are perfect squares or perfect cubes (or both). Another way to write this is using an indexed large union symbol:

$$\bigcup_{n=1}^{\infty} S_n.$$

This is the union version of the Σ notation that we will see in Chapter 2.

The next set operation is the *intersection*. Given two sets S and T, we can form a new set $S \cap T$, which consists of all the elements that are in both S and T. If $S = \{1, 2\}$ and $T = \{1, 4\}$, then $S \cap T = \{1\}$. Just like with the union, we can take the intersection of more than two sets: this

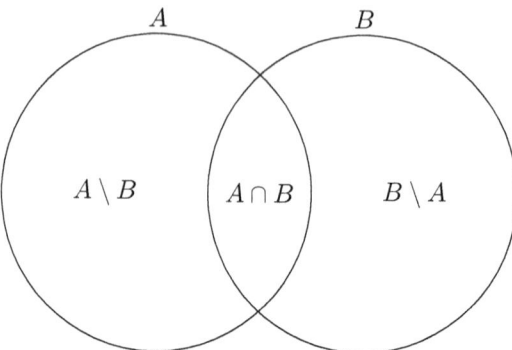

Figure −1.1. A Venn diagram showing how to form new sets from old ones. The set $A \cup B$ consists of the stuff inside at least one of the circles.

intersection consists of all the elements that are in all the sets. We can use a large intersection symbol to describe the intersection of many sets:

$$\bigcap_{n=1}^{\infty} S_n,$$

for instance.

Next up is the *set difference*: we write $S \setminus T$ for the set of elements that are in S but not in T. Another common notation is $S - T$. We also have the *symmetric difference* $S \triangle T$, which consists of all the elements that are in exactly one of S and T. Alternatively, we can write the symmetric difference as the union of two set differences:

$$S \triangle T = (S \setminus T) \cup (T \setminus S).$$

Next up is the *complement*: the complement S^c of S consists of all elements that are not in S. This is a little misleading, because we must first specify our underlying universe of all the elements under consideration. For example, if S is the set of even positive integers, what exactly do we want to consider to be in S^c? A tree, for instance, is not an even positive integer, so should it be an element of S^c? Perhaps under some circumstances, but not others. Thus, when talking about the complement of a set, it is important to consider it as a subset of a universal set. If we consider S to be a subset of the universal set U, then $S^c = U \setminus S$. In the case of the set S of even positive integers, if we're considering the universal set to be $\mathbb{Z}$, for instance, then

$$S^c = \{\ldots, -4, -3, -2, -1, 0, 1, 3, 5, 7, 9, \ldots\},$$

i.e. all the nonpositive integers together with the positive odd integers.

−1.3   Products of sets

Another way we can form a new set out of two sets S and T is to construct the set consisting of all *ordered pairs* (s,t), where $s \in S$ and $t \in T$. We call this new set the *product* (or *Cartesian product*, or *direct product*) of S and T, and we denote it by $S \times T$.

Example. Let $S = \{a,b,c\}$ and $T = \{x,y\}$. Then

$$S \times T = \{(a,x),(a,y),(b,x),(b,y),(c,x),(c,y)\}.$$

We may also take the product of more than two sets. If we have three sets S, T, and U, then $S \times (T \times U)$ and $(S \times T) \times U$ are not quite the same, because a typical element of $S \times (T \times U)$ has the form $(s,(t,u))$, whereas a typical element of $(S \times T) \times U$ has the form $((s,t),u)$, so the parentheses are in different places. However, this is the only difference, and if we drop the inner parentheses and instead write (s,t,u), then the two sets would be the same. We write $S \times T \times U$ for the set of all such ordered triples. Of course, we can apply the same construction to the product of more than two sets, even infinitely many of them. To take a product of many sets, we use a Π. For instance, we write

$$\prod_{n=1}^{\infty} S_n$$

for a product of the sets S_n.

A special case is the product of a set with itself, possibly multiple times. We write S^n for the n-fold product of S with itself, so that for instance

$$S^2 = S \times S, \qquad S^3 = S \times S \times S, \qquad S^4 = S \times S \times S \times S,$$

and so forth. Other important special cases are S^1, which formally speaking ought to be the set consisting of all ordered 1-tuples of elements of S, so that a typical element of S^1 would be (s). However, in this instance, we typically remove the parentheses and think of s as being a typical element of S^1. In this way, $S^1 = S$. Similarly, S^0 is formally equal to $\{()\}$, the set with one element, which is denoted $()$. However, we usually just think of S^0 as a set with one element without too much concern for what the element is named.

−1.4   Functions

Let S and T be two sets. A *function* from S to T is a way of assigning an element of T to each element of S. A function could be defined by a convenient formula, such as the function $f : \mathbb{Z} \to \mathbb{Z}$ given by $f(n) = n^2 + 2$, so that we have (for instance) $f(3) = 11$ and $f(-5) = 27$. But a function can also be something much more random-looking. For instance, suppose that $S = \{a, b, c, d\}$ and $T = \{v, w, x, y, z\}$. Then one example of a function is the function f defined by

$$f(a) = x, \quad f(b) = v, \quad f(c) = x, \quad f(d) = y.$$

We can picture this function graphically by drawing an arrow from s to $f(s)$ for each $s \in S$, as in Figure -1.2.

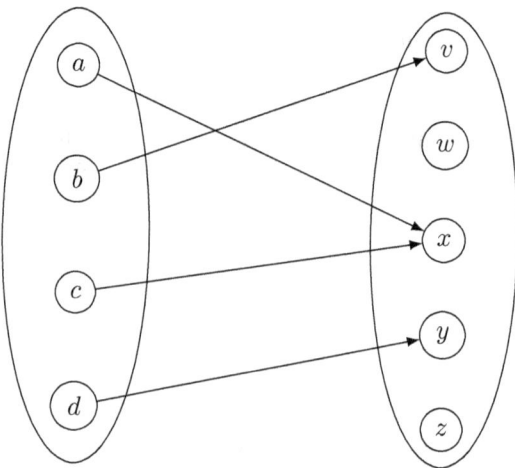

Figure -1.2. A graphical representation of a function.

The data of a function $f : S \to T$ consist of three parts. The first part is the set S, known as the *domain* of the function. The second part is the set T, known as the *codomain*. The third part is the rule f that tells us what $f(s)$ is for each $s \in S$. Changing any one of these three parts changes the function. So, for instance, the functions $f : \mathbb{Z} \to \mathbb{Z}$ defined by $f(n) = n^2 + 2$ and $g : \mathbb{Z} \to \mathbb{N}$ defined by $g(n) = n^2 + 2$ are different functions, even though for each $n \in \mathbb{Z}$, we have $f(n) = g(n)$. This is because the codomains are different.

Remark -1.1. We can also define functions in terms of products of sets. A function from S to T is a subset $F \subseteq S \times T$ such that for each $s \in S$, there is exactly one element $(s, t) \in F$. The connection with the previous definition is that $t = f(s)$.

Another key set associated to a function $f : S \to T$ is the *image*, or *range*.

Definition -1.2. Let S and T be two sets, and let $f : S \to T$ be a function. The *image* (or *range*) of f is $\mathrm{im}(f) = \{t \in T : t = f(s) \text{ for some } s \in S\}$.

In the function f shown in Figure -1.2, $\mathrm{im}(f) = \{v, x, y\}$.

Definition -1.3. Let S and T be two sets, and let $f : S \to T$ be a function.

- We say that f is *surjective* or *onto* if, for every $t \in T$, there is some $s \in S$ such that $f(s) = t$.
- We say that f is *injective* or *one-to-one* if, whenever $s \neq s'$, we have $f(s) \neq f(s')$.
- We say that f is *bijective* if it is both injective and surjective.

See Figure -1.3 for pictures of surjective and injective functions.

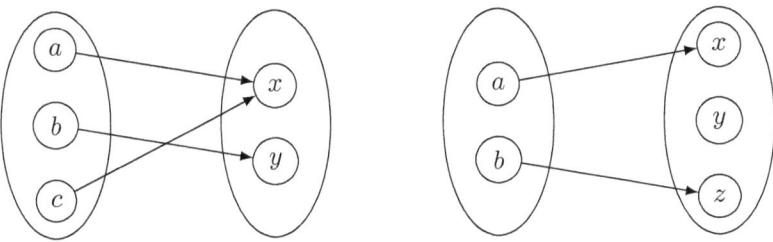

Figure -1.3. Left: a surjective function. Right: an injective function.

If S, T, and U are sets, and $f : S \to T$ and $g : T \to U$ are functions, then we can combine them into a single function from S to U.

Definition -1.4. Let S, T, and U be sets, and let $f : S \to T$ and $g : T \to U$ be functions. Then there is a function $g \circ f : S \to U$, defined by $(g \circ f)(s) = g(f(s))$ for all $s \in S$. We call $g \circ f$ the *composition* of g and f.

Remark -1.5. Note that $g \circ f$ means that we apply f first, then g second, and not the other way around!

Chapter 0

Logic

0.1 Statements

Logic is about the analysis of *statements*. A statement in mathematics is a mathematical expression or sentence that is either definitely true or definitely false. Here are some examples of true statements:

- 8 is a positive integer.
- All positive integers are real numbers.
- If a and b are the legs of a right triangle and the hypotenuse is c, then $a^2 + b^2 = c^2$.
- There exist positive integers x and y such that $x^2 - 5y^2 = 1$.

And here are some examples of false statements:

- $1 + 2 = 6$.
- If n is a positive integer, then $n + 100 < n^2$.
- All quadrilaterals are squares.
- Some squares are triangles.

Note that some of the false statements are true for *certain* inputs. For example, there are positive integers n such that $n + 100 < n^2$; $n = 11$ is an example of this. But it isn't true for *all* values of n, which is what the statement claims.

More subtly, there are also statements for which neither I—nor anyone else—knows whether they are true or false. Nonetheless, they are either definitely true or definitely false; our lack of knowledge does not change the fact that they have definitive truth values. Here are some examples of statements with unknown truth values:

- $\pi^{\pi^{\pi^{\pi^{\pi}}}}$ is an integer.

- There are infinitely many prime numbers p such that $p + 2$ is also prime.[1]
- It is possible to color every point in the plane with one of five colors in such a way that, whenever x and y are two points at a distance of 1 from each other, x and y have different colors.[2]
- 3003 occurs more times than any other number besides 1 in Pascal's triangle.[3]

Determining, with proof, whether these statements are true or false would be extremely interesting—and probably very challenging.

Finally, there are expressions that are not statements at all, because they do not have definitive truth values. Here are a few examples of non-statements:

- $x > 0$.
- $x + y = z$.
- p is a prime.

These are not statements, because they are sometimes true and sometimes false: their truth value depends on the value of some variable. For example, in the non-statement $x > 0$, this is true if $x = 2$, for instance, but not if $x = -3$. Since it is sometimes true and sometimes false, it does not have a definitive truth value.

0.2   Logical operations

There are various operations we can perform on statements to make new statements. For example, if P and Q are statements, we can form a new statement $P \wedge Q$, where $\wedge$ is read as "and." For instance, if P is the statement "8 is a positive integer," and Q is the statement "All positive integers are real numbers," then $P \wedge Q$ is the statement "8 is a positive integer, and all positive integers are real numbers." In order for $P \wedge Q$ to be true, both P and Q must be true; if at least one of them is false, then $P \wedge Q$ is a false statement.

Similarly, we have "or" statements. If P and Q are statements, their "or" is $P \vee Q$. With the statements P and Q as above, $P \vee Q$ is the statement "8 is a positive integer, or all positive integers are real numbers."

---

[1]This is a famous problem known as the *Twin Prime Conjecture*.
[2]This is part of the *Hadwiger–Nelson problem*.
[3]This is a version of *Singmaster's Conjecture*.

In mathematics, "or" is always inclusive, unless explicitly stated to the contrary. Thus $P \vee Q$ is true if at least one of P and Q is true, and it's only false when both P and Q are false.

The next sorts of statements we will consider are "not" statements. We write $\neg P$ for the statement "Not P." With P as above, $\neg P$ is the statement "It is not the case that 8 is a positive integer," or, in less stilted language, "8 is not a positive integer." The statement $\neg P$ is true exactly when P is false.

0.3 Truth tables

It is often helpful to make tables of truth values of various templates of statements, based on the truth values of their inputs. Let's see an example in the case of $\wedge$. Recall that $P \wedge Q$ is true exactly when both P and Q are true. We can represent this with the following truth table:

P	Q	$P \wedge Q$
T	T	T
T	F	F
F	T	F
F	F	F

Here, T stands for "true," and F stands for "false."
Here are the truth tables for $\vee$ and $\neg$:

P	Q	$P \vee Q$
T	T	T
T	F	T
F	T	T
F	F	F

P	$\neg P$
T	F
F	T

We can also make truth tables for more complicated statements. These might combine several of $\vee$, $\wedge$, and $\neg$, and perhaps more than two different statements. Often, it is helpful to break down the final truth value by listing out various intermediate results. Here is an example, illustrating the truth table of $P \wedge (Q \vee \neg R)$:

P	Q	R	$\neg R$	$Q \vee \neg R$	$P \wedge (Q \vee \neg R)$
T	T	T	F	T	T
T	T	F	T	T	T
T	F	T	F	F	F
T	F	F	T	T	T
F	T	T	F	T	F
F	T	F	T	T	F
F	F	T	F	F	F
F	F	F	T	T	F

Truth tables can help us detect that certain templates of statements are equivalent to certain other ones. For example, one of *De Morgan's Laws* says that $\neg(P \vee Q)$ and $(\neg P) \wedge (\neg Q)$ are equivalent, in the sense that if we fix the truth values of P and Q, then the truth values of $\neg(P \vee Q)$ and $(\neg P) \wedge (\neg Q)$ are always the same. To check this, we can just write out the truth tables for both of these expressions:

P	Q	$P \vee Q$	$\neg(P \vee Q)$
T	T	T	F
T	F	T	F
F	T	T	F
F	F	F	T

P	Q	$\neg P$	$\neg Q$	$(\neg P) \wedge (\neg Q)$
T	T	F	F	F
T	F	F	T	F
F	T	T	F	F
F	F	T	T	T

Indeed, we see that both $\neg(P \vee Q)$ and $(\neg P) \wedge (\neg Q)$ are true exactly when both P and Q are false.

0.4 Conditional statements

There is another logical operation we can perform to create a new statement out of two other statements. This is a statement of the form "If P then Q." We call P the *hypothesis* and Q the *conclusion*, and the entire statement is called a *conditional statement*. A typical example is the following: If an integer n is divisible by 4, then n is divisible by 2. The hypothesis is that n is divisible by 4, and the conclusion is that n is divisible by 2. Symbolically, we write $P \Rightarrow Q$ for the statement "If P then Q."

Sometimes, a conditional statement won't be explicitly written in if–then form. An example of a conditional statement that isn't written in if–then form is the statement "All perfect squares are nonnegative." This statement has exactly the same meaning as the statement "If n is a perfect square, then $n \geq 0$," which is in if–then form.

One way to think about if–then statements is that the statement "If P then Q" is a guarantee that if P is true, then Q is automatically also true. Using this interpretation, we can write out a truth table for implications.

P	Q	$P \Rightarrow Q$
T	T	T
T	F	F
F	T	T
F	F	T

Notice that the statement $P \Rightarrow Q$ is true when P is false! This is because nothing has broken in our guarantee if the hypothesis is false: the only way to break our guarantee is if P is true but Q is false, and we see that reflected in the truth table.

We can immediately see from the truth table that there is a type of statement not involving the $\Rightarrow$ symbol that is equivalent to the statement "If P then Q." This is the statement $\neg P \vee Q$. That is, $P \Rightarrow Q$ is true if at least one of $\neg P$ and Q is true. This shows that we never really need implications, since we can always replace them with statements involving negations and ors. Nonetheless, implications are extremely common, as they are very often the most convenient way to express statements in mathematics.

One other very important type of conditional statement is one of the form "If not Q then not P." To see why this is important, let's look at its truth table.

P	Q	$\neg Q$	$\neg P$	$\neg Q \Rightarrow \neg P$
T	T	F	F	T
T	F	T	F	F
F	T	F	T	T
F	F	T	T	T

Thus we find that the truth table for "If P then Q" is exactly the same as the truth table for "If not Q then not P." In other words, these two statements are *always equivalent*. The statement "If not Q then not P" is called the *contrapositive* of the statement "If P then Q."

Frequently, conditional statements have free variables in them, such as the n in our first example: If an integer n is divisible by 4, then n is divisible by 2. For any *particular* integer n, we get a statement about that particular number. For example, n could be 20, in which case the statement reads "If 20 is divisible by 4, then 20 is divisible by 2," which is clearly true. If, on the other hand, n were equal to 9, then the statement would read "If 9 is divisible by 4, then 9 is divisible by 2." This is also true, since the hypothesis is false! When we have a conditional statement that has a free variable—or perhaps multiple free variables—in it, then we say that the statement is true if it holds for all values of the free variable. This is the case in that example: if n is *any* integer that is divisible by 4, then n is divisible by 2.

There are other examples where the implication sometimes holds but sometimes doesn't hold. For instance, consider the statement "All integers are perfect squares," which can be turned into if–then form as "If n is an integer, then n is a perfect square." There are values of n for which this is true, such as $n = 16$. But there are other values of n for which it is false, such as -2. Since the implication does not *always* hold, the conditional statement is false.

One important special case of a conditional statement is one in which the hypothesis is *always* false. For example, consider the statement "If n is a negative perfect square, then $n \geq 10$." This seems at first glance as though it might be false: if we have a negative perfect square n, then $n < 0$, so it should not be the case that $n \geq 10$. However, this statement is actually true because the hypothesis is always false: there are no negative perfect squares. It is very important to remember that statements in which the hypothesis can never be satisfied are always automatically true. We call such statements *vacuous statements*, and we say that they are *vacuously true*.

PART 1
Number Theory

Chapter 1

Proof by contradiction

1.1 What is a proof?

Among the aspects and objectives that set mathematics apart from other fields of study is our insistence on proving things. It is not enough for us to *believe* or *suspect* that something might be true. Rather, we have to be completely certain that it is true, and to enable others to become equally certain by means of our arguments and explanations. We do this by means of proofs: in order to *prove* that some statement is true, we give a logically sound argument leading to that conclusion.

The statements we wish to prove are usually called *theorems*, although there are other terms such as *corollary*, *lemma*, and *proposition* that can serve in place of *theorem*. There is no clear rule about when to call something a theorem versus a proposition, and so forth, but a rough guide is this:

- A *theorem* is an important result. Of course, whether something is important may depend on your perspective or standards.
- A *corollary* is a result that follows easily from a more difficult or general result.
- A *lemma* is an intermediate result proven on the way to proving a larger theorem.
- A *proposition* is a less significant version of a theorem.

In contrast with theorems and their friends, we also have *definitions*. Definitions are not statements we prove; rather they introduce new terminology in order to make communication easier. For instance:

Definition 1.1. A nonnegative integer n is said to be a *triangular number* if $n = \frac{k(k+1)}{2}$ for some integer k.

There isn't anything we have to verify here—it's just naming something. In theory, we could do without definitions; but then we would need to rewrite the entire rest of the definition every time we wished to talk about triangular numbers. That would be inefficient and hard to understand.

Typically, the sorts of theorems (and so on) we wish to prove have the following form: they start with a hypothesis and end with a conclusion. Often, they will be expressed in the form "if P then Q," where P is some condition and Q is some other condition. For example, we can express the celebrated *Fermat's Last Theorem* in this form:

Theorem 1.2 (Fermat's Last Theorem). *If n is a positive integer greater than 2, and x, y, and z are positive integers, then $x^n + y^n \neq z^n$.*

Sometimes theorems (and so forth) are not written in exactly the "If P then Q" form, but they can generally be converted to this form, so as to have the same content. Let's look at an example:

Proposition 1.3. *The sum of two even numbers is even.*

This isn't written in the form "if P then Q," but we can rewrite the statement to put it into this form:

Proposition 1.4 (Restatement of Proposition 1.3). *If x and y are even numbers, then $x + y$ is even.*

Proposition 1.4 carries precisely the same content as Proposition 1.3, but now it is written in a form with a hypothesis and a conclusion.

Now, let's give a proof. We'll actually give two versions of the same proof: first a somewhat chatty version, and then a more concise one.

Proof 1. Recall that an *even number*, by definition, is a number that is two times an integer. If x and y are even, then *by definition* this means that there exist integers a and b such that $x = 2a$ and $y = 2b$. We need to show that their sum $x + y$ can be written as $x + y = 2c$, where c is some integer. To do this, we simply produce a suitable c: we have

$$x + y = 2a + 2b = 2(a + b),$$

so we let $c = a + b$. We have written $x + y$ as two times an integer, so that means $x + y$ is even. ∎

Proof 2. If x and y are even, then there exist integers a and b such that $x = 2a$ and $y = 2b$. Let $c = a + b$. Then $x + y = 2a + 2b = 2(a + b) = 2c$, so $x + y$ is even. ∎

This is a proof because it is completely general: we have explained that *any* numbers satisfying the hypothesis (the "if" part) also satisfy the conclusion (the "then" part). It would not be a proof to give one example, or even several examples, satisfying the hypothesis. After all, we can do that even for false statements. Imagine, for instance, wondering if three times an integer is always even. One can give many examples where this is true: $3 \times 2 = 6$, which is even, $3 \times 8 = 24$, which is even, $3 \times 18 = 54$, which is even, and so forth. Giving a proof means dealing with *all* possible examples, rather than picking and choosing some.[1]

While Proposition 1.3 is obvious, note that we treat it with respect: all theorems, propositions, and so forth, need to be proven. Use the proof above as a model for how to write your own proofs of theorems and propositions. Since Proposition 1.3 is a proposition about even numbers, we need to use the definition of even numbers, both in our hypothesis (x and y are even numbers) and in our conclusion ($x + y$ is even).

Also, note that we ended our proof with a ■. This is a common thing to do: it signifies to the reader that the proof is finished. Some people prefer to use a hollow square: □. Either one is fine. It is also possible to end a proof by writing "QED," an abbreviation of the Latin phrase *quod erat demonstrandum*, meaning "which was to be demonstrated." Other popular, but less formal, endings include "and we're done" and similar phrases. How you choose to end a proof, within reason, is a matter of personal preference.

1.2 Direct proofs

The most basic type of proof is the *direct proof*. Let us suppose we have a statement of the form "if P then Q" that we would like to prove. One way of doing this is to start by assuming P, then performing a sequence of logical deductions such that each statement follows from previous statements, and eventually arriving at the conclusion Q. Assuming that all the logical deductions are valid, this is a proof.

Let's see a simple example of how a direct proof goes.

Proposition 1.5. *If x is an odd number, then x^2 is also an odd number.*

To give a direct proof, we first *assume* that x is an odd number, then

---

[1] It would be remiss of me not to mention that, on occasion, it *is* actually possible to give a general proof by means of one or a small number of examples. See for instance [PWZ96] and [Mat19]. However, this should only be done in special cases, and only when you are absolutely certain that you know when this approach is valid.

follow some logical steps, then finally arrive at the conclusion that x^2 is an odd number. Let's see how this might be written.

Proof. Suppose that x is an odd number. This means that there exists an integer n such that $x = 2n + 1$. Then

$$x^2 = (2n + 1)^2 = 4n^2 + 4n + 1.$$

Letting $m = 2n^2 + 2n$, we have

$$x^2 = 4n^2 + 4n + 1 = 2(2n^2 + 2n) + 1 = 2m + 1.$$

Thus x^2 is an odd number.   ∎

Once again, we started by assuming the hypothesis (x is odd), did a bunch of logical steps, reached the conclusion (x^2 is odd), and then stopped. That is the standard structure of a direct proof.

1.3   Proof by contradiction

A more subtle way of proving theorems is called *proof by contradiction*, which also goes by other names such as *reductio ad absurdum* (reduction to the absurd). When using this technique, we assume that the statement we wish to prove is false and then show that something terrible happens. The general form this takes is to assume we have found a counterexample, then to perform some logical steps from there, and then to show that we violate the hypothesis. That is, given a statement of the form "if P then Q," we actually end up proving "if not Q, then not P." The two statements

$$\boxed{\text{If } P \text{ then } Q}$$

and

$$\boxed{\text{If not } Q \text{ then not } P}$$

are logically equivalent, as we discussed in §0.4. That is, regardless of what P and Q are, if one of those two statements is true, then so is the other one. Similarly, if one of them is false, then so is the other one. So, if we wish to prove "if P then Q," it is just as good to prove "if not Q then not P."

Definition 1.6. The *contrapositive* of the statement "if P then Q" is the statement "if not Q then not P."

Why is the contrapositive always equivalent to the original statement? We can see this with just a bit of thought. Suppose the original statement "if P then Q" is true. Then what happens if Q is false? Well, if P is true, then we know that Q must also be true from the original statement. Thus if Q is false, then P must also be false. Thus we see that the original statement "if P then Q" *implies* the contrapositive "if not Q then not P." But the contrapositive of the contrapositive is the original statement, so by exactly the same reasoning, the statement "if not Q then not P" implies the statement "if P then Q." Thus the two are equivalent.

But there's another, perhaps easier, way of seeing that a statement and its contrapositive must always be equivalent: they have the same truth tables! This is easier to see exactly because *it doesn't require any thinking*: it follows *effortlessly* from the formalism of truth tables. I generally prefer not to think about things that don't require thinking. Thinking is a great tool when it is necessary, but *thinking is the last resort*. Don't start thinking until you've first done all the things that don't require thinking. It's amazing how often a problem will solve itself in this way, before you ever reach the thinking stage.

In the case of Proposition 1.4 above, we can write down the contrapositive as follows:

Proposition 1.7 (Equivalent form of Propositions 1.3 and 1.4)**.** *If $x + y$ is not even, then it is not true that both x and y are even numbers.*

The phrasing here is a little awkward, because this is just a straight translation of Proposition 1.4. We can rephrase it to make it sound more natural. For example:

Proposition 1.8. *If $x + y$ is not even, then either x or y is not even.*

Several remarks are in order here:

Remark 1.9.

- Whenever we use the word "or" in mathematics, unless explicitly stated otherwise, we mean "inclusive or." For instance, when we say that either x or y is not even, then it *might* be the case that neither x nor y is even; that still counts. If we had meant to exclude this possibility, we would have made sure to say so, e.g. "either x is not even, or y is not even, but not both."
- You might wonder why we insist on saying "not even" rather than "odd." The reason is that we haven't stated anywhere that x and

y are supposed to be integers. If they are, then "not even" is the same as "odd." But what if x is some other sort of number, like $\frac{1}{2}$? This number is neither even nor odd, so "not even" is more general than "odd." Thus, the context determines how exactly we may negate a statement, or indeed do anything with it: is some variable supposed to be an integer? A real number? Something else? In this case, it is reasonable to believe that we are making the tacit assumption that x and y are integers, in which case we could freely replace "not even" with "odd." Ideally, the author will make it clear where each variable lives, either by explicitly saying so or by making it clear from context.

Now, let's return to the topic of doing proofs by contradiction. As mentioned above, we wish to prove a statement of the form "if P then Q" by instead proving "if not Q, then not P." Thus, we start by assuming that Q is false, then we perform some logical steps from there, and we end up showing that P is also false. Then we're done.

Often it's not so easy to reach the conclusion "not P" directly, so it is easier to start by assuming both P and not Q together, and then showing that universe has to explode: for example, maybe by assuming both of these, one can conclude that $1 = 0$ or some other catastrophe.

Okay, let's have a look at a simple example of proofs by contradiction.

Proposition 1.10. *If $7x - 3$ is even, then x is odd.*

We could prove this using a direct proof, but let's see how to do it by contradiction for pedagogical purposes. For the purpose of this proposition, let us assume that x is required to be an *integer*, so "not odd" is the same as "even," and "not even" is the same as "odd."

Proof. Suppose, on the contrary, that x is even. We will show that $7x - 3$ is odd, which will complete the proof of the contrapositive. Since x is even, there exists an integer n such that $x = 2n$. Then

$$7x - 3 = 7(2n) - 3 = 2(7n - 2) + 1.$$

Letting $m = 7n - 2$, we have

$$7x - 3 = 2m + 1,$$

so $7x - 3$ is odd, as desired.   ∎

1.4   Irrationality of $\sqrt{2}$

Let us now see how to use proof by contradiction in one of the most famous and celebrated proofs in all of mathematics. In order to state the theorem, we need a preliminary definition:

Definition 1.11. A number x is said to be *rational* if there exist integers a and b, with $b \neq 0$, such that $x = \frac{a}{b}$. A number that is *not* rational is said to be *irrational*.

For example, $\frac{2}{5}$ is a rational number, because we can let $a = 2$ and $b = 5$. There are also other choices for a and b, like $a = 8$ and $b = 20$, but one choice of a and b suffices.

It is not instantly obvious whether irrational numbers exist at all: might *every* number be rational? This turns out not to be the case, and there are irrational numbers, in fact lots and lots and lots of them. But it always requires some work to *prove* that a number is irrational. The first example that anyone came up with, and proved, is $\sqrt{2}$.

Theorem 1.12. $\sqrt{2}$ *is irrational.*

First, note that this theorem isn't stated as an "if P then Q" statement. We can fix this, using the definition of rational (and irrational) numbers: If a and b are integers with $b \neq 0$, then $\frac{a}{b} \neq \sqrt{2}$. That is what we will prove, by means of a contradiction.

Proof. Let us suppose, seeking a contradiction, that $\sqrt{2}$ is rational, with

$$\sqrt{2} = \frac{a}{b}, \tag{1.1}$$

chosen in such a way that b is positive and as small as possible. Note that $a > b$, since $\sqrt{2} > 1$. We may rewrite (1.1) as $2 = \frac{a^2}{b^2}$, or

$$2b^2 = a^2. \tag{1.2}$$

Now, the left side of (1.2) is an even number, and since the two sides are equal, the right side must also be an even number. So, is a even or odd? If we square an odd number, we get another odd number, so a must be even.

Now, since a is even, we may write $a = 2c$ for some integer c. Substituting $a = 2c$ into (1.2), we get

$$2b^2 = (2c)^2 = 4c^2.$$

We may divide by 2 to get

$$b^2 = 2c^2.$$

Thus $2 = \frac{b^2}{c^2}$, or $\sqrt{2} = \frac{b}{c}$, and $b > c > 0$. But this means that we have written $\sqrt{2}$ as a rational number whose denominator is smaller than b, contrary to our assumption that b is as small as possible. This is a contradiction, so our initial assumption, that $\sqrt{2}$ is rational, must be false. Thus $\sqrt{2}$ is irrational. ∎

Note that we ended up assuming that b is as small as possible, and then showing that it isn't. At the beginning of the proof, we might not have anticipated that *that* was where the contradiction was going to show up. When proving theorems, keep an open mind, and be flexible about where you get a contradiction: it might be in a most unexpected place!

1.5   Quantifiers

Many mathematical statements can be expressed in terms of one of two quantifiers: "there exists" and "for all." In fancy terms, these are known as *existential* and *universal* quantifiers, respectively. For example, our theorem that says that $\sqrt{2}$ is irrational is expressible using a universal quantifier:

$$\boxed{\text{For all integers } a \text{ and } b \text{ with } b \neq 0, \frac{a}{b} \neq \sqrt{2}.}$$

In general, we can consider statements of the form

$$\boxed{\text{For all } x, P(x).}$$

Here $P(x)$ is some statement about x.

When proving theorems by contradiction, we have to negate statements, so let's discuss how to negate existential and universal quantifiers. Let's suppose we have a statement

$$\boxed{\text{There exists an } x \text{ such that } P(x)}$$

with an existential quantifier. The negation of this statement is

$$\boxed{\text{There does not exist an } x \text{ such that } P(x).}$$

This is equivalent to saying that for every x, it is *not* the case that $P(x)$ is true, or

$$\boxed{\text{For all } x, \text{ not } P(x).}$$

Similarly, we can negate a universal quantifier: the negation of

$$\boxed{\text{For all } x, \; P(x)}$$

is

$$\boxed{\text{There exists an } x \text{ such that not } P(x).}$$

A more thorough discussion on negations of statements, including a general grammar for negating arbitrary types of statements, can be found in the Interlude at the end of this chapter, beginning on page 35.

There is a common shorthand for much of this stuff. It shouldn't be used in formal writing (i.e. don't use it when writing up your proofs), but it is convenient for taking notes or for developing a schematic picture of what is going on, just for your own purposes. It is also used "officially" in mathematical logic, where it has more precise meaning.

Here is a dictionary of shorthand symbols:

Symbol	Meaning
$\exists$	There exists
$\forall$	For all
$\Rightarrow$	Implies
$\Longleftrightarrow$	If and only if [2]
$\neg$	Not
$\vee$	Or
$\wedge$	And

The statement "$P \implies Q$" is the same as "if P then Q." In the future, we will sometimes say "P implies Q" rather than "if P then Q," which we have avoided doing so far. The statement "$P \iff Q$" is the same as "if P then Q, and if Q then P."

When we say P iff Q (or $P \iff Q$), that means P is true if and only if Q is true. We can unpack that into two different statements: "P if Q" and "P only if Q." These are somewhat awkwardly phrased. The statement "P if Q" can be translated to "if Q then P." The statement "P only if Q" requires more thought to unpack: it means that P is only true if Q is, i.e. if Q is false, then P is false, or "not Q implies not P." As we mentioned above when introducing proof by contradiction, this is equivalent to "if P then Q."

---

[2] Or iff, which is a real word in mathematics and can be used in proofs. We pronounce "iff" as "if and only if."

In shorthand, it is common to write things like

$$\exists x : P(x),$$

which translates to "there exists an x such that $P(x)$." While it is tempting to write as much as possible with symbols and few or no words, this temptation should be avoided at all costs. No one wants to read and parse a block full of symbols with no explanations.

Finally, let's talk a bit about proving statements with quantifiers. The techniques used to prove statements of the form "for all x, $P(x)$" (or, in shorthand, $\forall x : P(x)$) and "there exists an x such that $Q(x)$" ($\exists x : Q(x)$) are rather different. In the first case, we have something to prove for *every* x. It isn't enough to check one value of x, or even many values of x. Recall how we did this when showing that $\sqrt{2}$ is irrational: we had to show that for *every* pair of integers a and b with $b \neq 0$, $\frac{a}{b} \neq \sqrt{2}$. It wouldn't be good enough to check that, say, $\frac{a}{b} \neq \sqrt{2}$ when $a = 7$ and $b = 5$.

By contrast, when proving an existential statement of the form "there exists an x such that $Q(x)$," we're done as soon as we have found such an x. (Sometimes, we might not actually be able to find an x but only show that one must exist, somewhere, but we might not know where.) It might not be easy to find an x, but it can be done. For instance:

Proposition 1.13. *There exist positive integers x and y such that $x^2 - 97y^2 = 1$.*

Proof. Let $x = 62809633$ and $y = 6377352$. Then $x^2 - 97y^2 = 1$. ∎

That's a complete proof, because we only had to show that there is at least one solution to the equation and we did it. How one *finds* such a solution to begin with is a different story, but we don't have to explain everything we tried when writing up a proof; a logically correct sequence of statements that leads to the statement we're trying to prove is sufficient.[3]

1.6   Tips on mathematical writing

Observe that all the proofs we've seen above are written in *complete sentences*. All mathematical proofs should be written in complete sentences,

---

[3]The standard way of finding such solutions uses the theory of continued fractions. An equation of the form $x^2 - dy^2 = 1$ is called a *Pell equation*, and Pell equations always have solutions in positive integers as long as d isn't a perfect square. The smallest solution might be quite large though! See [Len08] for a survey about Pell equations, how to solve them, and more.

with a subject, predicate, and proper punctuation. Proofs are meant to be read by people, and people like reading stuff that is well-written and aligned with the usual rules of grammar. It is worth noting, though, that sometimes mathematical symbols are the subjects or predicates of sentences. For example, here is an example of a complete sentence:

$$x = 3.$$

The subject of this sentence is "x," and the verb is "$=$." Here are some things you should keep in mind in your mathematical writing:

- Always end your sentences with a period or some other sentence-ending form of punctuation (like an exclamation mark). This includes sentences that end with equations. For example: the following might be a sentence in a proof: It follows that

$$x^2 + 4 = 9y^3.$$

  Note that, even though the sentence has a displayed equation at the end, it still ends with a period. This is always a good idea, unless it's typographically problematical. In those cases, it's often worth reworking the sentence to end differently. For example, a sentence could conceivably end with a diagram from a grammatical standpoint, but this tends to be a bad idea from the perspective of typography, so don't do that if it is conveniently possible to avoid doing so.
- Don't be stupid with punctuation. For example, if you are doing something with factorials, don't end a sentence with an exclamation mark after a number or mathematical symbol. This may necessitate strange forms of punctuation at the end of a sentence, such as: We have

$$120 = 5!.$$

  The exclamation mark is a factorial, and the period ends the sentence; they have completely different grammatical roles.
- Begin sentences with capital letters. It is best to avoid beginning a sentence with a variable, because those are case-sensitive: for instance $x^2 + 5x - 3$ and $X^2 + 5x - 3$ have different meanings. One trick to get around the issue of starting sentences with variables or equations is to use connecting words, like "thus," "so," "therefore," "it follows that," and so on. If you want to begin with an equation, you can often deal with this by inserting the words "The equation."

For instance, instead of writing

$$\boxed{x^2 + y^2 = -1 \text{ has no real solutions.}}$$

you can write

$$\boxed{\text{The equation } x^2 + y^2 = -1 \text{ has no real solutions.}}$$

It is conventional in mathematics to write in first-person plural, using "we" and "us." In mathematics, "we" is not used royally; rather, it stands for "the reader and I." The reader is part of the journey too. Using the first-person plural helps us avoid writing in the passive voice (e.g. "it is shown that"), which is frowned upon in mathematics.

1.7   Problems

(1) Which of the following statements are equivalent to which others?

(a) If P then Q.   (f) Whenever P, Q.
(b) If Q then P.   (g) P implies Q.
(c) P only if Q.   (h) Not P implies not Q.
(d) Q only if P.   (i) Not Q implies not P.
(e) P if Q.

(2) Give an example of statements P and Q, such that P implies Q, but Q does not imply P. Given a statement "P implies Q," we call the statement "Q implies P" the *converse*. Statements and their converses are not the same!

(3) Given a statement "if P then Q," the statement "if not P then not Q" is called the *inverse*. Explain why the inverse and converse are always equivalent: if one of them is true, then so is the other; and if one of them is false, then so is the other.

(4) Suppose that $P(x, y)$ is a statement about two numbers x and y. Are the following necessarily equivalent: $\forall x : \exists y : P(x, y)$ and $\exists y : \forall x : P(x, y)$? Either prove they are equivalent for all statements P, or else give an example of a statement P such that one of them is true and the other is false. (Make sure you understand why this problem has content! You cannot conclude that they are equivalent just because they contain the same symbols, albeit in a different order!)

(5) Prove the following statements:

    (a) For all integers x, $x^2 + 5x$ is even.

    (b) If x is odd, then $x^3 - 3x^2 + 5x + 7$ is even.

    (c) Suppose x is an integer. If $x^2 + 2x + 3$ is even, then x is odd.

    (d) For all integers n, either n^2 or $n^2 - 1$ is a multiple of 4.

(6) Find, with proof, the largest integer n such that, for all integers x, $x(x + 1)(x + 2)(x + 3)$ is a multiple of n.

(7) Mimic the proof of the irrationality of $\sqrt{2}$ to prove that $\sqrt{3}$ and $\sqrt[3]{4}$ are irrational. What goes wrong when you try to use the same argument to prove that $\sqrt{9}$ is irrational?

(8) Prove that the sum of a rational number and an irrational number is irrational. Prove that there exist two irrational numbers whose sum is rational.

(9) Prove that the product of a *nonzero* rational number and an irrational number is irrational. Prove that there exist two irrational numbers whose product is rational.

(10) Prove that $\log_2(3)$ is irrational. (In case you haven't seen logarithms yet, $\log_2(3)$ is the number x such that $2^x = 3$.)

(11) Note that $(\sqrt{2}^{\sqrt{2}})^{\sqrt{2}} = 2$. Explain how to use this fact to prove that there exist two irrational numbers x and y such that x^y is rational.

(12) Prove that there are no integers x and y such that $x^2 - 3y^2 = 2$.

Interlude: Some sample proofs

This interlude is here to give you more examples of what proofs look like, so that you can follow their style as you write your own proofs.

Proposition 1.14. *There are no integers a and b such that $6a + 15b = 191$.*

Proof. Suppose that a and b are integers such that $6a + 15b = 191$. Then, dividing by three,

$$2a + 5b = \frac{6a + 15b}{3}$$

is an integer, but $\frac{191}{3}$ is not. This is a contradiction to the assertion that a and b are integers such that $6a + 15b = 191$. ∎

Theorem 1.15. *There exist integers a, b, and c such that $a^3 + b^3 + c^3 = 42$.*

Proof. Let $a = -80538738812075974$, $b = 80435758145817515$, and $c = 12602123297335631$. Then $a^3 + b^3 + c^3 = 42$. ∎

Definition 1.16. Let n be a positive integer. A *divisor* of n is a positive integer d such that n is a multiple of d. We write $d(n)$ for the number of divisors of n.

Example. The divisors of 14 are 1, 2, 7, and 14, so we have $d(14) = 4$.

Theorem 1.17. *If n is a positive integer, then $d(n)$ is odd if and only if n is a perfect square.*

Proof. Suppose that $n = \alpha^2$ where α is positive, but not necessarily an integer. For any divisor d of n other than α, the number $\frac{n}{d}$ is another divisor of n, distinct from n, and $\frac{n}{d} > \alpha$ iff $d < \alpha$. Thus the number of divisors of n that are less than α is equal to the number of divisors of n

greater than α, so the total number of divisors of n that are not equal to α is even. If n is not a perfect square, then α is not an integer, so the total number of divisors of n is even. On the other hand, if n is a perfect square, then α is an integer, so the total number of divisors of n is odd. ∎

Theorem 1.18. *If a, b, and c are odd positive integers, then there is no rational number x such that $ax^2 + bx + c = 0$.*

Proof. Suppose that $x = \frac{p}{q}$ is a rational number, where p and q are integers and $q \neq 0$. By reducing to lowest terms if necessary, we may assume that p and q are not both even. If $ax^2 + bx + c = 0$, then we may multiply by q^2 to get

$$ap^2 + bpq + cq^2 = 0. \tag{1.3}$$

The right side of (1.3) is even, so the left side must be as well. If both p and q are odd, then all three terms on the left side of (1.3) are odd, so the sum is odd. If one of p and q is even and the other is odd, then exactly one of the terms on the left side of (1.3) is odd, so again the sum is odd. Thus there is no rational x such that $ax^2 + bx + c = 0$. ∎

Interlude: Negating statements

When we prove a statement of the form "if P then Q" by contradiction, we need to prove the contrapositive "if not P then not Q." That means we need to know how to negate the statements P and Q. Of course, the statements "not P" and "not Q" are already themselves statements, but they typically aren't written in a form that is conducive to saying much about them. Fortunately, there is a grammar of logic that allows us to negate statements in symbolic form nearly automatically. When writing an actual proof, one typically doesn't use these symbols. But they are helpful for scratchwork, especially when determining the negation of a statement.

A statement is typically built out of the following ingredients:

Symbol	Meaning
$\wedge$	And
$\vee$	Or
$\neg$	Not
$\Rightarrow$	Implies
$\forall$	For all
$\exists$	There exists

Here is how we negate each one:

Statement	Negation
$P \wedge Q$	$(\neg P) \vee (\neg Q)$
$P \vee Q$	$(\neg P) \wedge (\neg Q)$
$\neg(\neg P)$	P
$P \implies Q$	$P \wedge (\neg Q)$
$\forall x : P(x)$	$\exists x : \neg P(x)$
$\exists x : P(x)$	$\forall x : \neg P(x)$

Let's see how this grammar of negation applies to concrete examples.

Example. Consider the statement S, which is "The numbers x and y are both even." Let P be the statement "x is even," and Q the statement "y is even," so that our original statement S is $P \wedge Q$ (read "P and Q"). Its negation is $\neg(P \wedge Q)$, which is the same as $(\neg P) \vee (\neg Q)$. Now, $\neg P$ is the statement "x is not even," and $\neg Q$ is the statement "y is not even," so the full negation $\neg S$ is "x is not even or y is not even." From here, it is possible to reword the negation if desired, perhaps as "At least one of x or y is not even." Remember as usual that "or" in mathematics is always inclusive unless it is explicitly stated otherwise.

Example. Consider the statement S, which is "For all numbers x, there exists a number y such that $y > x$ and y is a perfect square." Let $P(x,y)$ be the statement "$y > x$," and let $Q(x,y)$ be the statement "y is a perfect square." Notice that Q really only depends on y and not x, but we can still write it as a statement about both of them. Symbolically, S is

$$\forall x : \exists y : (P(x,y) \wedge Q(x,y)).$$

Its negation is

$$\neg \forall x : \exists y : (P(x,y) \wedge Q(x,y)).$$

We need to propagate the negation throughout the statement, from the outside in. We begin by negating the outermost part, which is the $\forall$, so we get

$$\exists x : \neg \exists y : (P(x,y) \wedge Q(x,y)).$$

Next, we need to negate the second $\exists$, so we get

$$\exists x : \forall y : \neg(P(x,y) \wedge Q(x,y)).$$

Finally, we need to negate the $\wedge$, so we get

$$\exists x : \forall y : (\neg P(x,y) \vee \neg Q(x,y)).$$

We've now propagated the $\neg$ throughout the statement, so it's time to convert our symbolic statement back into an English statement. In literal form it reads "There exists a number x such that for all numbers y, either it is not true that $y > x$ or it is not true that y is a perfect square." This is awkwardly stated, so we can rephrase it as "There exists a number x such that for all numbers y, either $y \leq x$ or y is not a perfect square." Better still is "There exists a number x such that for all numbers $y > x$, y is not a perfect square." Or, even better: "There exists a number x such that there are no perfect squares greater than x." Of course, this negated statement $\neg S$ is a *false* statement, but that's okay: statements are allowed to be false.

Chapter 2

Induction

2.1 Inductive proofs

Suppose we wish to prove that a statement $P(n)$ is true for all positive integers n; that is, it is true when $n = 1, 2, 3$, and so on. We could try to prove them all at once, but under some circumstances that might be hard.

Another possibility is to start by proving that $P(n)$ is true when $n = 1$. Then, *using* the fact that P is true when $n = 1$, show that P is true when $n = 2$. After that, use the fact that P is true when $n = 2$ to prove that P is true when $n = 3$, and so forth. Ideally, we'll be able to do all these steps at once. That is, we will be able to prove that the truth of $P(n)$ implies the truth of $P(n + 1)$ for all n, simultaneously. That, together with the truth of $P(1)$, guarantees that $P(n)$ is true for all positive integers n. This technique is known as *proof by induction*, or simply *induction*.

Let's see an example of induction in action.

Theorem 2.1. *For all positive integers n, we have*

$$1 + 2 + \cdots + n = \frac{n(n + 1)}{2}. \tag{2.1}$$

Theorem 2.1 can be proven in many ways (for instance, by contemplating Figure 2.1), but one of those ways is by induction.

Proof of Theorem 2.1. We first prove the statement when $n = 1$. This is easy: both sides are equal to 1, so the statement is true when $n = 1$.

Next, suppose that we know the result is true for n; we then prove the statement for $n + 1$. Since it's true for n, we have

$$1 + 2 + \cdots + n + (n + 1) = (1 + 2 + \cdots + n) + (n + 1) = \frac{n(n + 1)}{2} + n + 1,$$

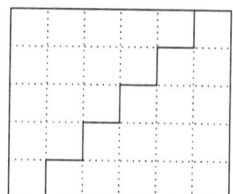

Figure 2.1. A pictorial "proof" of Theorem 2.1. Each half of the rectangle contains $1 + 2 + \cdots + n$ squares, and the entire rectangle contains $n(n+1)$ squares.

where in the last equality we have made use of the result for n. Now, we have

$$1 + 2 + \cdots + n + (n+1) = \frac{n(n+1)}{2} + n + 1$$

$$= \frac{n^2}{2} + \frac{n}{2} + n + 1$$

$$= \frac{n^2}{2} + \frac{3n}{2} + 1$$

$$= \frac{(n+1)(n+2)}{2},$$

which is what we get when we plug in $n+1$ into (2.1). This completes the proof. ∎

Why does this finish the proof? Well, we think of Theorem 2.1 as stating an infinite sequence of statements, one for each n. Giving a proof means explaining why each one of them is true. So let's take one of these statements, say the one with $n = 292$:

$$1 + 2 + \cdots + 292 = \frac{292 \cdot 293}{2}.$$

In our proof, we started by showing that the one with $n = 1$ is true. Then we used that to prove that the statement with $n = 2$ is true, then we used that to show that the one with $n = 3$ is true, and so forth. Eventually, this implies that the statement with $n = 291$ is true, and finally we use that to prove that the one with $n = 292$ is true. We accomplished our goal. We could do the same thing for *any* value of n.

2.2   The structure of an inductive proof

In a typical inductive proof, we want to prove a statement of the form

For all positive integers n, $P(n)$ is true.

In special cases, it might look a little bit different: it might only be true for all $n \geq 5$ for instance, or for $n \geq 0$, or perhaps only for even n. But these are only minor differences.

Let us suppose we have a standard statement, like the boxed one above. An inductive proof consists of two parts. The first part is proving the *base case*: we must show that $P(1)$ is true. Often this is straightforward: just checking that one *number* is equal to another, like in the proof of Theorem 2.1. On occasion, however, proving the base case might be challenging.

The other part of the proof is the *inductive step*. In the inductive step, we *use* the truth of the statement $P(n)$ to *prove* $P(n+1)$. This is what we did in the proof of Theorem 2.1: we proved that *if*

$$1 + 2 + \cdots + n = \frac{n(n+1)}{2},$$

then

$$1 + 2 + \cdots + n + (n+1) = \frac{(n+1)(n+2)}{2}.$$

But we already know that the statement is true for n thanks to the base case and previous inductive steps, so the statement for $n+1$ must also be true.

Note that the inductive proof didn't allow us to *notice* that $1 + 2 + \cdots + n = \frac{n(n+1)}{2}$. We had to know that in advance, either from being told, or by having guessed it by looking at a bunch of small cases, or by recognizing it in some other way. This is characteristic of inductive proofs: we can typically prove that a formula is always true once we know what we're looking for, but it is not possible to come up with the answer without having a guess about what it is first. There are, of course, other methods of proof that are better suited to coming up with the correct answer from scratch.

Okay, let's try another example, along the lines of Theorem 2.1.

Theorem 2.2. *For any positive integer n, we have*

$$1^2 + 2^2 + \cdots + n^2 = \frac{n(n+1)(2n+1)}{6}. \tag{2.2}$$

Proof. We begin by proving the base case, $n = 1$. When $n = 1$, the left side of (2.2) is 1, and the right side is $\frac{1 \cdot 2 \cdot 3}{6} = 1$. Thus (2.2) holds when $n = 1$.

Next, we do the inductive step. We assume that (2.2) holds for n, and we prove it for $n + 1$. That is, we assume that

$$1^2 + 2^2 + \cdots + n^2 = \frac{n(n+1)(2n+1)}{6},$$

and we must prove that

$$1^2 + 2^+ \cdots + n^2 + (n+1)^2 = \frac{(n+1)(n+2)(2n+3)}{6}.$$

We have

$$
\begin{aligned}
1^2 + 2^2 + \cdots + n^2 + (n+1)^2 &= (1^2 + 2^2 + \cdots + n^2) + (n+1)^2 \\
&= \frac{n(n+1)(2n+1)}{6} + (n+1)^2 \\
&= \frac{n+1}{6} \cdot (n(2n+1) + 6(n+1)) \\
&= \frac{n+1}{6} \cdot (2n^2 + 7n + 6) \\
&= \frac{(n+1)(n+2)(2n+3)}{6},
\end{aligned}
$$

as desired.   ■

2.3   Sigma notation

It is often desirable to write expressions like $1 + 2 + \cdots + n$ without the use of $\cdots$. To this end, we introduce Σ notation. Given a function $f(x)$ and integers $a \le b$, we write

$$\sum_{i=a}^{b} f(i)$$

to mean $f(a) + f(a + 1) + f(a + 2) + \cdots + f(b - 1) + f(b)$. Hence we can rewrite $1 + 2 + \cdots + n$ as $\sum_{i=1}^{n} i$. Note that the indexing variable, in this case i, does not matter: we can just as easily write $\sum_{j=1}^{n} j$, which means exactly the same thing. However, the indexing variable should not appear elsewhere in the statement in some other context. If we made the unfortunate choice of n as our index variable, we would end up with the nonsense expression $\sum_{n=1}^{n} n$.

In this new notation, we can rewrite the statements of the theorems we already proved: Theorem 2.1 now says that

$$\sum_{i=1}^{n} i = \frac{n(n+1)}{2},$$

and Theorem 2.2 now says that

$$\sum_{i=1}^{n} i^2 = \frac{n(n+1)(2n+1)}{6}.$$

This notation is actually more flexible. The subscript and superscript tell us which terms to include in the sum, but they can do so in more "literary" ways than just giving us the upper and lower bounds. For instance, we could rewrite the statement of Theorem 2.1 as

$$\sum_{1 \le i \le n} i = \frac{n(n+1)}{2}.$$

We can also imagine that we might have a function $g(x,y)$ of *two* variables, and we want to sum over those values of $g(i,j)$ where i and j are both between 1 and n. To do this, we write

$$\sum_{1 \le i,j \le n} g(i,j),$$

which can also be written as the iterated pair of sums

$$\sum_{i=1}^{n} \sum_{j=1}^{n} g(i,j) = \sum_{i=1}^{n} \left(\sum_{j=1}^{n} g(i,j) \right).$$

If we also want to require that $i < j$, then we would write

$$\sum_{1 \le i < j \le n} g(i,j).$$

If, on top of that, we require that i be odd, we can include that as part of our subscript:

$$\sum_{\substack{1 \le i < j \le n \\ i \text{ odd}}} g(i,j).$$

As you can see, this notation wields great power and can be used to say all sorts of things.

Sometimes we might wish to multiply the terms, instead of adding them. To signify this, we replace the Σ with a Π. For instance, if we wish to multiply the numbers from 1 to n, also known as $n!$ and read "n factorial," we can write

$$\prod_{i=1}^{n} i.$$

As you may have imagined, we call this Π notation.

2.4   A word on formality

You might be wondering why we need the formal proof structure that we have seen in our inductive proofs so far, that we will continue to require in all future induction proofs. Sometimes, it seems that an inductive proof can be explained perfectly satisfactorily with an argument that uses the phrase "and so on" or "et cetera" at some key moment. Consider, for instance, the following theorem, making use of our newly introduced Σ notation.

Theorem 2.3. *For every positive integer n, we have*

$$\sum_{k=1}^{n} \frac{1}{k(k+1)} = 1 - \frac{1}{n+1}.$$

Let's start with an argument that captures the right idea, but isn't written using the best practices for inductive proofs.

A Bad Proof. First, note that for every positive integer k, we have $\frac{1}{k(k+1)} = \frac{1}{k} - \frac{1}{k+1}$. Thus for $n = 1$, the left side is

$$1 - \frac{1}{2},$$

so the two sides agree. Next, for $n = 2$, the left side is

$$\left(1 - \frac{1}{2}\right) + \left(\frac{1}{2} - \frac{1}{3}\right) = 1 - \frac{1}{3},$$

so again the two sides agree. Next, for $n = 3$, the left side is

$$\left(1 - \frac{1}{2}\right) + \left(\frac{1}{2} - \frac{1}{3}\right) + \left(\frac{1}{3} - \frac{1}{4}\right) = 1 - \frac{1}{4},$$

so the two sides agree. For larger values of n, the same argument applies.

∎

Sure, some readers will be able to understand what is going on here, but there is an issue that make this a bad proof. What exactly is "the same argument"? The point is that there is a lot of cancellation when the parentheses are removed, and only the first and last terms remain. So, here is an improved, but still not great, proof. We'll continue using the fact that $\frac{1}{k(k+1)} = \frac{1}{k} - \frac{1}{k+1}$ in the next proof, as this is a key ingredient.

A Better, But Still Imperfect, Proof. We have

$$\sum_{k=1}^{n} \frac{1}{k(k+1)} = \left(1 - \frac{1}{2}\right) + \left(\frac{1}{2} - \frac{1}{3}\right) + \cdots + \left(\frac{1}{n} - \frac{1}{n+1}\right)$$

$$= 1 - \frac{1}{2} + \frac{1}{2} - \frac{1}{3} + \cdots + \frac{1}{n} - \frac{1}{n+1}$$

$$= 1 - \frac{1}{n+1},$$

because all the other terms cancel out. ∎

In this case, it's clear that indeed all the other terms do cancel out, so *in this particular instance,* a reader should not have trouble understanding what is going on. However, in more complicated examples, there may be subtleties going on that might make it harder to detect problems with this sort of argument. So, let's now see how to write this argument in the best possible way, using our standard template for an inductive proof.

A Good Proof. We begin with the base case $n = 1$. When $n = 1$, the left side is $\frac{1}{1\cdot2} = \frac{1}{2}$, and the right side is $1 - \frac{1}{2} = \frac{1}{2}$, so the two sides are equal.

Now, suppose that n is a positive integer, and the result is true for n. We'll show that it is true for $n + 1$ as well. Using the result for n, we have

$$\sum_{k=1}^{n+1} \frac{1}{k(k+1)} = \sum_{k=1}^{n} \frac{1}{k(k+1)} + \frac{1}{(n+1)(n+2)}$$

$$= 1 - \frac{1}{n+1} + \frac{1}{(n+1)(n+2)}$$

$$= 1 - \frac{1}{n+1}\left(1 - \frac{1}{n+2}\right)$$

$$= 1 - \frac{1}{n+1} \cdot \frac{n+1}{n+2}$$

$$= 1 - \frac{1}{n+2},$$

which is the result for $n + 1$. This completes the inductive step and thus the proof. ∎

Now there can't possibly be any guesswork left for the reader to do. All steps have been explained completely and thoroughly. Note that the good proof is longer in terms of the number of lines or characters, but this is because it takes more space to explain everything well. The proofs that take the fewest lines are not always the best proofs: sometimes they are artificially short because they are missing important steps or explanations. Note also that the middle argument has certain merits that the final, correct, proof does not: it demonstrates the crucial point that the terms cancel out in pairs. One may even reasonably argue that it is more insightful than the correct proof. But it is nonetheless not as good as the final argument *as a formal proof.* When you solve problems, try to notice things like the argument highlighted in the middle proof, but then write them up like the last one.

2.5   Recurrences

One place that induction is often useful is in finding closed formulae for sequences defined by recurrences. Let's try an example.

Example. Consider the sequence $a_0, a_1, a_2, \ldots$ defined by $a_0 = 3$ and $a_{n+1} = (a_n - 1)^2 + 1$. The first few terms are

$$a_0 = 3, \qquad a_1 = 5, \qquad a_2 = 17, \qquad a_3 = 257, \qquad a_4 = 65537,$$

and so forth. With a bit of insight, we might happen to notice that it *appears* that $a_n = 2^{2^n} + 1$, also known as the nth *Fermat number.*

So far, this is just a guess, and we have no guarantee that $a_n = 2^{2^n} + 1$ for all n, only that this is true for the small cases we have checked. However, we can prove this formula by induction. Let's do it!

As usual, we start with the base case, which is $n = 0$. This amounts to checking that a_0, which is *defined* to be 3, is equal to $2^{2^0} + 1$, which is also 3. Thus the base case holds.

Next, we perform the induction step. We assume that $a_n = 2^{2^n} + 1$, and we prove that this implies that $a_{n+1} = 2^{2^{n+1}} + 1$. We have

$$\begin{aligned}
a_{n+1} &= (a_n - 1)^2 + 1 \\
&= ((2^{2^n} + 1) - 1)^2 + 1 \\
&= (2^{2^n})^2 + 1 \\
&= 2^{2 \cdot 2^n} + 1 \\
&= 2^{2^{n+1}} + 1.
\end{aligned}$$

This completes the inductive step and the proof of the formula.

A similar argument can be used to prove many other formulae for recurrences. See problem 6 for a closed form for the Fibonacci numbers, known as *Binet's formula.*

2.6 Strong induction

In the examples we have seen so far, it is possible to prove a statement $P(n+1)$ directly from $P(n)$. Sometimes, however, just using $P(n)$ isn't enough, and we need to know values further back. Perhaps we need to use $P(n-1)$ as well, or $P(n-2)$, or perhaps even various $P(m)$'s, for unknown values of $m \leq n$.

We may prove theorems in this way using the technique of *strong induction*. In strong induction, we prove a base case, say $P(0)$, and then we derive the truth of $P(n+1)$ from knowing $P(0), P(1), \ldots, P(n)$. The setup of a strong induction proof is similar to that of a normal induction proof: we prove the base case $P(0)$, and then we prove the inductive step, which says that if $P(0), P(1), \ldots, P(n)$ are all true, then $P(n+1)$ is true as well. This implies that $P(n)$ is true for all nonnegative integers n.

Let's see an example of how this works.

Theorem 2.4. *Every positive integer has a binary representation. That is, if n is a positive integer, we may write n in the form*

$$n = a_0 \cdot 2^0 + a_1 \cdot 2^1 + \cdots + a_k \cdot 2^k \tag{2.3}$$

for some nonnegative integer k, where each a_i is either 0 or 1.

Remark 2.5. Using Σ notation, we can rewrite equation (2.3) as

$$n = \sum_{i=0}^{k} a_i \cdot 2^i.$$

Proof. We first prove the base case, which is the case of $n = 1$. We must show that the number 1 has a binary representation. Indeed, it does: let $k = 0$ and $a_0 = 1$, so that the binary representation of 1 is $1 \cdot 2^0$.

Next, suppose that, for all m with $1 \leq m < n$, the number m has a binary representation. We will show that n also has a binary representation. To do this, we split the problem into two cases, depending on whether n is even or odd.

Case 1: n is even. Since $1 \leq \frac{n}{2} < n$ and $\frac{n}{2}$ is a positive integer, we may assume that $\frac{n}{2}$ has a binary representation, say

$$\frac{n}{2} = \sum_{i=0}^{k} a_i \cdot 2^i.$$

Multiplying by 2, we get

$$n = \sum_{i=0}^{k} a_i \cdot 2^{i+1} = \sum_{i=1}^{k+1} a_{i-1} \cdot 2^i.$$

This is a binary representation for n.

Case 2: n is odd. We have already done the case $n = 1$, so we assume that $n \geq 3$. Thus $\frac{n-1}{2}$ is an integer, and $1 \leq \frac{n-1}{2} < n$. Thus, by induction, $\frac{n-1}{2}$ has a binary representation, say

$$\frac{n-1}{2} = \sum_{i=0}^{k} a_i \cdot 2^i.$$

If we multiply by 2, we get

$$n - 1 = \sum_{i=1}^{k+1} a_{i-1} \cdot 2^i.$$

Now, adding 1 (or rather $1 \cdot 2^0$), we get

$$n = 1 \cdot 2^0 + \sum_{i=1}^{k+1} a_{i-1} \cdot 2^i,$$

a binary representation for n. That was the last case, so this completes the inductive step and the proof. ∎

2.7   Problems

(1) We will prove that all horses are the same color. To do this, we will show that if we have any collection of $n \geq 1$ horses, then all n of these horses are the same color.

We begin with the base case, $n = 1$. If we have only one horse, then all the horses clearly have the same color.

Next, we do the inductive step. Suppose that n horses are always the same color. We will show that $n + 1$ horses are always the same color. Let us number the horses $H_1, \ldots, H_{n+1}$. Since we are assuming that *any* set of n horses are the same color, then $H_1, \ldots, H_n$ are the same color. Similarly, $H_2, \ldots, H_{n+1}$ are the same color. Now, any horse in the middle is the same color as both H_1 and H_{n+1}, so H_1 and H_{n+1} are the same color. Thus all $n + 1$ horses are the same color. What is the mistake in this argument?

(2) What is the sum of the first n odd numbers? Give a proof using induction.

(3) Prove that
$$1^3 + 2^3 + \cdots + n^3 = \frac{n^2(n+1)^2}{4} = (1 + 2 + \cdots + n)^2.$$
Then express this statement using Σ notation, without the $\cdots$.

(4) Prove that for every nonnegative integer n, the number $n^5 - n$ is a multiple of 5. What about for negative integers?

(5) Prove that if $x > -1$ and n is a nonnegative integer, then $(1 + x)^n \geq 1 + nx$.

(6) Let F_n be the nth Fibonacci number, defined by $F_0 = 0$, $F_1 = 1$, and $F_{n+2} = F_n + F_{n+1}$ for $n \geq 0$. Prove that
$$F_n = \frac{1}{\sqrt{5}}\left(\left(\frac{1 + \sqrt{5}}{2}\right)^n - \left(\frac{1 - \sqrt{5}}{2}\right)^n\right).$$

(7) Show that every positive integer n can be written in the form
$$n = \sum_{i=1}^{k} a_i \cdot i!,$$
where k is a positive integer, each a_i is a nonnegative integer, and for all i with $1 \leq i \leq k$, we have $0 \leq a_i \leq i$.

(8) Prove that for all nonnegative integers n, a number consisting of 3^n equal digits is divisible by 3^n.

(9) Prove that for every positive integer n, there exists an n-digit number divisible by 5^n whose digits are all odd.

(10) Prove that every positive integer $n \geq 1$ can be expressed as a sum of pairwise nonconsecutive Fibonacci numbers. That is, for every positive integer n, there is some positive integer $k \geq 2$ and numbers $a_2, \ldots, a_k$ which are each either 0 or 1 such that
$$n = \sum_{i=2}^{k} a_i F_i,$$
and if $a_i = 1$, then a_{i-1} and a_{i+1} must both be 0. This result is known as *Zeckendorf's Theorem*.

(11) The *Ackermann function* $A(m, n)$ is a function of two nonnegative integer variables, defined by
$$A(m, n) = \begin{cases} n + 1 & m = 0, \\ A(m - 1, 1) & m > 0 \text{ and } n = 0, \\ A(m - 1, A(m, n - 1)) & m > 0 \text{ and } n > 0. \end{cases}$$
For $m = 0, 1, 2, 3, 4$, find and prove a formula for $A(m, n)$. What about when $m = 5$?

(12) A *graph* consists of a collection of vertices, drawn as dots, together
with edges, which connect pairs of vertices. (See Figure 2.2 for an
example.) For a graph G, let $C_G(n)$ denote the number of ways of
coloring each vertex of G using one of n colors, such that if v and w
are vertices connected by an edge, then they are colored differently.
The goal of this problem is to prove some facts about the function
$C_G(n)$.

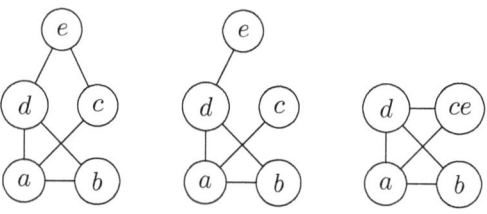

Figure 2.2. Left: A graph G. Middle: The deletion $G \setminus \{c, e\}$ of the edge
$\{c, e\}$ between c and e. Right: The contraction $G/\{c, e\}$ of the edge $\{c, e\}$
between c and e.

(a) Let e be an edge of a graph G. We define the *deletion* of e to be the
graph $G \setminus e$, which is the same as G except with edge e deleted. We
also define the *contraction* of e to be the graph G/e, which is the
same as G except that the edge e is deleted and the two endpoints
of e are glued together. Prove that $C_G(n) = C_{G \setminus e}(n) - C_{G/e}(n)$.
This is known as the *deletion-contraction formula*. This formula
will be helpful for the later parts of this problem.

(b) Prove that, for any graph G, $C_G(n)$ is a polynomial function of n.
(Hint: Use induction on the number of vertices plus the number
of edges of G.)

(c) Prove that the degree of the polynomial $C_G(n)$ is equal to the
number of vertices of G.

(d) Suppose that $C_G(n) = \sum_{i=0}^{k} a_i n^i$. Prove that the signs of the
coefficients a_i alternate: if k is the degree of $C_G(n)$, then $a_k > 0$,
$a_{k-1} \leq 0$, $a_{k-2} \geq 0$, $a_{k-3} \leq 0$, and so forth.

Interlude: More proofs by induction

In this interlude, we'll give a few more examples of proof by induction.

Definition 2.6. The *Fibonacci numbers* are the numbers F_n defined by $F_0 = 0$, $F_1 = 1$, and $F_{n+2} = F_n + F_{n+1}$ for all $n \geq 0$.

The first few Fibonacci numbers, starting from F_0, are 0, 1, 1, 2, 3, 5, 8, 13, 21, 34.

Theorem 2.7. *For every positive integer n, we have $\sum_{i=1}^{n} F_i^2 = F_n F_{n+1}$.*

Proof. We first prove the base case, $n = 1$. The left side is $F_1^2 = 1$, and the right side is $F_1 F_2 = 1 \cdot 1 = 1$. Thus the base case holds.

Now we do the inductive step. Suppose that n is a positive integer, and we have $\sum_{i=1}^{n} F_i^2 = F_n F_{n+1}$. We must prove the result for $n + 1$; that is, we must show that

$$\sum_{i=1}^{n+1} F_i^2 = F_{n+1} F_{n+2}.$$

We have

$$\sum_{i=1}^{n+1} F_i^2 = \sum_{i=1}^{n} F_i^2 + F_{n+1}^2$$
$$= F_n F_{n+1} + F_{n+1}^2$$
$$= F_{n+1}(F_n + F_{n+1})$$
$$= F_{n+1} F_{n+2},$$

as desired. ■

Proposition 2.8. *If n is a positive integer greater than or equal to 8, then there exist nonnegative integers a and b such that $3a + 5b = n$.*

Proof. We use three base cases: $n = 8$, $n = 9$, and $n = 10$. For $n = 8$, we have $a = 1$ and $b = 1$. For $n = 9$, we have $a = 3$ and $b = 0$. For $n = 10$, we have $a = 0$ and $b = 2$.

Now, suppose that $n \geq 11$, and every positive integer k with $8 \leq k < n$ can be written in the form $3a + 5b$, where a and b are nonnegative integers. Because $n \geq 11$, $n - 3 \geq 8$, so by the inductive hypothesis, there exist nonnegative integers a and b such that $3a + 5b = n - 3$. Thus we have $3(a + 1) + 5b = n$, so n can be written in this form as well. $\blacksquare$

Proposition 2.9. *For any positive integer n,*

$$\sum_{i=1}^{n} \frac{1}{\sqrt{i}} \geq \sqrt{n}.$$

Proof. The base case is $n = 1$, when both sides are equal to 1. Thus the base case holds.

Now, suppose that $\sum_{i=1}^{n} \frac{1}{\sqrt{i}} \geq \sqrt{n}$. We must show that $\sum_{i=1}^{n+1} \frac{1}{\sqrt{i}} \geq \sqrt{n+1}$. We have

$$\sum_{i=1}^{n+1} \frac{1}{\sqrt{i}} = \sum_{i=1}^{n} \frac{1}{\sqrt{i}} + \frac{1}{\sqrt{n+1}} \geq \sqrt{n} + \frac{1}{\sqrt{n+1}},$$

so to prove the result, it suffices to show that $\sqrt{n} + \frac{1}{\sqrt{n+1}} \geq \sqrt{n+1}$. We have

$$\sqrt{n^2 + n} + 1 \geq \sqrt{n^2} + 1 = n + 1.$$

Dividing both sides by $\sqrt{n+1}$, we have $\sqrt{n} + \frac{1}{\sqrt{n+1}} \geq \sqrt{n+1}$, as claimed. This completes the inductive step, and thus the proof. $\blacksquare$

Chapter 3

Prime numbers

3.1 Prime numbers

In number theory, we favor certain numbers over others. Among the numbers that number theorists are most excited by are the *prime numbers*. Let's define them.

Definition 3.1. A positive integer $n > 1$ is said to be *prime* if it is only divisible by 1 and itself. Equivalently, $n > 1$ is prime if whenever we factor $n = ab$ as a product of two positive integers a and b, then either $a = 1$ or $b = 1$.

Remark 3.2. Note that 1 is not considered a prime number. The reason for this is to make the Fundamental Theorem of Arithmetic hold; we will present this crucial theorem very shortly. The first few primes are 2, 3, 5, 7, 11, 13, 17, and 19.

Definition 3.3. A positive integer $n > 1$ that is *not* prime is said to be *composite*.

Mathematicians have been studying the prime numbers for thousands of years, and already the ancient Greeks knew a fair amount about them. Since then, we have learned many things about primes, but there are also many questions that we do not yet know the answer to. For example, the famous *Twin Prime Conjecture* asks if there are infinitely many pairs of primes, like 3 and 5, or 29 and 31, that differ by exactly 2. While some progress has been made in this direction in recent years (see for instance [Zha14, Pol14, May15]), the Twin Prime Conjecture remains open. Philosophically, it's easy to understand why problems like the Twin Prime Conjecture are hard: primes are meant to be multiplied, not added or subtracted, so it's hard to say what happens when you start adding or subtracting them.

We will now look at some of the most important theorems about primes. In order to prove these theorems, we will need to use the proof techniques we have discussed so far: contradiction and induction.

3.2   The Fundamental Theorem of Arithmetic

The primes are interesting because they are the multiplicative building blocks of all the positive integers: all positive integers can be expressed as a product of prime numbers. This is part of the content of a theorem so important that it is known as the *Fundamental Theorem of Arithmetic*.

Theorem 3.4 (Fundamental Theorem of Arithmetic). *Every positive integer can be expressed as a product of prime numbers. A decomposition of an integer n as a product of primes is unique, up to changing the order of the factors.*

Before we delve into the proof, let's discuss the ingredients that go into the proof. The theorem has two parts: the *existence* of the prime factorization, and the *uniqueness*. Theorems expressing the existence and uniqueness of some object or construction are very common in mathematics. Typically to prove these, we have to prove these two parts separately, and that is the case here. We start with the existence, which is the easier part in this case. To prove existence, we use strong induction: if all positive integers less than n have a prime factorization, then we show that n has one as well.

The uniqueness is a little more challenging here. To do this, we suppose that there exists some positive integer that doesn't have unique factorization, i.e. it contains more than one such factorization. In that case, there must be a smallest such number. We then get a contradiction by finding an even smaller number that must still have two prime factorizations. Let's now see how this works in detail!

Proof. There are two things to prove here. The first is to show that every positive integer *has* a prime factorization (at least one), and the second is to show that every positive integer has *at most one* prime factorization (uniqueness).

We'll start with the existence, i.e. showing that a number n has a prime factorization. We do this by (strong) induction. The base case is $n = 1$.

Now, it doesn't appear that we can write 1 as a product of primes. However, we can: if we do not multiply together any numbers at all, we end up with 1, so 1 is the product of *no primes at all!*

Next, we perform the inductive step. For this, we use strong induction. Suppose that every positive integer less than n has a prime factorization; we must use this to produce a prime factorization of n. We have two cases: n is prime and n is composite.

Case 1: n is prime. In this case, n is the product of just the single prime n, and we're done.

Case 2: n is composite. In this case, we can factor n as $n = ab$, where $a, b < n$. By the inductive hypothesis, both a and b can be written as products of primes, say $a = \prod_{i=1}^{k} p_i$ and $b = \prod_{j=1}^{\ell} q_j$. Then

$$n = ab = \prod_{i=1}^{k} p_i \cdot \prod_{j=1}^{\ell} q_j,$$

and we have just written n as a product of primes.

The next part is *uniqueness*. We must show that if n is a product of primes in two ways, say as $\prod_{i=1}^{k} p_i = \prod_{j=1}^{\ell} q_j$, then $k = \ell$, and the p_i's are a rearrangement of the q_j's. Again, we prove this by induction, but it is convenient to express our induction a bit differently.

Suppose that the prime factorization is *not* always unique. Then there is some *smallest* positive integer that has at least two prime factorizations. Let us call this smallest integer s, and suppose that s has two distinct prime factorizations

$$s = \prod_{i=1}^{k} p_i = \prod_{j=1}^{\ell} q_j.$$

Note that 1 only has a single prime factorization, so $s \neq 1$. If s were prime, then by definition it cannot be factored into a product of smaller primes, so we know that s is composite. Thus $k, \ell \geq 2$. Furthermore, if one of the p_i's were equal to one of the q_j's, say $p_i = q_j$, then $s/p_i = s/q_j$ would also have two prime factorizations: just omit the p_i term from the first factorization and the q_j term from the second one. Thus we may assume that for every i and j, $p_i \neq q_j$.

Since $p_1 \neq q_1$, we either have $p_1 < q_1$ or $p_1 > q_1$. Let us assume that $p_1 < q_1$; if not, then we switch the roles of the p_i's and the q_j's. Define t

to be

$$t = (q_1 - p_1) \prod_{j=2}^{\ell} q_j.$$

Note that $1 < q_2 \le t < s$. We have

$$t = q_1 \prod_{j=2}^{\ell} q_j - p_1 \prod_{j=2}^{\ell} q_j$$

$$= s - p_1 \prod_{j=2}^{\ell} q_j$$

$$= p_1 \left(\prod_{i=2}^{k} p_i - \prod_{j=2}^{\ell} q_j \right).$$

Since $t < s$, our assumption that s is *minimal* with respect to having at least two prime factorizations guarantees that t has a unique prime factorization. Thus p_1 appears in the (unique) prime factorization of t.

On the other hand, $(q_1 - p_1) \prod_{j=2}^{\ell} q_j$ is another factorization of t, so p_1 must divide one of the factors. It doesn't divide $\prod_{j=2}^{\ell} q_j$, so p_1 must divide $q_1 - p_1$. Let us say that $q_1 - p_1 = mp_1$. Then $q_1 = (m + 1)p_1$. Since p_1 is prime (and hence greater than 1) and $m + 1 > 1$, we have a factorization of q_1 into smaller integers. This contradicts q_1 being prime.

Our only assumption here was that the original number s had more than one prime factorization, and was the smallest positive integer with this property. The contradiction shows that this was false, so we have proven that all positive integers have unique prime factorizations, as desired. ∎

3.3 The infinitude of primes

Now that we have seen that every positive integer can be expressed uniquely as a product of primes, it is worth investigating more questions about primes. One of the most basic questions about primes is to determine how many there are.

Theorem 3.5 (Euclid). *There are infinitely many primes.*

There are many known proofs of this theorem; see [Meš12] for a historical survey of proofs. But Euclid's original proof is still a classic that everyone needs to know. The way it works is that we'll show that for any finite list of primes, there must be another prime not on the list. That implies that there must be infinitely many primes.

Proof. Suppose that we have a finite list of primes, say $p_1, p_2, \ldots, p_n$. Consider the number $M = \prod_{i=1}^{n} p_i + 1$: multiply them all together and add one.

By the Fundamental Theorem of Arithmetic, M must have a prime factorization. Since $M > 1$, it must be divisible by some prime q. Can q be one of the primes on our list? Let us suppose that it is divisible by p_i. By construction $M - 1 = \prod_{i=1}^{n} p_i$ is divisible by p_i, so M must not be divisible by p_i: two consecutive integers cannot both be divisible by the same integer > 1. Thus q must be a new prime not on our list.

We have shown that no finite list can include all the primes. Thus there are infinitely many primes. ∎

Remark 3.6. At this point, it is worth clearing up a common misconception about this proof. If we multiply the first n primes together and add one, the resulting number is sometimes prime and sometimes composite. The first time it is composite is

$$2 \cdot 3 \cdot 5 \cdot 7 \cdot 11 \cdot 13 + 1 = 59 \cdot 509.$$

We can carry out Euclid's proof a little further to show there are infinitely many primes of certain types:

Theorem 3.7. *There are infinitely many primes of the form $4n + 3$.*

Before we prove Theorem 3.7, let's first state and prove a lemma that we will need for the proof.

Lemma 3.8. *Let $a_1, a_2, \ldots, a_k$ be any integers. Then the product $\prod_{i=1}^{k}(4a_i + 1)$ has the form $4n + 1$ for some integer n.*

Proof. We prove this by induction on k. We need two base cases, namely $k = 1$ and $k = 2$. For the first base case $k = 1$, the product in question is $4a_1 + 1$, which has the form $4n + 1$, where $n = a_1$. For the second base case $k = 2$, we note that

$$(4a_1 + 1)(4a_2 + 1) = 4(4a_1 a_2 + a_1 + a_2) + 1,$$

which again has the form $4n + 1$ with $n = 4a_1 a_2 + a_1 + a_2$.

Now we perform the inductive step. Suppose the result is true for k, in that for *any* integers $a_1, \ldots, a_k$, there exists an integer n (which depends on $a_1, \ldots, a_k$) such that $\prod_{i=1}^{k}(4a_i + 1) = 4n + 1$. We wish to prove the result for $k + 1$. Let $a_1, a_2, \ldots, a_k, a_{k+1}$ be integers. We wish to prove that $\prod_{i=1}^{k+1}(4a_i + 1)$ has the form $4n + 1$ for some integer n. Because the result is true for k, we know that $\prod_{i=1}^{k}(4a_i + 1)$ can be written as $4m + 1$ for some integer m. Thus we have

$$\prod_{i=1}^{k+1}(4a_i + 1) = \prod_{i=1}^{k}(4a_i + 1) \cdot (4a_{k+1} + 1)$$

$$= (4m + 1)(4a_{k+1} + 1).$$

In other words, this product of $k + 1$ numbers of the form $4n + 1$ can be written as a product of only two such numbers. By the second base case, the product of two numbers of the form $4n+1$ again has the form $4n+1$, so this implies in particular that $\prod_{i=1}^{k+1}(4a_i + 1)$ again has the same form. ∎

In other words, a product of several numbers of the form $4n + 1$ is still of the form $4n + 1$. Thus if we have a number *not* of the form $4n + 1$, it *must* have a factor, and even a prime factor, that isn't of the form $4n + 1$. Let us now return to the proof of Theorem 3.7.

Proof of Theorem 3.7. Suppose we have a finite list of primes of the form $4n+3$, say $p_1, p_2, \ldots, p_r$. Consider the number $M = 4\prod_{i=1}^{r} p_i - 1$. Observe that M has the form $4n+3$, where $n = \prod_{i=1}^{r} p_i - 1$. Any factor of a number of the form $4n + 3$ must be odd, because $4n + 3$ is odd. Any odd number is of the form $4n + 1$ or $4n + 3$. So, since M is odd, either all prime factors are of the form $4n + 1$, or else there is at least one prime factor of M of the form $4n+3$. If all the prime factors of M had the form $4n+1$, then M would be a product of numbers of the form $4n + 1$, which again by Lemma 3.8 has the form $4n + 1$. But M has the form $4n + 3$, and not $4n + 1$. Thus it must be the case that M has at least one prime factor of the form $4n + 3$.

Now, M must have a prime factor q of the form $4n + 3$, and M is not divisible by any of the primes $p_1, \ldots, p_r$, since $M + 1$ is divisible by all of them. Thus in this case q is a new prime of the form $4n + 3$ not on our list. Thus our list did not contain all primes of the form $4n + 3$, which implies that there are infinitely many such primes. ∎

3.4   The sieve of Eratosthenes

Note that Euclid's proof does not give us a lot of insight into actually *finding* primes: it guarantees that there are infinitely many without actually producing any. They exist, but the proof doesn't tell us where they are.

There are many ways of finding primes. One of the oldest is the *sieve of Eratosthenes*, which can be used to find all the primes up to some fixed number n, all at once. The way the sieve of Eratosthenes works is as follows: we start by listing out all the numbers from 2 to n; here we take $n = 100$ for illustrative purposes, as shown in Figure 3.1. At each stage, starting with 2, we take the first number left. That number is prime, so we add it to our list of primes. Then we remove that number and all its multiples. Then we repeat the process. In Figure 3.1, we start by removing 2 and all its multiples, marked in blue. Once that is done, 3 is the next number remaining, so we remove 3 and all its multiples that haven't yet been removed; those are marked in green. Next, we remove 5 and all its multiples which haven't been removed yet; those are marked in purple. Next, we remove 7 and all its multiples which haven't been removed yet; those are marked in red. When we get to $\sqrt{n}$, we can stop: everything left is prime. (See problem 1.) In Figure 3.1, all the primes are marked in yellow.

3.5 The greatest common divisor

Definition 3.9. Given two nonzero integers a and b, their *greatest common divisor*, denoted $\gcd(a,b)$, is the largest positive integer that divides both a and b.

Example. The greatest common divisor of 12 and 18 is 6. The greatest common divisor of 9 and 14 is 1.

There is a fairly obvious way of computing the gcd of two numbers: write down their prime factorizations, and multiply together all the primes that are shared in these factorizations, the number of times they both appear. For instance, $12 = 2 \times 2 \times 3$ and $18 = 2 \times 3 \times 3$, so their gcd is 2×3, the primes that are shared. Note that we don't use the second 2 in 12, because there is no corresponding second 2 in 18.

While this is an easy way of computing gcds for small numbers, it isn't very good for larger numbers. The reason is that, in order to use this method, we need to be able to compute prime factorizations, which appears to be a difficult problem. However, there is a better method for computing gcds, which does not rely on knowing any prime factorizations. This method is called the *Euclidean algorithm*. The key input to the algorithm is this simple lemma:

Lemma 3.10. *If a and b are nonzero integers and k is any integer, then* $\gcd(a,b) = \gcd(a, b - ka)$.

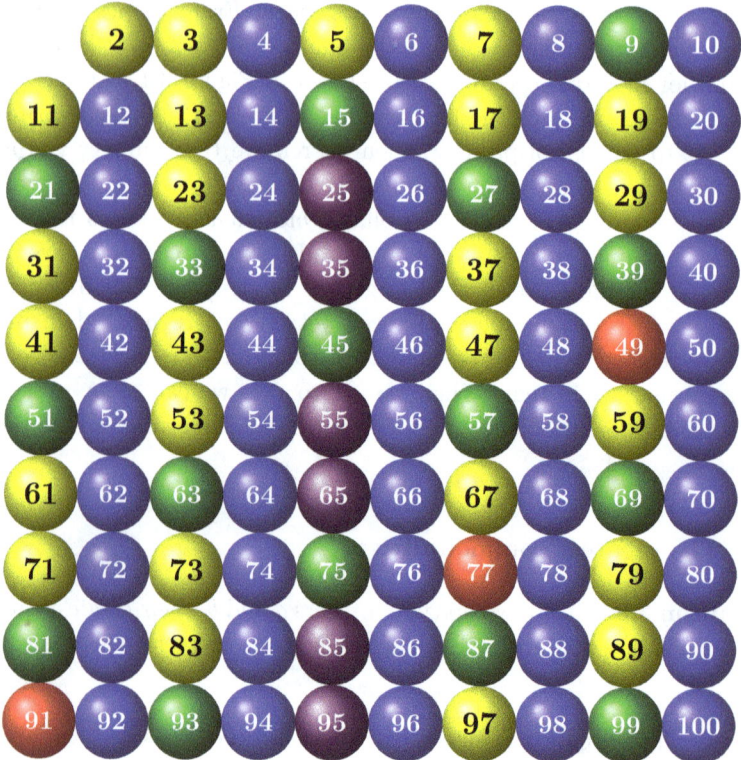

Figure 3.1. The sieve of Eratosthenes.

Proof. In order to show that $\gcd(a, b) = \gcd(a, b - ka)$, it suffices to show that *any* common divisor of a and b is also a common divisor of a and $b - ka$, and vice versa. So, let us suppose that d is a common divisor of a and b, say with $a = dx$ and $b = dy$. Then $b - ka = d(y - kx)$ is a multiple of d, and of course a is also a multiple of d. Thus d is a common divisor of a and $b - ka$.

On the other hand, if d is a common divisor of a and $b - ka$, then we can write $a = dx$ and $b - ka = dz$. Then we have $b = (b - ka) + ka = dz + dkx = d(z + kx)$, so d is a divisor of b and hence a common divisor of a and b. ∎

The way the Euclidean algorithm works is that we start with two numbers a and b, whose gcd we would like to find. Then, using Lemma 3.10, we replace a and b with smaller numbers until it is obvious what the gcd is. More precisely:

Theorem 3.11 (Euclidean Algorithm). *The following algorithm produces the gcd of two positive integers a and b:*

(1) *If $a > b$, then switch a and b. Thus we may assume that $a \leq b$. Let $b_0 = b$ and $b_1 = a$.*

(2) *If, at some stage, $b_n = 0$, then $\gcd(a, b) = b_{n-1}$. At this point, we may stop running the algorithm.*

(3) *If $b_n \neq 0$, then choose k such that $0 \leq b_{n-1} - kb_n < b_n$. Set $b_{n+1} = b_{n-1} - kb_n$.*

It is also rather common to express the Euclidean algorithm in terms of something that looks more like an algorithm we would code up in a programming language. The idea is that we express $\gcd(a, b)$ in terms of the gcd of other, smaller, numbers, which we can compute recursively. Let's see how to do that:

The Euclidean algorithm

```
function gcd(a,b)
    if a > b then
        swap a and b
    if a = 0 then
        return b
    k ← ⌊b/a⌋
    return gcd(a, b − ak)
```

Here the notation $\lfloor x \rfloor$ means the *floor* of x, i.e. the greatest integer $\leq x$, or equivalently "round down." For instance, $\lfloor 4 \rfloor = 4$ and $\lfloor \pi \rfloor = 3$.

Let's see how this algorithm works in practice.

Example. Let us compute $\gcd(35, 98)$. We have $b_0 = 98$ and $b_1 = 35$. We produce $b_2 = 98 - 2 \times 35 = 28$. Then we produce $b_3 = 35 - 1 \times 28 = 7$. Then we produce $b_4 = 28 - 4 \times 7 = 0$. Thus $\gcd(35, 98) = b_3 = 7$. Indeed, this is correct: the prime factorization of 35 is $35 = 5 \times 7$, and the prime factorization of 98 is $98 = 2 \times 7 \times 7$. Thus they share only a 7.

There is also an extension of the Euclidean algorithm.

Theorem 3.12 (Bézout's Lemma). *The greatest common divisor* $\gcd(a, b)$ *of two positive integers is the smallest positive integer that can be expressed in the form* $ax + by$, *where* x *and* y *are (not necessarily positive) integers.*

Example. In the example above, we have $7 = 3 \times 35 - 1 \times 98$.

Proof. If $d = \gcd(a, b)$, then any number of the form $ax + by$ must be a multiple of d because both a and b are multiples of d, so any positive integer of the form $ax + by$ must be $\geq d$. Thus, it suffices to show that d itself can be expressed in this form. To do that, we use the Euclidean algorithm, slightly more carefully. We prove by induction that each b_n that the Euclidean algorithm produces can be expressed in the form $ax + by$, and since d is some b_n, it too can be expressed in this form. The base cases, which in this case are $n = 0$ and $n = 1$, are clear: one of them is a and one of them is b, so both of these numbers can clearly be expressed in the form $ax + by$ by taking $(x, y) = (1, 0)$ and $(x, y) = (0, 1)$, respectively.

Now, we do the inductive step. Let us suppose that both b_{n-1} and b_n can be expressed in the form $ax + by$, say

$$b_{n-1} = ax_{n-1} + by_{n-1}, \qquad b_n = ax_n + by_n.$$

Then, since $b_{n+1} = b_{n-1} - kb_n$, we have

$$\begin{aligned} b_{n+1} &= b_{n-1} - kb_n \\ &= (ax_{n-1} + by_{n-1}) - k(ax_n + by_n) \\ &= a(x_{n-1} - kx_n) + b(y_{n-1} - ky_n). \end{aligned}$$

This completes the inductive step and the proof. ∎

While we stated this only as saying that there is *some* way of writing b_{n+1} in this form, the proof actually tells us a bit more: it tells us exactly *how to do it*.

Example. Let us compute $\gcd(76, 104)$ and write this number in the form $76x + 104y$. We just follow the algorithm. We have $b_0 = 104$ and $b_1 = 76$. Then we have $b_2 = 104 - 76 = 28$, and so we have written $b_2 = 28$ in the desired form. Next, we have $b_3 = 76 - 2 \times 28 = 20$, or

$$20 = 76 - 2 \times 28 = 76 - 2 \times (104 - 76) = 3 \times 76 - 2 \times 104.$$

Continuing on, we have $b_4 = 28 - 20 = 8$, and we have

$$8 = 28 - 20 = (104 - 76) - (3 \times 76 - 2 \times 104) = 3 \times 104 - 4 \times 76.$$

Next up, we have $b_5 = 20 - 2 \times 8 = 4$, and

$$4 = 20 - 2 \times 8 = (3 \times 76 - 2 \times 104) - 2(3 \times 104 - 4 \times 76) = 11 \times 76 - 8 \times 104.$$

Finally, we have $b_6 = 8 - 2 \times 4 = 0$, so $\gcd(76, 104) = 4$ and

$$4 = \gcd(76, 104) = 11 \times 76 - 8 \times 104.$$

A consequence of Bézout's Lemma is another characterization of prime numbers:

Corollary 3.13. *Let p be a prime number. Then, if a and b are integers such that ab is a multiple of p, then either a or b is a multiple of p.*

Proof. Suppose a is not a multiple of p. Then $\gcd(a, p) = 1$, so there exist integers x and y such that $ax + py = 1$. Multiply this equation by b to get

$$abx + pby = b. \qquad (3.1)$$

Since ab is a multiple of p, and pby is also a multiple of p, it follows that the left side of (3.1) is a multiple of p, and hence so too is the right side. We have shown that if a is *not* a multiple of p, then b *is*. In other words, it cannot be the case that *neither a nor b* is a multiple of p, as long as ab is a multiple of p. ∎

Of course, this is false for composite numbers; for instance, neither 2 nor 6 is a multiple of 4, yet their product 2×6 is a multiple of 4. In fact, Corollary 3.13 characterizes primes completely; see problem 3.

3.6 Problems

(1) Prove that if n is composite, then n has a prime factor less than or equal to $\sqrt{n}$.
(2) Prove that there are infinitely many primes of the form $6n + 5$.
(3) Prove that if n is a composite number, then there exist integers a and b, neither one a multiple of n, such that ab is a multiple of n.
(4) Generalize Corollary 3.13 as follows: if $a_1, \ldots, a_n$ are integers such that the product $\prod_{i=1}^{n} a_i$ is a multiple of a prime number p, then at least one of $a_1, \ldots, a_n$ is a multiple of p.
(5) As we showed, for a pair (a, b) of integers, there exist integers x and y such that $ax + by = 1$ if and only if $\gcd(a, b) = 1$. For which triples of integers (a, b, c) do there exist integers x, y, z with $ax + by + cz = 1$? For example, if $(a, b, c) = (6, 10, 15)$, then we can take $x = 1$, $y = 1$, $z = -1$, so that $ax + by + cz = 1$, so $(6, 10, 15)$ is an example of such a triple. On the other hand, $(a, b, c) = (2, 4, 6)$ is not, because for any integers x, y, z, $2x + 4y + 6z \neq 1$; the left side is always even. Prove that your answer is correct.

(6) Since $\gcd(5,8) = 1$, we know that there are integers x and y such that $5x + 8y = 1$; for instance, we can take $x = 5$ and $y = -3$. Similarly, there are integers x and y such that $5x + 8y = 21$, such as $x = 1$ and $y = 2$. Which integers can be expressed in the form $5x + 8y$, where x and y are *nonnegative* integers? (For example, 21 works, since we can take $x = 1$ and $y = 2$.) Give a proof.

(7) Prove that if a_1, a_2, b_1, b_2 are integers with $b_1 + b_2 \neq 0$ such that $a_1 b_2 - a_2 b_1 = \pm 1$, then

$$\frac{a_1 + a_2}{b_1 + b_2}$$

is in lowest terms.

(8) Let $f(x)$ be a nonconstant polynomial with integer coefficients. Prove that there is some integer n such that $|f(n)|$ is composite.

(9) Let F_n be the nth Fibonacci number, so that $F_0 = 0$, $F_1 = 1$, and $F_{n+2} = F_n + F_{n+1}$ for all $n \geq 0$.

(a) Prove that, for all $n \geq 1$, $\gcd(F_n, F_{n+1}) = 1$.

(b) Prove that if m and n are positive integers, then $\gcd(F_m, F_n) = F_{\gcd(m,n)}$.

(10) Prove that if n is a positive integer such that $2^n - 1$ is prime, then n must be a prime. (The converse is false: $2^{11} - 1 = 23 \times 89$ is the first counterexample.)

(11) Prove that if n is a positive integer $2^n + 1$ is prime, then n must be a power of 2. (The converse is false: $2^{2^5} + 1 = 641 \times 6700417$.)

Chapter 4

The pigeonhole principle

4.1 The pigeonhole principle

The *pigeonhole principle* is a simple technique, often used to show that something exists without actually finding it. It is often stated in terms of pigeons, and we shall continue that tradition here.

Theorem 4.1 (Pigeonhole principle). *Suppose there are n pigeons in m holes. If $m < n$, then there must be a hole containing at least two pigeons. More generally, if $km < n$, then there must be a hole containing at least $k + 1$ pigeons.*

This theorem is completely obvious, but let us give a proper proof of the general version, just for practice.

Proof. Suppose that $km < n$, but that each hole only contains at most k pigeons. Let us write a_i for the number of pigeons in hole i, so that $a_i \le k$ for all i. Thus the total number of pigeons is

$$\sum_{i=1}^{m} a_i \le mk < n,$$

i.e. there are fewer than n pigeons. But this contradicts our hypothesis that there are n pigeons. ∎

Now let's see something easy we can do with the pigeonhole principle.

Example. Suppose we have an urn[1] containing balls, each of which is colored red, blue, green, yellow, or black. If we pick any 21 balls from the urn, we must end up with five of the same color. To show this, we use the pigeonhole

---

[1] An urn is a natural habitat for balls in mathematics.

principle. The balls are the pigeons, and the colors are the pigeonholes. If we have 21 pigeons in 5 holes, then we must have at least 5 pigeons in one hole. Translating that back to colored balls, this means that we must have at least five balls of one color.

Here are some slightly more interesting applications of the pigeonhole principle:

Theorem 4.2. *Given five points in a square of sidelength 1, there must be two of them whose distance is at most $\frac{1}{\sqrt{2}}$.*

Proof. Divide the square into four smaller squares, as shown in Figure 4.1. Then two of the five points must lie in the same small square. For a point on an edge, we assign it to one of the squares containing that edge. It doesn't matter which small squares the edges are assigned to. The farthest distance those points could possibly be is at opposite corners of the small square, which is distance $\frac{1}{\sqrt{2}}$. ∎

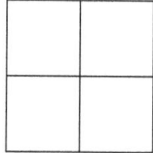

Figure 4.1. A square divided into four smaller squares. If the sidelength of the large square is 1, then the distance between two opposite corners of one of the small squares is $\frac{1}{\sqrt{2}}$.

Let's clarify exactly where we used the pigeonhole principle in this proof. The pigeons are the 5 points, and the holes are the 4 little squares. Since we have 5 pigeons in 4 holes, one of the holes must have at least 2 pigeons.

Theorem 4.3. *Given any six people, there must either be three people who all know each other, or three people none of whom know each other. We assume that "knowing" is mutual: if person A knows person B, then person B also knows person A.*

Proof. Pick any person among the six, say person A. There are five other people, and two possible relations to person A: knowing or not knowing them. Thus there must either be at least three people A knows, or else at least three people A does not know. We shall assume that there are three people A knows, but the other case is identical. Call these three people B, C, and D. If any of those two people know one another, then we have found three people who all know each other; for instance, if B and D know each other, then A, B, and D all know each other and we're done. The other case is that none of B, C, and D know one another. In that case, we have just found three people none of whom know one another, and again we're done. ∎

Remark 4.4. One way of picturing this is to construct a *graph* whose vertices are the people. We put a red edge between two people who know each other, and a blue edge between two people who do not know each other. The portion of the graph used in the proof is shown in Figure 4.2.

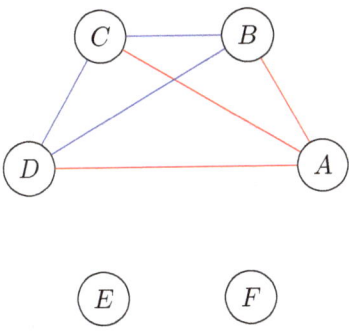

Figure 4.2. The graph associated to Theorem 4.3.

The pigeonhole principle was used in this proof to show that there must be three people A knows or three people A doesn't know. We think of the remaining five people as being the pigeons, and the pigeonholes as "knowing" and "not knowing." Since we have 5 pigeons in 2 holes, one of the holes must have at least 3 pigeons.

4.2 The Dirichlet Approximation Theorem

The pigeonhole principle was first introduced by Dirichlet in 1834 to prove the following theorem in number theory. At first glance, it doesn't look like a pigeonhole problem. But with a bit of cleverness, we shall see that in fact it is.

Theorem 4.5 (Dirichlet Approximation Theorem). *Let α be any real number, and let N be a positive integer. Then there exists an integer p and a positive integer q with $1 \leq q \leq N$ such that*

$$\left| \alpha - \frac{p}{q} \right| < \frac{1}{qN}.$$

Remark 4.6. For a real number β, we will write $\lfloor \beta \rfloor$ for the *floor* of β, i.e. the unique integer n such that $n \leq \beta < n + 1$. For instance, $\lfloor 2 \rfloor = 2$, $\lfloor \pi \rfloor = 3$, and $\lfloor -1.5 \rfloor = -2$. In other words, the floor function means "round down." There is also the analogous notion of the *ceiling* function, denoted $\lceil \beta \rceil$. If β is not an integer, then $\lceil \beta \rceil = \lfloor \beta \rfloor + 1$, and if β is an integer, then $\lceil \beta \rceil = \lfloor \beta \rfloor$.

Proof. Consider the $N + 1$ numbers a_i with $0 \leq i \leq N$, defined by $a_i = i\alpha - \lfloor i\alpha \rfloor$. All the a_i's satisfy $0 \leq a_i < 1$. Now, consider the N intervals $I_1, \ldots, I_N$, where I_k is the interval of x such that $\frac{k-1}{N} \leq x < \frac{k}{N}$; these partition the interval $0 \leq x < 1$. Since there are $N + 1$ a_i's and only N intervals, there must be two a_i's in the same interval; say a_i and a_j are in interval I_k. It follows that

$$\frac{k-1}{N} \leq a_i, a_j < \frac{k}{N},$$

so $|a_i - a_j| < \frac{1}{N}$. Without loss of generality, suppose that $i > j$.

Now, $a_i = i\alpha - \lfloor i\alpha \rfloor$ and $a_j = j\alpha - \lfloor j\alpha \rfloor$. Thus

$$a_i - a_j = (i - j)\alpha + (\lfloor j\alpha \rfloor - \lfloor i\alpha \rfloor).$$

Let $p = \lfloor i\alpha \rfloor - \lfloor j\alpha \rfloor$, and note that p is an integer. We have

$$|a_i - a_j| = |(i - j)\alpha - p| < \frac{1}{N}.$$

Now, let $q = i - j$. Then we have

$$|q\alpha - p| < \frac{1}{N},$$

or

$$\left| \alpha - \frac{p}{q} \right| < \frac{1}{qN},$$

as desired. ∎

That may have been a bit abstract, so let's take a look at an example of the proof in action.

Example. We'll take $\alpha = \pi$ and $N = 10$. We first have to compute the a_i's. We have

$$5\pi \approx 15.71 \Rightarrow a_5 \approx .71,$$

$$0\pi = 0 \Rightarrow a_0 = 0, \qquad 6\pi \approx 18.85 \Rightarrow a_6 \approx .85,$$

$$1\pi \approx 3.14 \Rightarrow a_1 \approx .14, \qquad 7\pi \approx 21.99 \Rightarrow a_7 \approx .99,$$

$$2\pi \approx 6.28 \Rightarrow a_2 \approx .28, \qquad 8\pi \approx 25.13 \Rightarrow a_8 \approx .13,$$

$$3\pi \approx 9.42 \Rightarrow a_3 \approx .42, \qquad 9\pi \approx 28.27 \Rightarrow a_9 \approx .27,$$

$$4\pi \approx 12.57 \Rightarrow a_4 \approx .57, \qquad 10\pi \approx 31.42 \Rightarrow a_{10} \approx .42.$$

The I_k's are the intervals with $\frac{k-1}{10} \leq x < \frac{k}{10}$. There are 10 I_k's and 11 a_i's, so there must be two in the same I_k. In fact, there are several I_k's with two a_i's, so let's just pick one. Let's pick I_2, which contains a_1 and a_8. We have $a_1 = \pi - 3$ and $a_8 = 8\pi - 25$. Then we have

$$a_8 - a_1 = (8-1)\pi + (\lfloor \pi \rfloor - \lfloor 8\pi \rfloor) = 7\pi + (3 - 25) = 7\pi - 22.$$

We let $p = 22$, so that $a_8 - a_1 = 7\pi - p$. Since a_1 and a_8 are both in I_2, we have $|a_1 - a_8| < \frac{1}{10}$, i.e. $|7\pi - 22| < \frac{1}{10}$. Dividing both sides by 7, we get

$$\left| \pi - \frac{22}{7} \right| < \frac{1}{7 \times 10},$$

so $q = 7$ and $p = 22$ in our theorem.

Recall that we proved that $\sqrt{2}$ is irrational back in Chapter 1. The Dirichlet Approximation Theorem gives us a *criterion* for detecting when a number is irrational.

Corollary 4.7. *If α is irrational, then there are infinitely many rational numbers $\frac{p}{q}$ in lowest terms such that*

$$\left| \alpha - \frac{p}{q} \right| < \frac{1}{q^2}. \tag{4.1}$$

Proof. Suppose that α is irrational, and suppose that there were only finitely many rational numbers $\frac{p}{q}$ such that

$$\left| \alpha - \frac{p}{q} \right| < \frac{1}{q^2},$$

and call these rational numbers $\frac{p_1}{q_1}, \ldots, \frac{p_r}{q_r}$. Let $\frac{p}{q}$ be the one of these that is closest to α, and choose N such that

$$\left| \alpha - \frac{p}{q} \right| > \frac{1}{N}.$$

For this value of N, we have

$$\left| \alpha - \frac{p_i}{q_i} \right| > \frac{1}{N} > \frac{1}{q_i N}$$

for each i with $1 \leq i \leq r$, so those rational numbers do not satisfy the Dirichlet Approximation Theorem. For any other rational number $\frac{p}{q}$ with $1 \leq q \leq N$, we have

$$\left| \alpha - \frac{p}{q} \right| \geq \frac{1}{q^2} \geq \frac{1}{qN}.$$

Thus, under the assumption that there are only finitely many rational numbers satisfying (4.1), we have a contradiction to the Dirichlet Approximation Theorem. Thus, this assumption must have been incorrect, and we have completed the proof. ∎

This argument breaks down when α is rational, because we can take $\frac{p}{q} = \alpha$, so that $\left| \alpha - \frac{p}{q} \right| = 0$, and there is no N such that $\left| \alpha - \frac{p}{q} \right| > \frac{1}{N}$. In fact, the analogous statement is *false* when α is rational, for a very simple reason:

Proposition 4.8. *If $\frac{p}{q}$ and $\frac{r}{s}$ are two distinct rational numbers with $q, s > 0$, then*

$$\left| \frac{p}{q} - \frac{r}{s} \right| \geq \frac{1}{qs}.$$

Proof. We have

$$\frac{p}{q} - \frac{r}{s} = \frac{ps - qr}{qs}. \tag{4.2}$$

Since $\frac{p}{q} \neq \frac{r}{s}$, the numerator of the right side of (4.2) is nonzero, and hence at least 1 in absolute value. Thus

$$\left| \frac{p}{q} - \frac{r}{s} \right| = \frac{|ps - qr|}{qs} \geq \frac{1}{qs},$$

as desired. ∎

It then follows that for $N \geq s$, there are no rational numbers $\frac{p}{q}$ *other than $\frac{r}{s}$, possibly in nonreduced form*, such that

$$\left| \frac{r}{s} - \frac{p}{q} \right| < \frac{1}{qN}.$$

Thus, there are only finitely many rational numbers such that

$$\left| \frac{r}{s} - \frac{p}{q} \right| < \frac{1}{q^2}.$$

Putting all of this together, we get the following result:

Theorem 4.9. *A real number α is irrational if and only if, for any $\varepsilon > 0$, there is a positive integer q and an integer p such that*

$$0 < |q\alpha - p| < \varepsilon.$$

For proofwriting practice, let's give a complete and clean proof of Theorem 4.9 here, using the results we have already proven.

Proof of Theorem 4.9. First, suppose that α is irrational. By Corollary 4.7, we know that there are infinitely many rational numbers $\frac{p}{q}$ such that

$$\left| \alpha - \frac{p}{q} \right| < \frac{1}{q^2}.$$

We may assume for each such rational approximation $\frac{p}{q}$ that $q > 0$. Multiplying both sides by q, we know that there are infinitely many $\frac{p}{q}$ such that

$$|q\alpha - p| < \frac{1}{q}.$$

Since there are infinitely many such numbers, and there can only be finitely many of them with any fixed value of q, there must exist such rational numbers $\frac{p}{q}$ with q arbitrarily large. In particular, we can find one such that $q > \frac{1}{\varepsilon}$, or $\frac{1}{q} < \varepsilon$. With such a q, we have

$$|q\alpha - p| < \frac{1}{q} < \varepsilon,$$

as desired.

Now, suppose that α is rational, say with $\alpha = \frac{r}{s}$. Then if $\frac{p}{q} \neq \frac{r}{s}$, and $q, s > 0$, then

$$\left| \frac{r}{s} - \frac{p}{q} \right| \geq \frac{1}{qs}.$$

Multiplying both sides by q, we have

$$|q\alpha - p| \geq \frac{1}{s}.$$

Thus if $0 < \varepsilon < \frac{1}{s}$, there are no rational numbers $\frac{p}{q}$ such that $0 < |q\alpha - p| < \varepsilon$, which is what we wanted to prove. ∎

We can actually use what we have just done to prove that the number e is irrational. There are many possible definitions of e, but the one that is useful to us at the moment is

$$e = \sum_{k=0}^{\infty} \frac{1}{k!} = \frac{1}{0!} + \frac{1}{1!} + \frac{1}{2!} + \frac{1}{3!} + \frac{1}{4!} + \cdots.$$

Theorem 4.10. *The number e is irrational.*

Proof. We use Theorem 4.9. Pick any $\varepsilon > 0$, and let n be a positive integer, depending on ε, that we will determine later. Then let

$$\frac{p}{q} = \sum_{k=0}^{n} \frac{1}{k!} = \frac{1}{0!} + \frac{1}{1!} + \frac{1}{2!} + \cdots + \frac{1}{n!}$$

in lowest terms, so that $q = n!$ and $p = n! \sum_{k=0}^{n} \frac{1}{k!}$. We have

$$
\begin{aligned}
|qe - p| &= \left| n!e - \sum_{k=0}^{n} \frac{n!}{k!} \right| \\
&= \sum_{k=n+1}^{\infty} \frac{n!}{k!} \\
&= \frac{1}{n+1} + \frac{1}{(n+1)(n+2)} + \frac{1}{(n+1)(n+2)(n+3)} + \cdots \\
&< \frac{1}{n+1} + \frac{1}{(n+1)^2} + \frac{1}{(n+1)^3} + \cdots \\
&= \frac{1/(n+1)}{1 - 1/(n+1)} \\
&= \frac{1}{n}.
\end{aligned}
$$

Thus, as long as $\frac{1}{n} < \varepsilon$, or equivalently $n > \frac{1}{\varepsilon}$, we have $|qe - p| < \varepsilon$. Thus Theorem 4.9 tells us that e is irrational. ∎

There is a stronger version of Corollary 4.7, due to Hurwitz.

Theorem 4.11 (Hurwitz). *If α is irrational, then there are infinitely many rational numbers $\frac{p}{q}$ such that*

$$\left| \alpha - \frac{p}{q} \right| < \frac{1}{\sqrt{5}q^2}.$$

We will not prove Hurwitz's Theorem here, as the proof is somewhat difficult. If you would like to read a proof, see [BE02]. However, we can show that $\frac{1}{\sqrt{5}}$ cannot be replaced by any smaller number.

Theorem 4.12. *Let $\alpha = \frac{1+\sqrt{5}}{2}$, and let $c < \frac{1}{\sqrt{5}}$. Then there are only finitely many rational numbers $\frac{p}{q}$ such that*

$$\left| \alpha - \frac{p}{q} \right| < \frac{c}{q^2}.$$

Proof. Let $f(x) = x^2 - x - 1$, and note that $f(\alpha) = 0$. The other root of $f(x)$ is $\frac{1-\sqrt{5}}{2}$, which is also irrational. Thus $f(x)$ has no rational roots. Thus if $\frac{p}{q}$ is rational, then $f\left(\frac{p}{q}\right) \neq 0$, and in fact $\left| f\left(\frac{p}{q}\right) \right| \geq \frac{1}{q^2}$. Now, rewrite $f(x)$ as

$$f(x) = \sqrt{5}(x - \alpha) + (x - \alpha)^2. \tag{4.3}$$

If $\frac{p}{q}$ is such that $\left| \alpha - \frac{p}{q} \right| < \frac{c}{q^2}$, then, by plugging $\frac{p}{q}$ in for x in (4.3), we have

$$\frac{1}{q^2} \leq \left| f\left(\frac{p}{q}\right) \right|$$

$$= \left| \sqrt{5}\left(\frac{p}{q} - \alpha\right) + \left(\frac{p}{q} - \alpha\right)^2 \right|$$

$$\leq \sqrt{5}\left| \frac{p}{q} - \alpha \right| + \left| \frac{p}{q} - \alpha \right|^2$$

$$< \frac{\sqrt{5}c}{q^2} + \frac{c^2}{q^4}.$$

Solving for q, we have

$$q < \frac{c}{\sqrt{1 - c\sqrt{5}}},$$

so there are only finitely many values of q that work, and hence only finitely many such rational numbers $\frac{p}{q}$. ∎

4.3 Further reading

We mentioned Hurwitz's Theorem (Theorem 4.11). Hurwitz's Theorem guarantees that for *any* irrational number α, we can approximate α by rational numbers $\frac{p}{q}$ such that

$$\left| \alpha - \frac{p}{q} \right| < \frac{1}{\sqrt{5}q^2}.$$

Furthermore, in Theorem 4.12, we saw that we can't do any better for $\alpha = \frac{1+\sqrt{5}}{2}$. However, it turns out that for most irrational numbers, we can do better.

Definition 4.13. If α is an irrational number, its *Lagrange number* $L(\alpha)$ is the smallest number such that if $C > L(\alpha)$, then there are only finitely many rational numbers $\frac{p}{q}$ such that

$$\left| \alpha - \frac{p}{q} \right| < \frac{1}{Cq^2},$$

assuming that such a number exists. If no such number exists, then $L(\alpha) = \infty$.

Thus we have $L(\frac{1+\sqrt{5}}{2}) = \sqrt{5}$. Furthermore, if we let $\phi = \frac{1+\sqrt{5}}{2}$ and choose integers a, b, c, d such that $ad - bc = \pm 1$, then $L\left(\frac{a\phi+b}{c\phi+d}\right) = \sqrt{5}$ as well. However, for any irrational number α not of this form, $L(\alpha) \geq \sqrt{8}$. The set of numbers of the form $L(\alpha)$ for some irrational number α is called the *Lagrange spectrum*.

Good places to read more about this and related mathematics are [SS07, Chapter 4] and [Bak19].

After that, we can wonder about increasing the exponent in the denominator. What does it say about a number if it can be approximated to within the cube of the denominator, say, or a higher power? There's a beautiful theorem due to Liouville on that:

Theorem 4.14 (Liouville). *Let α be an irrational number that is a root of a degree-n polynomial with integer coefficients. Then there exists a constant $C > 0$ (depending on α) such that if $\frac{p}{q}$ is any rational number, then*

$$\left| \alpha - \frac{p}{q} \right| \geq \frac{C}{q^n}.$$

But there are some numbers that can be approximated to within $\frac{C}{q^n}$ for *any* positive integer n. It follows, then, that such numbers cannot be roots of any nonzero polynomial with integer coefficients, i.e. they are *transcendental*. A popular example is

$$\alpha = \sum_{n=1}^{\infty} 10^{-n!} = .110001000000000000000001000\ldots,$$

where the 1's are in the factorial positions. Liouville's Theorem is the starting point of the fascinating subject of transcendental number theory. A good place to start learning about it is [BT04].

4.4   Problems

(1) Suppose there are nine points (a_i, b_i, c_i) with $1 \leq i \leq 9$ in three-dimensional Euclidean space, also known as $\mathbb{R}^3$, all of whose coordinates are integers (i.e. a_i, b_i, c_i are integers). Show that there are two points among these nine such that the midpoint of the segment connecting them has integer coordinates. For instance, if two of the points are $(1, 2, 3)$ and $(4, 6, 8)$, then their midpoint is $(\frac{5}{2}, 4, \frac{11}{2})$. So

this midpoint does not have integer coordinates, but some pair must if there are 9 points.

(2) Let S be a set consisting of 51 positive integers from 1 to 100. Prove that S must contain two consecutive integers.

(3) Find, with proof, the maximum number of knights one can place on a chessboard such that no two of them attack each other. Do the same for bishops.

(4) Prove that if S is any subset consisting of 10 distinct integers from 1 to 100, there are two nonempty disjoint subsets $T, U \subseteq S$, such that the sum of the integers in T is equal to the sum of the integers in U.

(5) Prove that if n is any positive integer, there is a positive multiple of n whose decimal expansion contains only 0's and 1's. For instance, 1001 is a multiple of 7 consisting of only 0's and 1's.

(6) Prove that among any set S of $n \geq 2$ people, there are two of them who have the same number of friends in S. We assume the friendship is mutual: if person A is friends with person B, then person B is also friends with person A.

(7) Prove that, given any five points on the surface of a sphere, there is some closed hemisphere (i.e. including the boundary) that contains at least four of them.

(8) Let $a_0, a_1, \ldots, a_n$ be any integers. Prove that

$$\prod_{0 \leq i < j \leq n} (a_i - a_j)$$

is a multiple of $n!$.

(9) (a) Prove that for any positive integer k, we can find k consecutive positive integers that are all composite.

(b) Prove that the infinite sum

$$\sum_{p \text{ prime}} \frac{1}{2^p}$$

is an irrational number.

Chapter 5

Equivalence relations

5.1 Relations

In mathematics, it is often crucial to specify what it means for two things to be related to one another, or equal to each other, in a certain sense. In the case of numbers, we might say that two numbers are related when they are equal, or perhaps a and b are related when $a < b$, or when the numbers are related based on some other sort of rule. In the case of shapes, perhaps we say that two shapes (triangles, perhaps) are related if they are congruent, or if they are similar, or if they have the same area. Similarly, we might say that two quadrilaterals are related if they have the same angles, or the same sidelengths. The point of this chapter is to understand what sorts of properties various kinds of relatedness ought to have.

Definition 5.1. Let X be a set. A *binary relation* (or simply *relation*) R on X is a collection of pairs of elements of X, i.e. it consists of various pairs (x, y) with $x, y \in X$. If (x, y) is one of these pairs, write xRy.

Example. One possible relation is the empty relation, consisting of no pairs. Thus it is never true, in this case, that xRy.

Example. Another very simple relation is equality. This is the relation consisting of the pairs (x, x), so that xRy if and only if $x = y$. Note that if we replace the "R" with an "$=$," then xRx becomes $x = x$. In the definition, R is the name of the relation, but it doesn't have to be called R: it could just as easily be called $=$, in which case we recover the usual notation.

Example. Let $X = \mathbb{Z}$ be the set of integers. We have the relation $<$ on $\mathbb{Z}$ that means the usual thing: $x < y$ if, well, $x < y$. A similar relation is $\leq$.

Example. Let $X = \mathbb{Z}$. We have the relation $\mid$ on $\mathbb{Z}$, where $x \mid y$ means that x divides y, i.e. that y is a multiple of x.

Example. Relations can be rather arbitrary-looking, since they can be completely general sets of pairs of elements of X. For instance, suppose $X = \{a, b, c, d\}$. Then we have a relation R on X, where aRc, bRb, bRd, cRb, dRa, dRb, and dRc. This isn't generated by any logical rule: it's just a random-looking collection of pairs. Generally, the relations of interest will be more natural, but it's important to remember that relations can be arbitrary.

There are several properties that a relation might or might not have.

Definition 5.2. Let X be a set, and let R be a relation on X.

- We say that R is *reflexive* if xRx for all $x \in X$.
- We say that R is *irreflexive* if it is never true that xRx.
- We say that R is *symmetric* if whenever xRy, then yRx.
- We say that R is *antisymmetric* if whenever xRy and yRx, then $x = y$.
- We say that R is *asymmetric* if whenever xRy, then it is not the case that yRx.
- We say that R is *transitive* if whenever xRy and yRz, then xRz.
- We say that R is *trichotomous* if for any $x, y \in X$, exactly one of xRy, $x = y$, and yRx is true.

Let's consider the $<$ relation on $\mathbb{Z}$. Which of those properties is satisfied? Let's go through them one by one.

- Reflexivity is not satisfied, because there exists an $x \in \mathbb{Z}$ such that $x \not< x$. (In fact, $x \not< x$ for all $x \in \mathbb{Z}$.)
- On the other hand, $<$ is irreflexive.
- $<$ is not symmetric; for instance, $0 < 1$ but $1 \not< 0$.
- $<$ is antisymmetric, because the hypothesis $x < y$ and $y < x$ never holds. Thus the conclusion that $x = y$ whenever $x < y$ and $y < x$ is vacuously true.
- $<$ is asymmetric: if $x < y$, then $y \not< x$.
- $<$ is transitive: if $x < y$ and $y < z$, then $x < z$.
- $<$ is trichotomous, because given any x, y, exactly one of the statements $x < y$, $x = y$, and $y < x$ is true.

There's a term for a relation that satisfies the same sorts of structural properties as $<$.

Definition 5.3. Let X be a set and $<$ a binary relation on X. We say that $<$ is a *total ordering* on X if $<$ is transitive and trichotomous.

In other words, $<$ is a total ordering if the following hold:

- If $x < y$ and $y < z$, then $x < z$.
- Given any two elements $x, y \in X$, either $x < y$, $x = y$, or $y < x$.

One of the other relations we saw was $|$. There's a generalization of that one as well.

Definition 5.4. Let X be a set and $\leq$ a binary relation on X. We say that $\leq$ is a *partial ordering* on X if $\leq$ is reflexive, antisymmetric, and transitive.

In other words, $\leq$ is a partial ordering if the following hold:

- $x \leq x$ for all $x \in X$.
- If $x \leq y$ and $y \leq x$, then $x = y$.
- If $x \leq y$ and $y \leq z$, then $x \leq z$.

The relation $|$ is a partial ordering on $\mathbb{Z}$ because it satisfies all these properties. It's called a "partial" ordering because we might only be allowed to compare some of the elements. Given certain pairs of numbers like 6 and 18, we can say that $6 \mid 18$, so we can compare them. On the other hand, some pairs of elements may be incomparable: with 4 and 7 for instance, $4 \nmid 7$ and $7 \nmid 4$, but $4 \neq 7$. If all pairs of elements are comparable, then we end up with a total ordering again, but in its $\leq$ version rather than the $<$ version. (This is still called a "total" ordering, and people don't fuss much about the difference between the $<$ version and the $\leq$ version.)

5.2 Equivalence relations

For our purposes, the most important type of binary relation is the equivalence relation.

Definition 5.5. A binary relation $\sim$ on a set X is said to be an *equivalence relation* if it satisfies the following three properties:

- Reflexivity: $x \sim x$ for all $x \in X$.
- Symmetry: if $x \sim y$, then $y \sim x$.
- Transitivity: if $x \sim y$ and $y \sim z$, then $x \sim z$.

Remark 5.6. It might seem as though reflexivity follows from symmetry and transitivity. However, this is not the case, as you will show in problem 3.

The simplest equivalence relation on any set X is just equality: $x \sim y$ if and only if $x = y$. Let's verify that this is indeed an equivalence relation:

- Reflexivity: we must check that $x \sim x$, i.e. that $x = x$ for all $x \in X$. Well, yes...
- Symmetry: we must check that if $x \sim y$, then $y \sim x$. If $x \sim y$, that means that $x = y$, so $y = x$ as well, which means that $y \sim x$.
- Transitivity: we must check that if $x \sim y$ and $y \sim z$, then $x \sim z$. If $x \sim y$, that means that $x = y$, and if $y \sim z$, that means that $y = z$. Thus we have $x = y = z$. Thus $x \sim z$.

We've now checked that all the properties hold, so $\sim$ (or $=$) is an equivalence relation.

Example. Let X be the set of all triangles in the plane. We say that two triangles $\triangle ABC$ and $\triangle DEF$ are equivalent if they are similar; we write $\sim$ for this relation, and $\sim$ is an equivalence relation on X. The relation $\cong$ of congruence ($\triangle ABC \cong \triangle DEF$ if and only if they are congruent) is also an equivalence relation.

Example. Let X and Y be two sets, and let $f : X \to Y$ be a function. Then we can define an equivalence relation $\sim$ on X by declaring that $x_1 \sim x_2$ if and only if $f(x_1) = f(x_2)$.

You will verify that these two are in fact examples of equivalence relations in problem 4.

5.3   Congruence modulo m

The reason we study equivalence relations as we learn number theory is because one of the most powerful tools in number theory is a certain equivalence relation, known as *congruence modulo m*. It is reminiscent of what we do when reading a clock. If it's currently 9:00, then in five hours we won't say that it's 14:00, but rather that it's 2:00. How do we get 2 instead of 14? After 12, we don't continue on with 13 but rather go back to 1. What we're doing here is keeping track of *remainders* upon division by 12.

There is a small difference between the way mathematicians usually like to think about modular arithmetic and the way we read clocks: instead of going from 1 to 12, mathematicians instead usually prefer to go from 0 to 11. We also don't have to divide by 12; the same technique works if we replace 12 with any positive integer.

Here is how modular arithmetic works at a more formal level. We select a modulus m, which serves the same role as 12 does in the case of clocks. We say that two integers a and b are *congruent* modulo m if they leave the same remainder upon division by m. Alternatively, and often more conveniently, a and b are congruent modulo m if and only if $a - b$ is a multiple of m. When this happens, we write $a \equiv b \pmod{m}$.

Example. $14 \equiv 2 \pmod{12}$ because both 14 and 2 leave a remainder of 2 upon division by 12. Alternatively, $14 \equiv 2 \pmod{12}$ because $14 - 2 = 12$, which is a multiple of 12.

There are many important properties of congruences that get used all the time. Let's start with a few very basic ones.

Proposition 5.7. *Let a, b, c, and m be integers.*

(1) $a \equiv a \pmod{m}$.
(2) If $a \equiv b \pmod{m}$, then $b \equiv a \pmod{m}$.
(3) If $a \equiv b \pmod{m}$ and $b \equiv c \pmod{m}$, then $a \equiv c \pmod{m}$.

Proof. In all three of these, we will use the criterion that $a \equiv b \pmod{m}$ if and only if $a - b$ is a multiple of m.

(1) $a - a = 0 = 0 \cdot m$ is a multiple of m, so $a \equiv a \pmod{m}$.
(2) Suppose $a - b = km$. Then $b - a = -km$, which is also a multiple of m.
(3) Suppose $a - b = km$ and $b - c = \ell m$. Then $a - c = (a - b) + (b - c) = km + \ell m = (k + \ell)m$, which is a multiple of m. ∎

If we translate back into the language of equivalence, then all that Proposition 5.7 is saying is that $\equiv \pmod{m}$ is an equivalence relation. The relation of congruence modulo m on $\mathbb{Z}$ consists of all pairs (a, b), where a and b are integers and $a \equiv b \pmod{m}$.

Whenever we have an equivalence relation $\sim$ on a set X, we can find a *set of representatives*: a subset $Y \subseteq X$ such that every $x \in X$ is equivalent to exactly one element $y \in Y$. In the case of congruence modulo m, we can take our set of representatives to be the integers $\{0, 1, \ldots, m - 1\}$. Of course, there are other choices too. For instance, we could use $\{1, 2, \ldots, m\}$ or $\{2, 3, \ldots, m+1\}$ or $\{-2, -1, 0, \ldots, m-3\}$, or even more exotic sets where the representatives are not all consecutive.

Also closely related to equivalence relations are equivalence *classes*.

Definition 5.8. Let X be a set and $\sim$ an equivalence relation on X. If $x \in X$, then the *equivalence class* of x is the set of everything in X equivalent to x: $[x] = \{y \in X : y \sim x\}$. We write $X/\sim$ for the set of all the equivalence classes.

Example. Let $X = \mathbb{Z}$ and the equivalence relation be congruence modulo 7. Then $[3] = \{\ldots, -11, -4, 3, 10, 17, \ldots\}$.

Note that finding a set of representatives just means picking one element out of each equivalence class.

When $X = \mathbb{Z}$ and the equivalence relation is congruence modulo m, we write $\mathbb{Z}/m\mathbb{Z}$ for the set of equivalence classes. Note that each element of $\mathbb{Z}/m\mathbb{Z}$ is an infinite set of integers, all of which are congruent to each other modulo m. In the case where the equivalence relation is $\equiv \pmod{m}$, then we call an equivalence class a *congruence class* or *residue class*; we'll use these two terms interchangeably.

The equivalence classes do something neat to the set X: every element of X is contained in exactly one equivalence class. That is, we have the following:

Theorem 5.9. *Let X be a set and $\sim$ an equivalence relation on X. Let $[x]$ and $[y]$ be two equivalence classes. Then either $[x] = [y]$ or $[x] \cap [y] = \varnothing$.*

So, as soon as two equivalence classes $[x]$ and $[y]$ have a single element in common, they are automatically forced to be identical.

Proof. Suppose that $[x] \cap [y] \neq \varnothing$. This means that there is some element $z \in [x] \cap [y]$. We must now show that $[x] = [y]$. Suppose $w \in [x]$. Then we have $w \sim x$ and $x \sim z$, so by transitivity $w \sim z$. Now we have $w \sim z$ and $z \sim y$, so $w \sim y$. Thus $w \in [y]$. Since an arbitrary element $w \in [x]$ also lies in $[y]$, this shows that $[x] \subseteq [y]$. The reverse direction follows by symmetry. ∎

Thus, the equivalence classes partition up X into a bunch of disjoint sets. In the case of $\equiv \pmod{5}$, for instance, on $\mathbb{Z}$, this means that for every integer n, exactly one of the following five statements holds:

$$n \equiv 0 \pmod{5}, \qquad n \equiv 1 \pmod{5}, \qquad n \equiv 2 \pmod{5},$$
$$n \equiv 3 \pmod{5}, \qquad n \equiv 4 \pmod{5},$$

because those are all the equivalence classes. Because of this, one way to think of the set $X/\sim$ of equivalence classes under $\sim$ on X is that the elements are the same as the elements of X, except that we consider two elements $x, y \in X$ to be equal if $x \sim y$.

5.4 Problems

(1) Define a relation R on $\mathbb{R}$ by saying that xRy if $|x - y| < 1$. Is R an equivalence relation? Which of the properties of a relation are satisfied by R?

(2) Is the relation R on $\mathbb{R}$, given by xRy if $x - y \in \mathbb{Z}$, an equivalence relation? What about the relation S given by xSy if $x - y \in \mathbb{Q}$?

(3) In Definition 5.5, it seems as though reflexivity should follow from symmetry and transitivity: $a \sim b$ implies $b \sim a$, so we have $a \sim b$ and $b \sim a$ and thus $a \sim a$. Why is this wrong? What is the flaw in the argument? Also, give an example of a binary relation on a set X that satisfies symmetry and transitivity but not reflexivity.

(4) Prove that the two examples of equivalence relations (similarity of triangles and equivalence of function inputs) given right after Definition 5.5 are actually equivalence relations, by checking that they satisfy the three properties in the definition.

(5) Suppose $\sim$ and $\equiv$ are two equivalence relations on a set X. Define a new relation R on X by saying that xRy if either $x \sim y$ or $x \equiv y$. Is R necessarily an equivalence relation? Prove or find a counterexample.

(6) Suppose $\sim$ and $\equiv$ are two equivalence relations on a set X. Define a new relation R on X by saying that xRy if both $x \sim y$ and $x \equiv y$. Is R necessarily an equivalence relation? Prove or find a counterexample.

(7) Suppose $\sim$ is an equivalence relation on a finite set X with n elements. Suppose, furthermore, that every equivalence class has exactly k elements. How many equivalence classes are there?

(8) Suppose that R is a reflexive and symmetric relation on a set X. Define a new relation $\sim$ on X by saying that $x \sim y$ if there exists a positive integer n and elements $x_1, x_2, x_3, \ldots, x_n \in X$ such that $x_1 = x$, $x_n = y$, and for $1 \leq i \leq n-1$, we have $x_i R x_{i+1}$. Prove that $\sim$ is an equivalence relation on X.

(9) Let $X = \mathbb{Z} \times (\mathbb{Z} \backslash \{0\})$ be the set of pairs of integers, where the second one cannot be zero. Define a relation $\sim$ on X by declaring that $(a, b) \sim (c, d)$ if $ad = bc$. Prove that $\sim$ is an equivalence relation. Explain why the equivalence classes are related to the rational numbers $\mathbb{Q}$.

Chapter 6

Congruences and modular arithmetic

6.1 Arithmetic in $\mathbb{Z}/m\mathbb{Z}$

Recall from Chapter 5 that, for each positive integer m, we put an equivalence relation $\equiv \pmod m$ on $\mathbb{Z}$ by declaring that $a \equiv b \pmod m$ if $a - b$ is a multiple of m, in which case we say that a is *congruent* to b modulo m. We write $\mathbb{Z}/m\mathbb{Z}$ for the set of equivalence classes modulo m.

The reason that congruences are useful is that they are well-behaved with respect to addition and multiplication. What this means is that we can add and multiply two elements of $\mathbb{Z}/m\mathbb{Z}$. More precisely:

Proposition 6.1.

(1) Suppose that $a \equiv b \pmod m$ and $c \equiv d \pmod m$. Then $a + c \equiv b + d \pmod m$.

(2) Suppose that $a \equiv b \pmod m$ and $c \equiv d \pmod m$. Then $ac \equiv bd \pmod m$.

Proof. Suppose that $a - b = km$ and $c - d = \ell m$.

(1) We have
$$(a + c) - (b + d) = (a - b) + (c - d) = km + \ell m = (k + \ell)m,$$
which is a multiple of m. Thus $a + c \equiv b + d \pmod m$.

(2) We have
$$
\begin{aligned}
ac - bd &= ac - ad + ad - bd \\
&= a(c - d) + d(a - b) \\
&= a\ell m + dkm \\
&= (a\ell + dk)m,
\end{aligned}
$$
which is also a multiple of m. ∎

We can also subtract in $\mathbb{Z}/m\mathbb{Z}$; we save the proof for problem 5. However, don't start believing that we can perform just *any* arithmetical operation modulo m.

Example. Exponentiation is not well-defined in $\mathbb{Z}/m\mathbb{Z}$, at least not in the most obvious way. For instance $1 \equiv 4 \pmod 3$, yet $2^1 \not\equiv 2^4 \pmod 3$. There *are* ways of exponentiating modulo m, but the rules are a bit more subtle. We will return to this point in Chapter 8. However, a certain kind of exponentiation *does* work in $\mathbb{Z}/m\mathbb{Z}$: If $a \equiv b \pmod m$ and e is any nonnegative integer, then $a^e \equiv b^e \pmod m$. Note that we require the exponents to be *exactly* the same, rather than just the same modulo m, which as we have just seen is not good enough.

What we have done in Proposition 6.1 is to show that addition and multiplication are *well-defined* modulo m—or, if you prefer, in $\mathbb{Z}/m\mathbb{Z}$. A function is said to be *well-defined* if it always gives the same answer when handed the same inputs. More precisely:

Definition 6.2. Let S and T be two sets, and suppose $f : S \to T$. Then f is *well-defined* if, whenever $s_1 = s_2 \in S$, then $f(s_1) = f(s_2) \in T$.

This statement is a little bit unfortunate, because how could we even define f if that weren't the case? But let's analyze the difference between what happens with addition and with exponentiation in $\mathbb{Z}/m\mathbb{Z}$. In this case, our set S will be $(\mathbb{Z}/m\mathbb{Z}) \times (\mathbb{Z}/m\mathbb{Z})$, or ordered pairs (a, b) of elements in $\mathbb{Z}/m\mathbb{Z}$, and T will just be $\mathbb{Z}/m\mathbb{Z}$.

We can try to define addition in $\mathbb{Z}/m\mathbb{Z}$, as follows. Let us write $[a]$ for the congruence class of a modulo m, i.e. the set of numbers congruent to $a \pmod m$. Then we can add two elements $[a]$ and $[c]$ of $\mathbb{Z}/m\mathbb{Z}$ by setting $[a] + [c] = [a + c]$. The issue that we need to resolve, in order to make sure that this is a well-defined definition, is to determine whether it depends on how we write our congruence classes. If $[a] = [b]$ and $[c] = [d]$, i.e. $a \equiv b \pmod m$ and $c \equiv d \pmod m$, is it true that $[a] + [c] = [b] + [d]$? Well, $[a] + [c] = [a + c]$ and $[b] + [d] = [b + d]$, so we need to ensure that $[a + c] = [b + d]$. But that's exactly what we showed in part (1) of Proposition 6.1.

By contrast, if we try to define $[a]^{[c]}$ by setting it equal to $[a^c]$, then we run into problems, because $[2^4] = [1]$, whereas $[2^1] = [2]$.

We can do all this more generally in terms of equivalence classes and equivalence relations. Let us suppose that X and Y are sets, and $f : X \to Y$ is a function. Let us also suppose that $\sim$ is an equivalence relation on X.

Then we can define f on $X/\sim$ by setting $f([x]) = f(x)$. This is a well-defined function on $X/\sim$ if and only if $f(x_1) = f(x_2)$ whenever $x_1 \sim x_2$. In the setting of addition modulo m, we have $X = \mathbb{Z} \times \mathbb{Z}$, $Y = \mathbb{Z}/m\mathbb{Z}$, and $f(x_1, x_2) = (x_1 + x_2) \pmod{m}$. The equivalence relation $\sim$ is such that $(a, b) \sim (c, d)$ if $a \equiv c \pmod{m}$ and $b \equiv d \pmod{d}$.

Now that we have spent all this time showing that addition and multiplication are well-defined in $\mathbb{Z}/m\mathbb{Z}$, let's see how to use that information to solve some problems.

Example. Let us work out the remainder when 52^{2000} is divided by 7. In other words, we want to know what 52^{2000} is modulo 7, i.e. find a representative of $[52^{2000}]$ in $\{0, 1, 2, 3, 4, 5, 6\}$. To do this, we note that, because $52 \equiv 3 \pmod{7}$ and multiplication is well-defined in $\mathbb{Z}/7\mathbb{Z}$, we have $52^{2000} \equiv 3^{2000} \pmod{7}$. Next, let's look at the powers of 3 modulo 7: we have

$$3^0 \equiv 1, \qquad 3^1 \equiv 3, \qquad 3^2 \equiv 2, \qquad 3^3 \equiv 6,$$

$$3^4 \equiv 4, \qquad 3^5 \equiv 5, \qquad 3^6 \equiv 1,$$

where all congruences are modulo 7. After that, they repeat. For instance, $3^7 \equiv 3^6 \times 3^1 \equiv 1 \times 3^1 \equiv 3 \pmod{7}$, and so on. Thus we find that the remainder depends only on the exponent modulo 6. Since $2000 \equiv 2 \pmod{6}$, we have

$$52^{2000} \equiv 3^{2000} \equiv 3^2 \equiv 2 \pmod{7}.$$

Now, recall that we *aren't* allowed to reduce the exponent modulo 7, so we may *not* say that since $2000 \equiv 5$, we have $52^{2000} \equiv 3^5 \equiv 5 \pmod{7}$: we get the wrong answer that way. Once again, that is because exponentiation is not well-defined in $\mathbb{Z}/m\mathbb{Z}$, whereas addition and multiplication are.

Let's try another example.

Example. Let us show that 999999 cannot be written as the sum of squares of two integers. That is, we wish to show that for all integers x and y, $x^2 + y^2 \neq 999999$. To do this, let us suppose that there *are* integers x and y such that $x^2 + y^2 = 999999$, and then we'll see what goes wrong. The trick here is to work modulo 4, i.e. in $\mathbb{Z}/4\mathbb{Z}$. Because addition and multiplication are well-defined in $\mathbb{Z}/4\mathbb{Z}$, we would have $[x]^2 + [y]^2 = [999999] = [3]$. So, now we need to know what happens with squares modulo 4. Since multiplication is well-defined modulo 4, we only have to check what happens with x^2 for $x \in \{0, 1, 2, 3\}$. We have

$$[0^2] = [0], \qquad [1^2] = [1], \qquad [2^2] = [0], \qquad [3^2] = [1].$$

In other words, all squares are 0 or 1 modulo 4. Thus it is not possible to add two squares and get a number that is 3 (mod 4), since the sum of two elements in $\{0, 1\}$ cannot be 3 (mod 4).

More generally:

Theorem 6.3. *If $n \equiv 3$ (mod 4), then n cannot be written as a sum of two squares.*

See problem 2 for a version for sums of three squares, which can be proven similarly. See also Chapter 10, where we will determine which numbers can be written as sums of two squares.

Remark 6.4. How did we know to work modulo 4? This comes from experience: squares are very nice modulo 4, and even nicer modulo 8, so these are common moduli to try. There is much more to be said about squares and moduli, culminating in the celebrated *Quadratic Reciprocity Theorem*, which unfortunately we won't get to in this book. For a detailed look at quadratic reciprocity, see for instance [Sil14, Chapters 20–23].

6.2   The Chinese Remainder Theorem

One of the most useful theorems about modular arithmetic is the *Chinese Remainder Theorem*. The Chinese Remainder Theorem allows us to turn two (or more congruences) into a single congruence, or vice versa: if $\gcd(m, n) = 1$, then a congruence modulo mn is exactly the same as a congruence modulo m together with a congruence modulo n.

Theorem 6.5 (Chinese Remainder Theorem). *Let a, b, m, n be integers with $m, n > 0$ and $\gcd(m, n) = 1$. Then there exists an integer x satisfying*

$$x \equiv a \pmod{m}, \qquad x \equiv b \pmod{n}. \tag{6.1}$$

Furthermore, all such x's form a congruence class modulo mn.

Proof. We begin by solving the case $a = 1$ and $b = 0$. That is, we want to find some y such that

$$y \equiv 1 \pmod{m}, \qquad y \equiv 0 \pmod{n}. \tag{6.2}$$

In order to do this, we use Bézout's Lemma from Chapter 3. As we recall, Bézout's Lemma says that if m and n are relatively prime, then there exist integers α and β such that $\alpha m + \beta n = 1$. With these values of α and β, we observe that $y = \beta n$ satisfies (6.2): $y \equiv 0 \pmod{n}$ because $y = \beta n$ is a multiple of n. On the other hand, since $\alpha m + \beta n = 1$, we have

$\beta n = 1 - \alpha m \equiv 1 \pmod{m}$. This proves that there is such a y when $a = 1$ and $b = 0$. By symmetry, there is also a z satisfying

$$z \equiv 0 \pmod{m}, \qquad z \equiv 1 \pmod{n}.$$

To solve the more general problem of showing that there is an x satisfying (6.1), we simply take $x = ay + bz$. Then

$$x \equiv ay + bz \equiv a \pmod{m},$$

and similarly $x \equiv b \pmod{n}$.

Now, we must prove that if x satisfies (6.1), then any $y \equiv x \pmod{mn}$ also satisfies (6.1) and vice versa. Suppose that $x \equiv y \pmod{mn}$, so that $x - y = cmn$ for some integer c. Then $y = x - cmn$, so $y \equiv x \equiv a \pmod{m}$ and $y \equiv x \equiv b \pmod{n}$, so y does indeed satisfy (6.1). Finally, if y satisfies (6.1), then $x \equiv y \pmod{m}$ and $x \equiv y \pmod{n}$, so $x - y$ is a multiple of both m and n. Since m and n are relatively prime, this means that $x - y$ is a multiple of mn, i.e. $x \equiv y \pmod{mn}$, as desired. ∎

Remark 6.6. Note that this proof tells you not only that such an x exists, but also how to find it. In order to do that, you'll have to use the Euclidean algorithm to give you the actual values of α and β in Bézout's Lemma. You will do this in problem 4.

We can also extend the Chinese Remainder Theorem to the case of more than two moduli.

Theorem 6.7 (Chinese Remainder Theorem)**.** *Let r be a positive integer, and let $a_1, a_2, \ldots, a_r$ be integers and $m_1, m_2, \ldots, m_r$ be positive integers such that for any distinct i, j with $1 \leq i, j \leq r$, we have $\gcd(m_i, m_j) = 1$. Then there exists an integer x satisfying*

$$x \equiv a_i \pmod{m_i} \qquad \text{for all } i \text{ with } 1 \leq i \leq r.$$

Furthermore, all such x's form a congruence class modulo $\prod_{i=1}^{r} m_i$.

Let's see what we can do with the Chinese Remainder Theorem. In practice, the way it's usually used is to split up a congruence modulo a number with several prime factors into a bunch of smaller congruences with prime or prime power moduli, which may be easier to work with.

Example. Let us find the remainder when 19^{659} is divided by 30. Thus, we want to know $19^{659} \pmod{30}$ and express the answer as an element of $\{0, 1, \ldots, 29\}$. By the Chinese Remainder Theorem, it suffices to determine the answer modulo 2, 3, and 5, and then put the pieces back together again.

First, $19^{659} \equiv 1^{659} \equiv 1 \pmod{2}$. Next, $19^{659} \equiv 1^{659} \equiv 1 \pmod{3}$. Finally, $19^{659} \equiv 4^{659} \equiv (-1)^{659} \equiv -1 \equiv 4 \pmod{5}$. To finish the problem off, we need to find a number that is 1 (mod 2), 1 (mod 3), and 4 (mod 5). We could do this systematically, but it's easier not to. By inspection, 4 is a number that is 1 (mod 3) and 4 (mod 5), and any other such number is congruent to 4 (mod 15). That leaves 4 and 19 as possibilities. Since only 19 is 1 (mod 2), that's the answer.

The point here is that we can solve this problem not by working modulo a large number directly, but by working with smaller numbers, where it's easier to see what's going on. In fact, we may always restrict to the case where congruences are modulo prime powers. For instance, when working modulo $120 = 2^3 \times 3 \times 5$, it suffices to work modulo 8, 3, and 5.

Here is another thing we can do with the Chinese Remainder Theorem:

Example. Find all integer solutions to $x^3 - x + 1 \equiv 0 \pmod{35}$. Let us write $f(x)$ for $x^3 - x + 1$. To solve this, we note that the statement $y \equiv 0 \pmod{35}$ is equivalent to the pair of statements $y \equiv 0 \pmod{5}$ and $y \equiv 0 \pmod{7}$. The latter two are easier. To determine those x for which $f(x) \equiv 0 \pmod{5}$, we just try out the possibilities $x = 0, 1, 2, 3, 4$; we find that $f(x) \equiv 0 \pmod{5}$ if and only if $x \equiv 3 \pmod{5}$. To determine those x for which $f(x) \equiv 0 \pmod{7}$, we try out $x = 0, 1, 2, 3, 4, 5, 6$; we find that $f(x) \equiv 0 \pmod{7}$ if and only if $x \equiv 2 \pmod{7}$. Thus $f(x) \equiv 0 \pmod{35}$ if and only if both $x \equiv 3 \pmod{5}$ and $x \equiv 2 \pmod{7}$. We can use the Chinese Remainder Theorem to see that this is equivalent to $x \equiv 23 \pmod{35}$. Thus $f(x) \equiv 0 \pmod{35}$ if and only if $x \equiv 23 \pmod{35}$.

6.3   Problems

(1) What is

$$9 \times 99 \times 999 \times \cdots \times 999 \cdots 9 \quad (\bmod\ 1000),$$

where the last number contains 999 9's? That is, determine the remainder of the product when divided by 1000.

(2) Prove that if $n \equiv 7 \pmod{8}$, then n cannot be written as a sum of three squares. (Hint: What are the squares modulo 8?)

(3) Prove that for any nonnegative integer n, $3^n + 13 \cdot 10^n$ is divisible by 7.

(4) Describe all integers x satisfying *all* of the following congruences:

$$x \equiv 2 \pmod 3, \qquad\qquad x \equiv 5 \pmod 7,$$
$$x \equiv 3 \pmod 5, \qquad\qquad x \equiv 7 \pmod{11}.$$

(The answer should be a congruence class modulo some number, such as "all $x \equiv 4 \pmod 9$," although that is not the correct answer in this case.)

(5) Prove that subtraction is well-defined in $\mathbb{Z}/m\mathbb{Z}$, i.e. if $a \equiv b \pmod m$ and $c \equiv d \pmod m$, then $a - c \equiv b - d \pmod m$.

(6) Prove that a number in base 10 is divisible by 9 if and only if the sum of its digits is divisible by 9. Show that a number $d_n d_{n-1} \cdots d_1$ (in terms of its digits) is divisible by 11 if and only if $d_1 - d_2 + d_3 - \cdots + (-1)^{n-1} d_n$ is divisible by 11. Devise, with proof, a (somewhat more involved) test to determine if a number is divisible by 7 or 13.

(7) Let $f(x) = a_n x^n + a_{n-1} x^{n-1} + \cdots + a_1 x + a_0$ be a polynomial with integer coefficients. Show that if, for some integer $m > 0$, there are m consecutive integers $r, r+1, \ldots, r+m-1$ such that $f(r+i)$ is divisible by m for $0 \le i \le m-1$, then $f(x)$ is divisible by m for every integer x.

(8) A lattice point $(x, y) \in \mathbb{Z}^2$ (i.e. a point in $\mathbb{R}^2$ such that both of its coordinates are integers) is said to be *blocked* if $\gcd(x, y) > 1$. Show that, for any positive integer n, there is an $n \times n$ square consisting entirely of blocked points, i.e. there exist positive integers a and b such that all the points $(a + i, b + j)$ with $0 \le i, j \le n-1$ are blocked. For instance, the four points $(14, 20), (14, 21), (15, 20), (15, 21)$ form a 2×2 square of blocked points.

Chapter 7

Modular multiplication and division

7.1 Modular multiplication

We saw in Chapter 6 that we can multiply modulo m, in the sense that if $a \equiv b \pmod{m}$ and $c \equiv d \pmod{m}$, then $ac \equiv bd \pmod{m}$. However, a glance at modular addition and multiplication tables suggests that these two operations can behave quite differently from one another. Here are the tables for addition and multiplication in $\mathbb{Z}/9\mathbb{Z}$:

+	0	1	2	3	4	5	6	7	8		×	0	1	2	3	4	5	6	7	8
0	0	1	2	3	4	5	6	7	8		0	0	0	0	0	0	0	0	0	0
1	1	2	3	4	5	6	7	8	0		1	0	1	2	3	4	5	6	7	8
2	2	3	4	5	6	7	8	0	1		2	0	2	4	6	8	1	3	5	7
3	3	4	5	6	7	8	0	1	2		3	0	3	6	0	3	6	0	3	6
4	4	5	6	7	8	0	1	2	3		4	0	4	8	3	7	2	6	1	5
5	5	6	7	8	0	1	2	3	4		5	0	5	1	6	2	7	3	8	4
6	6	7	8	0	1	2	3	4	5		6	0	6	3	0	6	3	0	6	3
7	7	8	0	1	2	3	4	5	6		7	0	7	5	3	1	8	6	4	2
8	8	0	1	2	3	4	5	6	7		8	0	8	7	6	5	4	3	2	1

In the addition table, every row and column contains each residue class exactly once. This is pretty straightforward to understand: it's just saying that for every $a, x \in \mathbb{Z}/m\mathbb{Z}$, there is a unique $b \in \mathbb{Z}/m\mathbb{Z}$ such that $a + b \equiv x \pmod{m}$. Indeed, $b \equiv x - a \pmod{m}$, which makes sense because subtraction is a well-defined operation in $\mathbb{Z}/m\mathbb{Z}$.

On the other hand, this property is not true in the multiplication table. There are some rows and columns in which every element of $\mathbb{Z}/m\mathbb{Z}$ appears exactly once, but then there are others where this doesn't happen: some elements are repeated, and some are missing. For example, in $\mathbb{Z}/9\mathbb{Z}$, every element appears exactly once in the 2 row, but not in the 3 row. This says that for every $x \in \mathbb{Z}/9\mathbb{Z}$, there is a unique $b \in \mathbb{Z}/9\mathbb{Z}$ such that $2b \equiv x$ (mod 9), but not such that $3b \equiv x$ (mod 9). For instance, this doesn't work when $x = 4$: no multiple of 3 can ever be 4 (mod 9).

We would like to identify those rows (or columns) of the multiplication table in $\mathbb{Z}/m\mathbb{Z}$ in which every element of $\mathbb{Z}/m\mathbb{Z}$ appears exactly once. Let's look at the a-row, i.e. the row of multiples of a. Note that there are m entries in the a-row, and also m elements of $\mathbb{Z}/m\mathbb{Z}$. There are two ways that the a-row can fail to have the property that every element appears exactly once. It could be the case that some element doesn't appear at all, and it could also be the case that some element appears more than once. But the pigeonhole principle says that these are exactly the same: if we are missing some element (say x), then there are m elements to fill, using only $m - 1$ elements of $\mathbb{Z}/m\mathbb{Z}$. So some element must be repeated. For much the same reason, if some element is repeated, then we don't have enough space to put all m elements.

Now, note that if every element of $\mathbb{Z}/m\mathbb{Z}$ appears in the a-row, then 1 appears. On the other hand, if 1 appears—say as $ab \equiv 1$ (mod m)—then every x appears, because $a(bx) \equiv x$ (mod m). So it suffices to determine when 1 appears in the a-row.

Definition 7.1. An element $a \in \mathbb{Z}/m\mathbb{Z}$ is said to be a *unit* if there is some b such that $ab \equiv 1$ (mod m). We write $(\mathbb{Z}/m\mathbb{Z})^\times$ (pronounced "Z mod mZ star") for the set of units of $\mathbb{Z}/m\mathbb{Z}$.

In other words, we're trying to identify the units in $\mathbb{Z}/m\mathbb{Z}$.

Proposition 7.2. *An element $a \in \mathbb{Z}/m\mathbb{Z}$ is a unit iff* $\gcd(a, m) = 1$.

Proof. We have two things to prove: if a is a unit, then $\gcd(a, m) = 1$; and if $\gcd(a, m) = 1$, then a is a unit.

Let us begin by showing that if a is a unit, then $\gcd(a, m) = 1$. Since a is a unit, there is some b such that $ab \equiv 1$ (mod m). By the definition of modular arithmetic, this means that there is some integer k such that $ab = 1 + km$, or $ab - km = 1$. By Bézout's Lemma, this means that $\gcd(a, m) = 1$.

For the other direction, we assume that $\gcd(a, m) = 1$, and we must show that a is a unit. By Bézout's Lemma, since $\gcd(a, m) = 1$, there are integers b and k such that $ab + km = 1$. This means that $ab \equiv 1$ (mod m), so a is a unit.   ∎

What we have just shown is that if $a \in (\mathbb{Z}/m\mathbb{Z})^{\times}$, and b and c are such that $ab \equiv ac \pmod{m}$, then $b \equiv c \pmod{m}$. In other words, we may divide by units modulo m.

One peculiar thing we can do with this modular division is to define congruences of fractions. For example, we may say that $\frac{1}{2} \equiv 5 \pmod{9}$. There are several possible interpretations of this. First, we can clear denominators: $\frac{1}{2} \equiv 5 \pmod 9$ should be the same as $1 \equiv 2 \times 5 \pmod 9$, which is indeed true. Another interpretation is to say that—while 1 isn't divisible by 2—we can find some number congruent to 1 $\pmod 9$ that is divisible by 2, then divide that number by 2. Our previous work shows that our choice of even number congruent to 1 $\pmod 9$ doesn't matter: we'll get the same answer after dividing by 2 regardless of which one we choose. So, if we instead take 28 as our even number congruent to 1 $\pmod 9$, then $\frac{1}{2} \equiv \frac{28}{2} \equiv 14 \equiv 5 \pmod 9$, the same as we get when we choose 10.

On the other hand, we may *not* divide by non-units modulo m. For example, $3 \times 2 \equiv 3 \times 5 \pmod 9$, even though $2 \not\equiv 5 \pmod 9$. So we may not divide by 3 modulo 9.

Still, it turns out that we may partially divide by 3 in $\mathbb{Z}/9\mathbb{Z}$.

Proposition 7.3. *Suppose that $ab \equiv ac \pmod{m}$. Let $g = \gcd(a, m)$. Then $b \equiv c \pmod{\frac{m}{g}}$.*

First we'll prove a simple lemma.

Lemma 7.4. *Suppose $\gcd(a, m) = g$. Let $a = ga'$ and $m = gm'$. Then $\gcd(a', m') = 1$.*

Proof. By Bézout's Lemma, we may write g as $g = as + mt$ for some integers s and t. Replacing a with ga' and m with gm', we get

$$g = ga's + gm't.$$

If we divide both sides by g, we get

$$1 = a's + m't,$$

which by Bézout's Lemma again implies that $\gcd(a', m') = 1$. ∎

Let's now return to the proof of Proposition 7.3.

Proof of Proposition 7.3. Since $ab \equiv ac \pmod{m}$, there is some integer k such that $ab = ac + km$. Let us write $a = ga'$ and $m = gm'$. Then we have $ga'b = ga'c + gkm'$. Dividing by g, we have $a'b = a'c + km'$, so $a'b \equiv a'c \pmod m$. Since $\gcd(a', m') = 1$, by Lemma 7.4, we may divide by a' to obtain $b \equiv c \pmod{m'}$, or $b \equiv c \pmod{\frac{m}{g}}$, as desired. ∎

Recall that $3 \times 2 \equiv 3 \times 5 \pmod 9$ did not imply that $2 \equiv 5 \pmod 9$, but Proposition 7.3 at least allows us to say that $2 \equiv 5 \pmod 3$, which is true.

7.2   The totient function

An interesting question is to determine the number of units in $\mathbb{Z}/m\mathbb{Z}$, i.e. the size of $(\mathbb{Z}/m\mathbb{Z})^{\times}$.

Definition 7.5. The *totient* of m, denoted $\phi(m)$, is the number of units in $\mathbb{Z}/m\mathbb{Z}$. A number a with $1 \leq a \leq m$ and $\gcd(a, m) = 1$ is called a *totative* of m.

Remark 7.6. Another common notation is $\varphi(m)$. The symbols ϕ and φ are just different ways of writing the Greek letter phi.

By Proposition 7.2, $\phi(m)$ is the number of totatives of m.

Remark 7.7. No one knows what the word "totient" means. The word was coined by the 19th-century English mathematician James Joseph Sylvester, who is also responsible for many other fun names in mathematics, including "discriminant" and "matrix," and his motivation for choosing this particular word is unknown. Presumably there is some connection with "quotient."

To find a formula for $\phi(n)$, it will be helpful to introduce a preliminary definition and proposition.

Definition 7.8. A function $f : \mathbb{N} \to \mathbb{R}$ (or $\mathbb{C}$, or any other reasonable collection of numbers) is called *multiplicative* if, whenever $\gcd(m, n) = 1$, we have $f(mn) = f(m)f(n)$.

Proposition 7.9. *The totient function is a multiplicative function.*

Proof. Suppose that $\gcd(m, n) = 1$. We wish to figure out how many numbers up to mn are relatively prime to mn. Write the numbers from 1 to mn in a table, like this:

1	2	3	$\cdots$	m
$m + 1$	$m + 2$	$m + 3$	$\cdots$	$2m$
$2m + 1$	$2m + 2$	$2m + 3$	$\cdots$	$3m$
$\vdots$	$\vdots$	$\vdots$	$\ddots$	$\vdots$
$(n-1)m + 1$	$(n-1)m + 2$	$(n-1)m + 3$	$\cdots$	$nm.$

There are some columns in which there can't be any numbers relatively prime to mn. A typical column contains numbers of the form $km+\ell$, where ℓ is fixed and k is variable. By the Euclidean algorithm, $\gcd(km+\ell, m) = \gcd(m, \ell)$, and $\gcd(km+\ell, m) \le \gcd(km+\ell, mn)$. Hence, if $\gcd(m, \ell) > 1$, then there is no way that $km + \ell$ can be relatively prime to mn.

Hence, there are $\phi(m)$ columns that could conceivably contain numbers relatively prime to mn. In fact, all the numbers in all these columns are relatively prime to m, so we just have to check how many of them are relatively prime to n. Let us do this by focusing on one such column, say the one consisting of numbers of the form $km + \ell$. We want to show that no two numbers in this column are congruent modulo n. Suppose, to the contrary, that $rm + \ell \equiv sm + \ell \pmod{n}$. Then we can subtract ℓ from both sides to get $rm \equiv sm \pmod{n}$. Since $\gcd(m, n) = 1$, we can divide by m to get $r \equiv s \pmod{n}$. But $0 \le r, s \le n-1$, so if $r \equiv s \pmod{n}$, they are actually equal. Hence, no two numbers in this column are congruent modulo n, and since there are n numbers in the column, they fill up all the residue classes modulo n. We know that, of the residue classes modulo n, exactly $\phi(n)$ are relatively prime to n. Hence, in this column, $\phi(n)$ are relatively prime to n. Since there are $\phi(m)$ relevant columns, we have

$$\phi(mn) = \phi(m)\phi(n),$$

so ϕ is multiplicative. ∎

Now, if f is a multiplicative function and the prime factorization of n is $n = \prod_{i=1}^{r} p_i^{e_i}$, then we have

$$f(n) = \prod_{i=1}^{r} f(p_i^{e_i}).$$

Thus, in order to calculate $f(n)$ for any n, it suffices to calculate f at all the prime powers p^e. We now do this in the case of the totient function.

Proposition 7.10. *If p is a prime, then* $\phi(p^e) = (p-1)p^{e-1} = p^e - p^{e-1} = \left(1 - \frac{1}{p}\right)p^e.$

Proof. Suppose that $1 \le a \le p^e$. Since the only divisors of p^e are powers of p, it follows that $\gcd(a, p^e) > 1$ if and only if a is a multiple of p. Hence the number of units in $\mathbb{Z}/p^e\mathbb{Z}$ is simply the number of elements that are *not* multiples of p. There are p^{e-1} multiples of p between 1 and p^e, so $p^e - p^{e-1}$ non-multiples of p. Hence $\phi(p^e) = p^e - p^{e-1}$. ∎

Putting all this together, we have the following theorem:

Theorem 7.11. *For every positive integer n, we have*

$$\phi(n) = n \prod_{\substack{p|n \\ p \text{ prime}}} \left(1 - \frac{1}{p}\right).$$

Remark 7.12. The notation $p \mid n$ reads "p divides n," and it just means that n is a multiple of p. More generally, we write $m \mid n$ to mean that n is a multiple of m. We also write $m \nmid n$ when n is not a multiple of m.

Proof. Suppose that the prime factorization of n is $n = \prod_{i=1}^{r} p_i^{e_i}$. Then we have

$$\phi(n) = \prod_{i=1}^{r} \phi(p_i^{e_i})$$

$$= \prod_{i=1}^{r} \left[p_i^{e_i} \left(1 - \frac{1}{p_i}\right) \right]$$

$$= \prod_{i=1}^{r} p_i^{e_i} \cdot \prod_{i=1}^{r} \left(1 - \frac{1}{p_i}\right)$$

$$= n \prod_{\substack{p|n \\ p \text{ prime}}} \left(1 - \frac{1}{p}\right),$$

as desired. ∎

Given some number a with $1 \leq a \leq n$, it might or might not be relatively prime to n, so it might or might not be counted by $\phi(n)$. If $\gcd(a, n) = g$, then $\gcd\left(\frac{a}{g}, \frac{n}{g}\right) = 1$, so $\frac{a}{g}$ is a totative of $\frac{n}{g}$, i.e. it is counted by $\phi\left(\frac{n}{g}\right)$. A careful analysis of this line of reasoning yields the following formula.

Theorem 7.13. *Let n be a positive integer. Then*

$$\sum_{d|n} \phi(d) = n. \tag{7.1}$$

Proof. Let $d_1, \ldots, d_k$ be the divisors of n. We partition the numbers $1, 2, \ldots, n$ into disjoint subsets $C_{d_1}, \ldots, C_{d_k}$ such that, for each i, $|C_{d_i}| = \phi(d_i)$. What this means is that every integer from 1 to n will be placed into exactly one of the sets $C_{d_1}, \ldots, C_{d_k}$, and each C_{d_i} has $\phi(d_i)$ elements. To do this, we let

$$C_{d_i} = \left\{ j : 1 \leq j \leq n, \gcd(j, n) = \frac{n}{d_i} \right\}.$$

If $\gcd(j, n) = \frac{n}{d_i}$, then $k = \frac{j}{n/d_i}$ is relatively prime to d_i and is in the range $1 \leq k \leq d_i$, i.e. k is a totative of d_i. The number of such elements k is $\phi(d_i)$, by definition, so C_{d_i} also has $\phi(d_i)$ elements: take all these values of k and multiply them by $\frac{n}{d_i}$. Since every element j with $1 \leq j \leq n$ is in exactly one C_{d_i}, it follows that the sum of the number of elements in the C_{d_i}'s is equal to n, which is simply (7.1).   ∎

Example. Let's do this partitioning when $n = 14$. The divisors of 14 are 1, 2, 7, and 14. The set C_1 consists of those numbers j from 1 to 14 with $\gcd(j, 14) = \frac{14}{1} = 14$, so just 14. Similarly, C_2 consists of those numbers j with $\gcd(j, 14) = \frac{14}{2} = 7$, which is just 7. Continuing on this way, we get

$$C_1 = \{14\}, \qquad C_2 = \{7\}, \qquad C_7 = \{2, 4, 6, 8, 10, 12\},$$

$$C_{14} = \{1, 3, 5, 9, 11, 13\}.$$

Note that for each divisor d of 14, the number of elements in C_d is $\phi(d)$, as claimed in the proof.

7.3   Problems

(1) Find, with proof, all integers x such that $6x \equiv 4 \pmod{28}$.
(2) Give a good algorithm for finding multiplicative inverses modulo m, provided they exist. That is, given a and m with $\gcd(a, m) = 1$, explain how to compute $\frac{1}{a} \pmod{m}$. Use your algorithm to compute $\frac{1}{14} \pmod{107}$. (Hint: Think about the Euclidean algorithm and Bézout's Lemma.)
(3) Prove that if p is prime, then the only elements of $(\mathbb{Z}/p\mathbb{Z})^\times$ that are congruent to their multiplicative inverses modulo p are $\pm 1 \pmod{p}$, i.e. these are the only $x \in (\mathbb{Z}/p\mathbb{Z})^\times$ such that $x \equiv \frac{1}{x} \pmod{p}$. Find an m such that there are more than two elements of $(\mathbb{Z}/m\mathbb{Z})^\times$ that are congruent to their own inverses. Show that for any positive integer n, there is an integer m (depending on n) such that at least n elements of $(\mathbb{Z}/m\mathbb{Z})^\times$ are congruent to their own inverses.
(4) Prove that if $\gcd(a, b) = g$, then
$$\phi(ab) = \frac{g\phi(a)\phi(b)}{\phi(g)}.$$
(5) Prove that for any integer $n > 1$, we have
$$\sum_{\substack{1 \leq i \leq n \\ \gcd(i, n) = 1}} i = \frac{n\phi(n)}{2}.$$

(6) Prove that if $d \mid n$, then $\phi(d) \mid \phi(n)$.

(7) Find all numbers n such that $\phi(n) = 36$. Prove that you found all of them.

(8) Prove that for every positive integer k, there are only finitely many positive integers n such that $\phi(n) = k$.

(9) Suppose f is a multiplicative function, and let

$$g(n) = \sum_{d \mid n} f(d).$$

Prove that g is a multiplicative function.

(10) Let ε be any positive real number. Prove that there is some positive integer n, depending on ε, such that $\frac{\phi(n)}{n} < \varepsilon$. (This is probably too hard in general, given what we have discussed so far. But you should be able to do it when $\varepsilon = \frac{1}{5}$, so do that if you can't do the general case. The construction for the general case isn't significantly different from what you'll probably do for $\frac{1}{5}$, but the problem is in proving that it gets arbitrarily small.)

Chapter 8

Fermat's Little Theorem

8.1 Modular exponentiation

Recall that we showed in Chapter 6 that addition, subtraction, and multiplication are well-defined modulo m. This means that if $a \equiv b \pmod{m}$ and $c \equiv d \pmod{m}$, then $a + c \equiv b + d \pmod{m}$, and similarly for subtraction and multiplication. The situation for division is a bit more subtle, as we saw in Chapter 7, but it too works in much the same way once the hypotheses are satisfied.

Yet, we saw that nothing of the sort works for exponentiation. That is, we can find numbers a, b, c, d such that $a \equiv b \pmod{m}$ and $c \equiv d \pmod{m}$, yet $a^c \not\equiv b^d \pmod{m}$. For instance, if we take $a = b = 2$, $c = 1$, $d = 6$, and $m = 5$, then we have $a^c = 2^1 \equiv 2 \pmod 5$, but $b^d = 2^6 = 64 \equiv 4 \pmod 5$.

Using the well-definedness of *multiplication* modulo m, it follows that if $a \equiv b \pmod{m}$, then $a^c \equiv b^c \pmod{m}$. The only thing we *can't* always get away with is modifying the *exponent* modulo m.

Still, it turns out that there is something we can say about changing the exponent in modular arithmetic; it just looks a bit different from the other operations. It's somewhat simpler in the case that m is a prime number, so we'll replace m with a p to remind ourselves that the modulus is prime. The version that works for composite m is given in problem 8.

8.2 Fermat's Little Theorem

Here is the key theorem about modular exponentiation:

Theorem 8.1 (Fermat's Little Theorem). *If p is a prime number and a is not a multiple of p, then $a^{p-1} \equiv 1 \pmod{p}$.*

Remark 8.2. An equivalent statement, which doesn't rely on a not being a multiple of p, is that $a^p \equiv a \pmod{p}$. (See problem 1.) We will present two proofs of Fermat's Little Theorem, one of which will prove the version in Theorem 8.1, and one of which will prove the equivalent statement.

Proof 1. Let a and p be as in Theorem 8.1. Consider the numbers $a, 2a, 3a, \ldots, (p-1)a$. We claim that no two of these numbers are congruent to each other modulo p, and that none of them are congruent to $0 \pmod{p}$. First, suppose that two of them are congruent to each other modulo p, say $ar \equiv as \pmod{p}$. Since p is a prime and a is not a multiple of p, $\gcd(a, p) = 1$, so we may divide by a modulo p, i.e. $r \equiv s \pmod{p}$. But since $1 \leq r, s \leq p-1$, there isn't enough room for r to be congruent to s modulo p unless $r = s$. Thus the numbers $a, 2a, 3a, \ldots, (p-1)a$ are all different modulo p.

Next, we need to show that none of $a, 2a, 3a, \ldots, (p-1)a$ can be $0 \pmod{p}$. If one of them, say ar, is $0 \pmod{p}$, then either a or r must be a multiple of p. But we're assuming that a isn't a multiple of p, and since $1 \leq r \leq p-1$, it's also not a multiple of p. Thus none of these numbers is $0 \pmod{p}$.

Now, we have $p-1$ numbers $a, 2a, 3a, \ldots, (p-1)a$, and they are all distinct and nonzero modulo p. It follows that they are congruent to $1, 2, 3, \ldots, p-1 \pmod{p}$, in some order. Now, let's multiply them all together. We have

$$(a)(2a)(3a) \cdots ((p-1)a) \equiv 1 \cdot 2 \cdot 3 \cdots (p-1) \pmod{p}.$$

Let's rewrite the left side by grouping all the a's and then everything else. The other stuff is simply $(p-1)!$. Thus we have

$$a^{p-1} \cdot (p-1)! \equiv (p-1)! \pmod{p}.$$

Since $(p-1)!$ is not a multiple of p, we can divide both sides by $(p-1)!$ to get $a^{p-1} \equiv 1 \pmod{p}$, which is what we were trying to prove all along. ∎

Let's now give a completely different proof, this time of the alternative version $a^p \equiv a \pmod{p}$.

Proof 2. Let a be a positive integer, and let S be the set of sequences of length p whose elements are integers from 1 to a. In other words, a typical element of S is something of the form $(x_1, x_2, \ldots, x_p)$, where $1 \leq x_i \leq a$ for each i with $1 \leq i \leq p$. There are a choices for each x_i, so there are a total of a^p elements of S.

Given a sequence $x = (x_1, x_2, \ldots, x_p) \in S$, consider the *shift* f of x, which moves the first term to the end, i.e. $f(x) = (x_2, x_3, \ldots, x_p, x_1)$. We can keep shifting, so $f^2(x)$ is the double shift $(x_3, x_4, \ldots, x_p, x_1, x_2)$, and so forth. Note that $f^p(x) = x$, since if we shift p times then we're just back to where we started.

Now let's divide S into several smaller sets, called *orbits*, based on f: given $x \in S$, let's define

$$Y_x = \{x, f(x), f^2(x), \ldots, f^{p-1}(x)\}$$

to be the set of all shifts of x. Note that $Y_x = Y_{f(x)} = Y_{f^2(x)}$, and so forth, because these just consist of the same elements—but in a different order. Furthermore, note that each $x \in S$ is contained in exactly one of these orbits Y_x, where we consider two of them to be the same if they have exactly the same elements.

Now, note that every Y_x must contain either 1 or p elements. If x is a constant sequence of the form $(b, b, \ldots, b)$, then $x = f(x) = f^2(x) = \cdots = f^{p-1}(x)$, so Y_x consists of just the element x. But otherwise Y_x has exactly p distinct elements, because all of $x, f(x), f^2(x), \ldots, f^{p-1}(x)$ are different from each other. The number of Y_x's with exactly 1 element is simply a, since there are a choices of constant sequences.

Now, let's take all the Y_x's. Some of them have 1 element, and some have p. In fact, we know that a of them have 1 element, and let's say that n of them have p elements. In total, they contain a^p elements. Thus we have $a + pn = a^p$. Reducing modulo p, we get $a \equiv a^p \pmod{p}$, which is what we wanted to prove. ∎

See Figure 8.1 for an example of this proof when $a = 2$ and $p = 5$. For visual purposes, we have drawn red and blue dots instead of writing 1's and 2's. Each row consists of one complete orbit, i.e. one of the Y_x's.

Remark 8.3. Proof 2 only works when a is a positive integer (or perhaps a nonnegative integer, if you can figure out how to handle the case 0 appropriately), because it is not possible to have a sequence whose elements are integers from 1 to some negative integer. See if you can figure out an easy way of deducing the result for negative integers a if you already know the result for positive integers. This can be done in several ways.

8.3   Using Fermat's Little Theorem

How does Fermat's Little Theorem help us to simplify modular exponentiation problems? Let's look at an example.

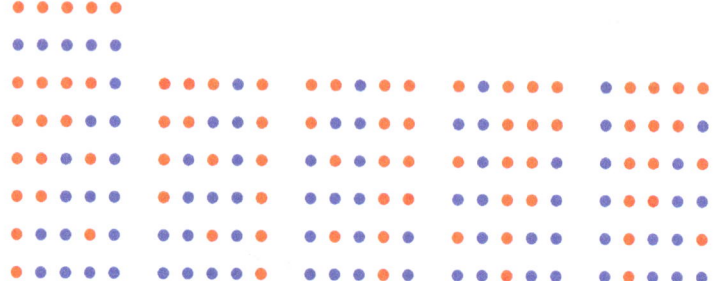

Figure 8.1. Proof 2 of Fermat's Little Theorem, in the case $a = 2$ and $p = 5$. Each row consists of one of the Y_x's.

Example. Let us compute 3^{100} (mod 7). By Fermat's Little Theorem, we know that $3^6 \equiv 1$ (mod 7). We can now write 3^{100} as $(3^6)^{16} \times 3^4$. Thus we have

$$3^{100} = (3^6)^{16} \times 3^4 \equiv 1^{16} \times 3^4 \equiv 3^4 \equiv 4 \quad (\text{mod } 7).$$

By the same logic, it follows that if p is prime, $a \equiv b$ (mod p), and $c \equiv d$ (mod $p - 1$), then $a^c \equiv b^d$ (mod p). Note that the congruence in the exponent is modulo $p - 1$, rather than modulo p. This is the initially-surprising consequence of Fermat's Little Theorem.

An even more surprising consequence of Fermat's Little Theorem is that we can use it to conclude that a number is composite, without ever finding a factor! Let's see how that works.

Example. Let us show that 57 is composite, without factoring it. If 57 were prime, then Fermat's Little Theorem would tell us that $2^{56} \equiv 1$ (mod 57). So, in order to conclude that 57 is composite, it suffices to check, somehow, that $2^{56} \not\equiv 1$ (mod 57). The easiest way of doing such computations is by repeated squaring: we have

$$2^{56} = 2^{32} \times 2^{16} \times 2^8,$$

since $56 = 32 + 16 + 8$. We don't need to know the exact values of 2^{32}, 2^{16}, and 2^8 though: we only have to know congruences modulo 57. To work those out, we just keep squaring, reducing modulo 57 whenever possible. We have:

$$2^1 = 2, \qquad 2^2 = 4, \qquad 2^4 = 16,$$

$$2^8 = 16^2 = 256 \equiv 28 \quad (\text{mod } 57).$$

So, that's the first reduction. Continuing on, we have

$$2^{16} = (2^8)^2 \equiv 28^2 \equiv 43 \pmod{57}.$$

Finally, we have

$$2^{32} = (2^{16})^2 \equiv 43^2 \equiv 25 \pmod{57}.$$

Now, we just have to compute $2^{56} \pmod{57}$:

$$2^{56} = 2^{32} \times 2^{16} \times 2^8 \equiv 28 \times 43 \times 25 \equiv 7 \times 25 \equiv 4 \pmod{57}.$$

Since that's not 1, we may conclude that 57 is composite, even without ever finding a factor.

This method doesn't always work to conclude that a number is composite. For instance, we may notice that 341 is composite, because $341 = 11 \times 31$. However, $2^{340} \equiv 1 \pmod{341}$. But we can pick some value of a other than 2 in Fermat's Little Theorem for this test. For instance, if we choose $a = 3$, then we may conclude that 341 is composite, because $3^{340} \equiv 56 \pmod{341}$.

Remark 8.4. Let us note that the *easy* way to check that $2^{340} \equiv 1$ $\pmod{341}$ is to use the Chinese Remainder Theorem, at least if we already know the factorization of 341. The Chinese Remainder Theorem tells us that $x \equiv 1 \pmod{341}$ if and only if both $x \equiv 1 \pmod{11}$ and $x \equiv 1 \pmod{31}$. So we have to check that $2^{340} \equiv 1 \pmod{11}$ and $2^{340} \equiv 1$ $\pmod{31}$. The first of these is very easy using Fermat's Little Theorem, for we have

$$2^{340} = (2^{10})^{34} \equiv 1^{34} \equiv 1 \pmod{11}.$$

For the second one, we note (just by pure calculation) that $2^5 \equiv 1$ $\pmod{31}$, so

$$2^{340} = (2^5)^{68} \equiv 1^{68} \equiv 1 \pmod{31}.$$

Thus $2^{340} \equiv 1 \pmod{341}$. Now, we can see what is special about $a = 2$: for any a that isn't a multiple of 11, we have $a^{340} \equiv 1 \pmod{11}$. However, usually $a^{340} \not\equiv 1 \pmod{31}$: it was something of a coincidence that $2^5 \equiv 1$ $\pmod{31}$.

Sometimes, however, a number n is composite even though when we choose *any* a—or at least any a that is relatively prime to n—we have $a^{n-1} \equiv 1 \pmod{n}$. Such numbers are called *Carmichael numbers*, and the first one is 561.

Theorem 8.5. *561 is a Carmichael number.*

Proof. The prime factorization of 561 is $561 = 3 \times 11 \times 17$. We must show that if $\gcd(a, 561) = 1$, then $a^{560} \equiv 1 \pmod{561}$. By the Chinese Remainder Theorem, this amounts to showing that $a^{560} \equiv 1 \pmod 3$, $a^{560} \equiv 1 \pmod{11}$, and $a^{560} \equiv 1 \pmod{17}$. We have

$$a^{560} = (a^2)^{280} \equiv 1^{280} \equiv 1 \pmod 3,$$

$$a^{560} = (a^{10})^{56} \equiv 1^{56} \equiv 1 \pmod{11},$$

and

$$a^{560} = (a^{16})^{35} \equiv 1^{35} \equiv 1 \pmod{17}.$$

This completes the proof. ∎

Remark 8.6. The first three Carmichael numbers are 561, 1105, and 1729. There are infinitely many Carmichael numbers, as shown in [AGP94].

So, we see that Fermat's Little Theorem can be used as a *compositeness test*: given an integer n, Fermat's Little Theorem can be used to prove that a number is composite. It cannot be used to prove that a number is prime, but there is a modified version called the *Miller–Rabin test* that can actually be used to prove primality, more or less. This is actually one of the main ways that primality testing is done on computers today, as it is faster than other known primality tests. See, for instance, Chapter 10 of [Sho09] for a detailed discussion of the Miller–Rabin test.

8.4 Wilson's Theorem

Another theorem of a similar flavor to Fermat's Little Theorem is Wilson's Theorem.

Theorem 8.7 (Wilson). *If p is a prime number, then $(p - 1)! \equiv -1 \pmod p$.*

The proof we'll see of Wilson's Theorem should remind you of the proofs we gave of Fermat's Little Theorem. Although the two proofs we saw of Fermat's Little Theorem were quite different, our proof of Wilson's Theorem will blend some ingredients of each.

Proof. By definition, $(p-1)! = 1 \times 2 \times 3 \times \cdots \times (p-2) \times (p-1)$. For any a with $1 \le a \le p - 1$, there is a unique b in this range with $ab \equiv 1 \pmod p$. Furthermore, $a = b$ if and only if $a^2 \equiv 1 \pmod p$, meaning that $a^2 - 1 \equiv 0 \pmod p$, or $(a - 1)(a + 1) \equiv 0 \pmod p$. Since p is prime, this means that either $a - 1$ or $a + 1$ is a multiple of p, i.e. $a \equiv \pm 1 \pmod{p}$. In the range of such a, this means that $a = b$ if and only if $a = 1$ or $a = p - 1$.

We next divide the elements of $\{1, 2, \ldots, p-1\}$ into sets $S_a = \{a, b\}$ such that $ab \equiv 1 \pmod{p}$. By the above analysis, we know that S_a consists of two elements, which multiply to 1 $\pmod{p}$, unless $a = 1$ or $a = p-1$. Thus, modulo p, we can compute $(p - 1)!$ by multiplying all the pairs modulo p, together with the two remaining elements 1 and $p - 1$, which are not part of any pair. Since each pair multiplies to 1 $\pmod{p}$, we have

$$(p - 1)! \equiv \prod 1 \times 1 \times (p - 1) \equiv -1 \pmod{p},$$

as desired. ∎

8.5 Further reading

Primality testing and factoring are both very important problems in cryptography. Thanks to the Miller–Rabin test and others like it, it is possible to test whether large numbers are prime on a computer very quickly. On the other hand, it does not appear to be possible to solve the related problem of actually factoring a (composite) number quickly, and the difficulty of factoring a number of around 1000 digits that is the product of two primes of around 500 digits each is crucial to making one of the most popular cryptosystems (the Rivest–Shamir–Adleman, or RSA, cryptosystem) work. It is possible to factor a large number significantly more quickly than just dividing by one prime after another, but still not fast enough to factor 1000-digit numbers. See for instance my book [Rub18] on cryptography for a discussion of RSA together with some of the clever methods of factoring.

8.6 Problems

(1) Explain why the two statements of Fermat's Little Theorem, i.e. Theorem 8.1 and the statement in Remark 8.2, are equivalent, by explaining how to deduce each one of them from the other one easily.

(2) Give another proof of Fermat's Little Theorem using induction and the identity

$$(x+1)^p = 1 + px + \frac{p(p-1)}{2!}x^2 + \frac{p(p-1)(p-2)}{3!}x^3 + \cdots + px^{p-1} + x^p,$$

which comes from the Binomial Theorem (see Chapter 17).

(3) Let $p_1, \ldots, p_k$ be distinct primes, and let d be the least common multiple of $p_1 - 1, p_2 - 1, \ldots, p_k - 1$. Prove that for any integer a, $a^{d+1} \equiv a \pmod{p_1 p_2 \cdots p_k}$.

(4) Prove that if p is prime, then

$$(p-1)! \equiv (p-1) \quad \left(\bmod \sum_{i=1}^{p-1} i \right).$$

(5) State and prove the converse of Wilson's Theorem.

(6) Let p be a prime. Prove that p is the *smallest* prime dividing $(p-1)!+1$.

(7) Prove that if p is a prime and a is not a multiple of p, then $a^{(p-1)p^{n-1}} \equiv 1 \pmod{p^n}$.

(8) Prove Euler's generalization of Fermat's Little Theorem, which says that if $\gcd(a, n) = 1$, then $a^{\phi(n)} \equiv 1 \pmod{n}$, where $\phi(n)$ is the totient function of n.

(9) Prove that if $6k + 1$, $12k + 1$, and $18k + 1$ are all primes, then their product $n = (6k + 1)(12k + 1)(18k + 1)$ is a Carmichael number.

(10) For nonnegative integers n and k with $n \geq k$, let $\binom{n}{k}$ denote the number of ways of selecting k objects from a set of size n; it is not too hard to check that

$$\binom{n}{k} = \frac{n!}{k!(n-k)!}.$$

Prove that if p is a prime, then $\binom{2p}{p} \equiv 2 \pmod{p^2}$. (Hint: Use a variant of the sequence-shifting proof we gave of Fermat's Little Theorem. How will you make all the orbits have size either 1 or p^2?)

Chapter 9

Pythagorean triples

9.1 Parametrizing Pythagorean triples

The Pythagorean Theorem is among the earliest mathematical theorems known to human civilization. It says that if a, b, and c are the sidelengths of a right triangle, with c being the hypotenuse, then $a^2 + b^2 = c^2$. (See Figure 9.1 for a hint of a proof of the Pythagorean Theorem.) A natural quest, then, is to find right triangles all of whose sides are integers, which is equivalent to searching for positive integer solutions to $a^2 + b^2 = c^2$. The most famous example is $(3, 4, 5)$, and indeed $3^2 + 4^2 = 5^2$. Other well-known examples are $(5, 12, 13)$, $(8, 15, 17)$, $(7, 24, 25)$, and $(9, 40, 41)$. We call such a triple (a, b, c) a *Pythagorean triple*.

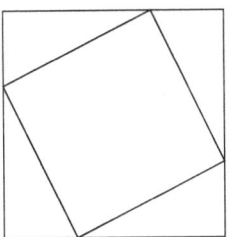

Figure 9.1. A pictorial proof of the Pythagorean Theorem.

Given any Pythagorean triple, there is an especially easy way to make more: just multiply all the numbers by some positive integer. For instance, if we start with the triple $(3, 4, 5)$ and multiply by 3, we get $(9, 12, 15)$, which is another Pythagorean triple. But, in some sense, this doesn't really feel like a *different* triple: it's just a scaled version of the original triple. For any such scaled family of triples, there is always one minimal one: the one where $\gcd(a, b, c) = 1$ (or $\gcd(a, b) = 1$, which is equivalent). We say that a Pythagorean triple (a, b, c) is *primitive* if $\gcd(a, b, c) = 1$.

An interesting problem is to determine all the primitive Pythagorean triples. What this means is that we want a function that, when we plug something in, we get a primitive Pythagorean triple as an output, in such a way that every primitive Pythagorean triple can be formed via this process. This is something of a lofty goal. Ordinarily, we cannot hope to write down all solutions to such an equation in a convenient manner. It is quite magical that in the case of primitive Pythagorean triples, there is a simple way of doing this.

A lot of deep mathematics from number theory and algebraic geometry goes into explaining, in some sense, why we can parametrize primitive Pythagorean triples—but not necessarily solutions to other sorts of equations. Or, if we can in theory parametrize solutions to other types of equation—because, for instance, there are only finitely many—it requires a lot of work and clever proofs in order to guarantee that we have found all of them. One very famous question along these lines is *Hilbert's Tenth Problem*, which asks whether there is an algorithm to determine whether a polynomial equation in several variables has a solution in integers. We now know (see [Mat70]) that there is no such algorithm, which can be interpreted casually as saying that number theory is hard: we can't write a computer program to solve number theory problems for us in an automatic manner.

Okay, so let's see how to parametrize primitive Pythagorean triples! Let's start with the equation $a^2 + b^2 = c^2$ and divide by c^2 to get

$$\left(\frac{a}{c}\right)^2 + \left(\frac{b}{c}\right)^2 = 1.$$

If we now let $x = \frac{a}{c}$ and $y = \frac{b}{c}$, we get $x^2 + y^2 = 1$, which is the equation of a circle. Because we obtained x and y by dividing two integers, we're no longer restricting to *integer* solutions for x and y, but rather *rational* solutions. Switching from integer solutions to rational solutions might not seem like an improvement, but on the plus side, we've reduced the number of variables from 3 to 2, and we're now working with a circle, which is a very familiar object.

In order to find rational points on the circle, we have to start with one rational point to get the process off the ground. It doesn't matter which one we choose, but the most popular choice for solving this problem is the point $(-1, 0)$, so that's what we'll use.

Now, given any *other* rational point (x_0, y_0) on the circle, consider the line through $(-1, 0)$ and (x_0, y_0). Because both of these points are rational, so is the slope $\frac{y_0}{x_0+1}$. Thus every rational point on the circle corresponds to a rational slope. Furthermore, because a line and a circle can only intersect in at most two points, every rational point gives us a different rational slope.

Now, what happens if we take an *arbitrary* line through $(-1, 0)$ with rational slope? Does this line intersect the circle at another rational point? It does! There are two ways of seeing this. The first way—the more conceptual and less computational way—is to note that we're solving a quadratic equation with rational coefficients in order to find the x-coordinate. This doesn't mean that the solutions are rational, because the quadratic formula has a square root in it. However, one of the solutions, namely $x = -1$, is rational, so the other one must be as well. Then, because this point (x_0, y_0) has rational x-coordinate, and the slope and x-intercept are rational, the y-coordinate is as well.

If you didn't like that explanation, that's okay, because we have a second explanation, which is that we simply find the point. We'll need to do this in order to parametrize primitive Pythagorean triples anyway, so let's get on with it. Suppose the line has slope t. Since it passes through the point $(-1, 0)$, this means its equation is $y = t(x + 1)$. We're interested in the second point, the one other than $(-1, 0)$, of intersection between this line and the circle $x^2 + y^2 = 1$. That is, we need to solve the simultaneous equations

$$y = t(x+1), \qquad x^2 + y^2 = 1.$$

To do this, substitute the first equation into the second to get

$$x^2 + t^2(x+1)^2 = 1,$$

or

$$(1 + t^2)x^2 + 2t^2 x + (t^2 - 1) = 0.$$

Applying the quadratic formula to solve for x, we get

$$x = \frac{-2t^2 \pm \sqrt{4t^4 - 4(t^4 - 1)}}{2(1 + t^2)} = \frac{-2t^2 \pm 2}{2(1 + t^2)} = -1, \frac{1 - t^2}{1 + t^2}.$$

The first solution, $x = -1$, corresponds to the point $(-1, 0)$, so we're only interested in the second solution $x = \frac{1-t^2}{1+t^2}$. Since $y = t(x+1)$, the y-coordinate of this point is

$$y = t\left(\frac{1 - t^2}{1 + t^2} + 1\right) = \frac{2t}{1 + t^2}.$$

Thus the second point of intersection is

$$\left(\frac{1-t^2}{1+t^2}, \frac{2t}{1+t^2}\right).$$

What we have shown is that there is a one-to-one correspondence between rational numbers (slopes) on the one hand, and points on the circle $x^2 + y^2 = 1$ with rational coordinates, if we exclude the point $(-1, 0)$, on the other hand. See Figure 9.2 for an illustration of this correspondence.

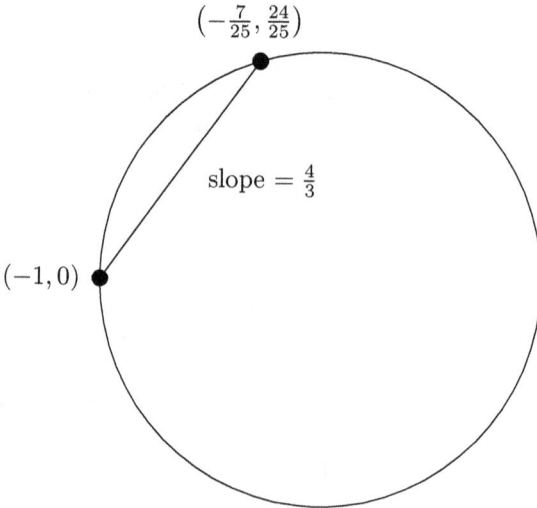

Figure 9.2. Parametrizing rational points on the circle in terms of rational slopes.

Since t is a rational number, we can write it as $t = \frac{m}{n}$, where m and n are relatively prime integers. With this notation, our rational point on the circle is

$$\left(\frac{1-m^2/n^2}{1+m^2/n^2}, \frac{2m/n}{1+m^2/n^2}\right),$$

or

$$\left(\frac{n^2-m^2}{n^2+m^2}, \frac{2mn}{n^2+m^2}\right).$$

Now, keep in mind that this is a point on the circle $x^2 + y^2 = 1$. We really want to know about primitive Pythagorean triples: relatively prime solutions to $a^2 + b^2 = c^2$. To convert from a rational point on the circle to a primitive Pythagorean triple, we can just clear denominators: if

$$\left(\frac{n^2 - m^2}{n^2 + m^2}\right)^2 + \left(\frac{2mn}{n^2 + m^2}\right)^2 = 1,$$

then

$$(n^2 - m^2)^2 + (2mn)^2 = (n^2 + m^2)^2. \tag{9.1}$$

We can easily check that if m and n are any integers, then (9.1) holds, so we get a Pythagorean triple. But two questions still remain. First, is this triple primitive? Second, do all primitive Pythagorean triples arise in this way?

Let us answer the first question first. We want to know if the triple $(n^2 - m^2, 2mn, n^2 + m^2)$ is primitive, meaning that we want to know whether the gcd of two of these numbers is equal to 1 or not. The easiest ones to work with are the first and third, so let's determine the gcd of $n^2 - m^2$ and $n^2 + m^2$. Suppose a prime p divides both $n^2 - m^2$ and $n^2 + m^2$, so that there are integers k and ℓ such that

$$n^2 - m^2 = pk,$$
$$n^2 + m^2 = p\ell.$$

Then we have

$$2n^2 = p(k + \ell), \qquad 2m^2 = p(\ell - k).$$

In other words, p divides both $2n^2$ and $2m^2$. Since m and n are assumed to be relatively prime, the only prime that can divide both $2n^2$ and $2m^2$ is 2. We check that 2 divides both $n^2 - m^2$ and $n^2 + m^2$ if and only if m and n are either both even or both odd. The both even case can't happen, because we assumed that m and n are relatively prime, and hence not both even. But the both odd case is something we have to watch out for: that case gives us Pythagorean triples that aren't primitive, because a, b, and c are all even. Thus we find that this construction gives us a primitive Pythagorean triple if and only if m and n are relatively prime, and one of m and n is even and the other is odd.

Now, let's answer the second question: Do all primitive Pythagorean triples arise in this way? We know that if (a, b, c) is a (primitive) Pythagorean triple, then $\left(\frac{a}{c}, \frac{b}{c}\right)$ is a rational point on the circle $x^2 + y^2 = 1$, so there exist m and n such that $\frac{a}{c} = \frac{n^2 - m^2}{n^2 + m^2}$ and $\frac{b}{c} = \frac{2mn}{n^2 + m^2}$. But this really means that, for some nonzero integer k, we have

$$ka = n^2 - m^2$$
$$kb = 2mn$$
$$kc = n^2 + m^2.$$

It turns out that we can't really guarantee that $k = 1$. For instance, consider the Pythagorean triple $(4, 3, 5)$ in that order. We cannot find integers m and n such that $n^2 - m^2 = 4$, $2mn = 3$, and $n^2 + m^2 = 5$, because (among other reasons) the middle equation $2mn = 3$ is problematical. But we can avoid that particular issue. If (a, b, c) is a primitive Pythagorean triple, then exactly one of a and b is even. Let's make a convention that b will always be the even one. Now, with that in mind, can we always find integers m and n such that $a = n^2 - m^2$, $b = 2mn$, and $c = n^2 + m^2$?

Yes, we can! To do so, let's just find them. Since we're assuming b is even, let's write $b = 2d$. Since $b^2 = c^2 - a^2$, we have

$$4d^2 = c^2 - a^2 = (c - a)(c + a),$$

and we want this to be equal to $(2mn)^2$. Since both a and c are odd, $c - a$ and $c + a$ are both even, so let's divide by 4 to get

$$d^2 = \frac{c - a}{2} \cdot \frac{c + a}{2}.$$

Since a and c are relatively prime, so are $\frac{c-a}{2}$ and $\frac{c+a}{2}$. (To see this, note that if g divides both $\frac{c-a}{2}$ and $\frac{c+a}{2}$, then g also divides $\frac{c+a}{2} - \frac{c-a}{2} = a$ and $\frac{c+a}{2} + \frac{c-a}{2} = c$.) Since their product is a perfect square, they must each be perfect squares individually, so let's set $m = \sqrt{\frac{c-a}{2}}$ and $n = \sqrt{\frac{c+a}{2}}$. Thus $d^2 = m^2 n^2$, or $d = mn$. Since $b = 2d$, we have $b = 2mn$, and m and n are relatively prime.

Let's check that this actually works. If $m = \sqrt{\frac{c-a}{2}}$ and $n = \sqrt{\frac{c+a}{2}}$, then

$$n^2 - m^2 = \frac{c+a}{2} - \frac{c-a}{2} = a,$$

$$2mn = 2\sqrt{\frac{c-a}{2}} \cdot \sqrt{\frac{c+a}{2}} = 2\sqrt{\frac{c^2-a^2}{4}} = 2\sqrt{\frac{b^2}{4}} = b,$$

and

$$n^2 + m^2 = \frac{c+a}{2} + \frac{c-a}{2} = c.$$

Putting all this together, we have proven the following theorem.

Theorem 9.1. *Let (a, b, c) be a primitive Pythagorean triple, where b is even. Then there exist relatively prime positive integers m and n such that*

$$a = n^2 - m^2, \qquad b = 2mn, \qquad c = n^2 + m^2.$$

Conversely, if m and n are relatively prime positive integers such that one of m and n is even and one of them is odd, and $n > m$, then

$$(n^2 - m^2, 2mn, n^2 + m^2)$$

is a primitive Pythagorean triple.

9.2 Descent and Fermat's Last Theorem for $n = 4$

One neat thing we can do with the parametrization of primitive Pythagorean triples is to prove one case of *Fermat's Last Theorem*.

Theorem 9.2 (Fermat's Last Theorem). *Let n be a positive integer greater than 2. Then there are no positive integers x, y, z such that $x^n + y^n = z^n$.*

Of course, when $n = 2$, there are many solutions, because these are just the Pythagorean triples. Fermat's Last Theorem is one of the most famous and difficult problems in mathematics, and it took over 350 years from the time Fermat stated it until Andrew Wiles finally proved it. Wiles's proof is extremely difficult and uses much of the machinery of modern number theory and algebraic geometry.

The only case that Fermat himself proved was the case $n = 4$ (and by extension, all multiples of 4, since if $x^{4k} + y^{4k} = z^{4k}$ is a solution with $n = 4k$, then $(x^k)^4 + (y^k)^4 = (z^k)^4$ is a solution with $n = 4$). Fermat showed that there are no triples (x, y, z) of positive integers such that $x^4 + y^4 = z^4$. In fact, he showed something a bit stronger: There are no triples of positive integers (x, y, z) such that $x^4 + y^4 = z^2$, which is better because all fourth powers are automatically squares.

The proof uses the method of *descent*, which is really a special kind of hybrid induction and contradiction. What Fermat did was to show that if we start with a solution to $x^4 + y^4 = z^2$, then we can produce another solution, which is smaller in the sense that z is smaller. But that process can't go on forever: there must be a smallest one. To make this more precise, we suppose that we start with a solution, and it's the smallest one possible. Then we produce a smaller one. But this contradicts our hypothesis that we started with the smallest one. This technique can also be phrased in terms of induction, where the induction hypothesis is that there are no solutions for a given value of z. Let's see how that proof goes.

Theorem 9.3. *There are no positive integers x, y, z such that $x^4 + y^4 = z^2$.*

Proof. Suppose that positive integers x, y, z satisfy $x^4 + y^4 = z^2$, and that this solution is minimal in the sense that there is no solution with a smaller value of z. Let us write $a = x^2$ and $b = y^2$, so that $a^2 + b^2 = z^2$. In other words, (a, b, z) is a Pythagorean triple. If both a and b are odd, then a^2 and b^2 are both 1 (mod 4), so $z^2 \equiv 2$ (mod 4), but squares can never be 2 (mod 4). Thus a and b can't both be odd. On the other hand, if a and b are both even, then x and y are also even, and z is a multiple of 4, so we can write $x = 2x'$, $y = 2y'$, and $z = 4z'$. Thus we have

$$16x'^4 + 16y'^4 = 16z'^2,$$

or

$$x'^4 + y'^4 = z'^2,$$

which means that (x', y', z') is a smaller solution than (x, y, z), contradicting our hypothesis.

Thus we may assume that one of a and b is even, and the other one is odd, and that (a, b, z) is a primitive Pythagorean triple. Without loss of generality, let's say that a is odd and b is even. Thus, by the parametrization of primitive Pythagorean triples, there are integers m and n such that

$$a = n^2 - m^2 \qquad b = 2mn \qquad z = n^2 + m^2.$$

Let us look at $a = n^2 - m^2$. Since $a = x^2$, we have $x^2 = n^2 - m^2$, so that (x, m, n) is another Pythagorean triple. It is also primitive, so that means there exist integers s and t such that

$$x = t^2 - s^2 \qquad m = 2st \qquad n = t^2 + s^2.$$

Recall that $b = y^2$, so we have

$$y^2 = b = 2mn = 4st(t^2 + s^2).$$

Now, $4st$ and $t^2 + s^2$ are relatively prime, so they are both perfect squares, say $4st = u^2$ and $t^2 + s^2 = v^2$.

Next, because $4st = u^2$, u is even, so we have $st = \left(\frac{u}{2}\right)^2$. Since s and t are relatively prime, they are both squares, say $s = k^2$ and $t = \ell^2$. Plugging these into the equation $t^2 + s^2 = v^2$, we have

$$\ell^4 + k^4 = v^2.$$

Thus we have produced another solution to $x^4 + y^4 = z^2$. It remains only to show that $v < z$. To see this, note that

$$v \le v^2 = t^2 + s^2 = n < n^2 + m^2 = z,$$

as desired. Thus we have reached a contradiction, and the proof is complete. ∎

9.3 Problems

(1) Let a and b be two positive integers that differ by 2. Consider the number $\frac{1}{a} + \frac{1}{b}$. For instance, if a and b are 4 and 6, you get $\frac{1}{4} + \frac{1}{6} = \frac{5}{12}$. What does this have to do with Pythagorean triples? Make a precise statement along these lines, and prove that it is correct.

(2) Prove that if (a, b, c) is a primitive Pythagorean triple where a is odd, then $2(c - a)$ is a perfect square.

(3) Find, with proof, a parametrization for all relatively prime positive integer solutions to $a^2 + 2b^2 = c^2$.

(4) For which primes p do there exist positive integers x, y, and n such that $p^n = x^3 + y^3$?

(5) Find two primitive Pythagorean triples (a, b, c) with $c = 65$. Find two other values of c for which there are at least two primitive Pythagorean triples (a, b, c) with the same value of c, like in the case of $c = 65$.

(6) Find a value of c for which there are at least three primitive Pythagorean triples of the form (a, b, c).

(7) Find, with proof, all integer solutions to $x^3 + 2y^3 = 4z^3$.

(8) Prove that there are no solutions in positive integers to $x^4 - y^4 = z^2$.

(9) Let $f : \mathbb{N} \to \mathbb{N}$ be a function such that $f(n + 1) > f(f(n))$ for all $n \in \mathbb{N}$. Prove that $f(n) = n$ for all n.

(10) Prove that it is not possible to divide a cube into a finite number of smaller cubes, in such a way that all the small cubes have a different sidelength.

(11) Find infinitely many triangular numbers that are also perfect squares. (A triangular number is a number of the form $\frac{n(n+1)}{2}$, where n is a positive integer.)

Chapter 10

Sums of two squares

10.1 Sums of two squares

Our main goal this chapter is to determine which integers can be represented as sums of two squares. That is:

Question 10.1. *For which integers n do there exist integers x and y such that $x^2 + y^2 = n$?*

We will be able to answer this question completely, but it will take a little while. Let us start with some observations. The first one, which is really obvious, is that any square is nonnegative, so if n can be represented as a sum of two squares, then $n \geq 0$.

The second observation is that, if we pick a modulus m, then only some of the residue classes $a \pmod{m}$ are squares. Thus, for good choices of m, we can learn something about which integers definitely cannot be written as sums of two squares. For example, one can check that any square is $0, 1 \pmod 4$, and adding two of these numbers gives a result that is $0, 1, 2 \pmod 4$. Thus if $n \equiv 3 \pmod 4$, then n cannot be written as a sum of two squares.

Theorem 10.2. *Suppose m and n are both sums of two squares. Then so is mn.*

To see this, we simply write down a representation of mn as a sum of two squares based on the representations of m and n.

Proof. If $m = a^2 + b^2$ and $n = c^2 + d^2$, then

$$mn = (ac - bd)^2 + (ad + bc)^2$$

is a sum of two squares. ∎

That proof was rather disappointing, wasn't it? The problem is that it lacks insight: we can easily verify that it is correct, but how would we think of it? Often, proofs are written in such a way as to hide the thought process behind them. This tradition of writing proofs that are completely correct but unenlightening goes back to Gauß.[1] Gauß was criticized for doing this, to which he responded that no architect leaves the scaffolding after completing the building.

What we learn from this is that there is a major difference between a correct argument and an insightful one. Sometimes correct proofs are insightful, and sometimes they aren't. When writing the solution to a homework problem, in which the goal is to show the reader that *you* understand what is going on, you do not necessarily have to provide any insight. But when writing a paper or book—in which the goal is to teach the reader what is going on—if your proof is not insightful, then it ought to be accompanied by some further remarks explaining what the insight is.

Indeed, the proof of Theorem 10.2 hides some very useful and beautiful insight. In fact, I can never remember how to write down the squares in the proof, so I have to rederive it every time. Here is how I derive it.

Let us suppose that $m = a^2 + b^2$ and $n = c^2 + d^2$. Then, assuming we allow ourselves the use of $i = \sqrt{-1}$, we can factor m and n:

$$m = (a + bi)(a - bi), \qquad n = (c + di)(c - di).$$

Thus, when we multiply them together, we get

$$mn = (a + bi)(c + di)(a - bi)(c - di).$$

Multiplying out the first two terms on the right and the last two terms, we get

$$mn = ((ac - bd) + (ad + bc)i)((ac - bd) - (ad + bc)i)$$
$$= (ac - bd)^2 + (ad + bc)^2.$$

We have just written mn as a sum of two squares!

Theorem 10.2 suggests that it sensible to start with the primes when determining which numbers can be written as a sum of two squares. That is, which primes can be written as sums of two squares? We know that if all the prime factors of n can be written as sums of two squares, then so can n. The converse is not exactly true, though: 9 can be written as a sum of two squares even though 3 cannot. However, as we will see, there is something close to a converse available.

---

[1] The letter "ß" is called an "Eszett." It is a German letter (or, more precisely, ligature) used for a double-s under certain circumstances. People less pretentious than I will write "Gauss" rather than "Gauß."

10.2   Primes congruent to 1 (mod 4)

We know from the above discussion that numbers congruent to 3 (mod 4) cannot be written as sums of two squares. The primes left to be considered are 2 (which can clearly be written as a sum of two squares, namely $2 = 1^2 + 1^2$) and the primes congruent to 1 (mod 4). We will show that if $p \equiv 1$ (mod 4) is prime, then p can be written as a sum of two squares. But first, we'll prove something weaker that will be handy later.

Lemma 10.3. *If $p \equiv 1$ (mod 4), then there is some number x such that $x^2 + 1$ is a multiple of p. If $p \equiv 3$ (mod 4), then there is no such x.*

Another way of stating this is to say that -1 is a *quadratic residue* modulo p if $p \equiv 1$ (mod 4), but -1 is a *quadratic nonresidue* modulo p if $p \equiv 3$ (mod 4). By definition, a number a is a quadratic residue modulo a prime p if there is some integer x such that $x^2 \equiv a$ (mod p). Similarly, a is a quadratic nonresidue modulo p if there is no integer x such that $x^2 \equiv a$ (mod p). Some people add to their definition of quadratic residue that a must not be a multiple of p; there is good reason for doing so, but it won't affect us here, so we'll not dwell on this point.

Proof. Let us look at the numbers $\{1, 2, \ldots, p - 1\}$, which are considered to be the nonzero residue classes modulo p. We break these numbers up into subsets $\{x, -x, \bar{x}, -\bar{x}\}$, where these numbers are taken modulo p. Here $\bar{x}$ denotes the *multiplicative inverse* of x, i.e. the number such that $x\bar{x} \equiv 1$ (mod p). Usually, these subsets consist of four distinct numbers, but sometimes two of them are equal. Let us investigate:

- If $x \equiv -x$ (mod p), then $2x \equiv 0$ (mod p), which means that $x \equiv 0$ (mod p) since p is odd. But 0 is not in our set. Thus it cannot happen that $x \equiv -x$ (mod p).
- If $x \equiv \bar{x}$ (mod p), then $x^2 \equiv 1$ (mod p), so that $x \equiv 1$ (mod p) or $x \equiv p - 1$ (mod p). When either of these are true, then the subset containing x becomes $\{1, p - 1\}$ and hence has size 2.
- If $x \equiv -\bar{x}$ (mod p), then $x^2 \equiv -1$ (mod p). When this happens, then $x \equiv -\bar{x}$ and $-x \equiv \bar{x}$, so the subset containing x becomes $\{x, -x\}$ and has size 2. Such a subset exists if and only if there is some x such that $x^2 \equiv -1$ (mod p).

So, we see that most of the subsets have size 4, and there is one subset $\{1, p - 1\}$ of size 2, and potentially one more $\{x, -x\}$ of size 2, if $x^2 \equiv -1$

(mod p) has a solution modulo p. Since the total number of elements is $p - 1$, we find that this additional subset of size 2 exists if $p \equiv 1$ (mod 4) and does not exist if $p \equiv 3$ (mod 4). In other words, $x^2 \equiv -1$ (mod p) has a solution if $p \equiv 1$ (mod 4) and does not have a solution if $p \equiv 3$ (mod 4). ∎

Recall that in Chapter 3 we proved that there are infinitely many primes congruent to 3 (mod 4). The technique we used there is not applicable for primes congruent to 1 (mod 4). But with the help of Lemma 10.3, we are now able to prove that there are also infinitely many primes congruent to 1 (mod 4).

Corollary 10.4. *There are infinitely many primes congruent to* 1 (mod 4).

Proof. Suppose we have any finite list of primes congruent to 1 (mod 4), say $p_1, \ldots, p_k$. Then consider the number

$$M = 4 \left(\prod_{i=1}^{k} p_i \right)^2 + 1.$$

By Lemma 10.3, if p is a prime dividing M, then $p \equiv 1$ (mod 4). Furthermore, none of $p_1, \ldots, p_k$ divides M. So there must be more primes congruent to 1 (mod 4) not among $p_1, \ldots, p_k$. ∎

We are now ready to show that if $p \equiv 1$ (mod 4), then p be can be written as a sum of two squares.

Theorem 10.5. *If $p \equiv 1$ (mod 4) is prime, then there are integers x and y with $x^2 + y^2 = p$.*

We will present two proofs. The first uses Lemma 10.3. The second one is a shorter "pure magic" proof.

Proof 1. Let $X = \{(x', y') : 0 \leq x', y' \leq \lfloor \sqrt{p} \rfloor\}$. Then $|X| = (1 + \lfloor \sqrt{p} \rfloor)^2 > p$. Thus by the pigeonhole principle, for any s, we can find two elements (x', y') and $(x'', y'') \in X$ such that

$$x' - sy' \equiv x'' - sy'' \pmod{p}.$$

Rewriting this, we have

$$x' - x'' \equiv s(y' - y'') \pmod{p}.$$

Let $x = |x' - x''|$ and $y = |y' - y''|$; thus $x \equiv \pm sy$ (mod p), or $x^2 \equiv s^2 y^2$ (mod p). Note also that x and y are not both zero. Now, by Lemma 10.3, there is some s such that $s^2 \equiv -1 \pmod{p}$; choose one. Then $x^2 \equiv -y^2$ (mod p), or $x^2 + y^2 \equiv 0 \pmod{p}$. Now, $0 \le x, y \le \lfloor \sqrt{p} \rfloor$, so $0 \le x^2, y^2 < p$. It cannot be that both x^2 and y^2 are 0, so $0 < x^2 + y^2 < 2p$. Since $x^2 + y^2$ is divisible by p and is strictly between 0 to $2p$, it must be exactly p. So, we have found a solution to $x^2 + y^2 = p$. ∎

We now give another proof of Theorem 10.5. This one, due to Zagier in [Zag90] and based on ideas of Heath-Brown and Liouville, is just a single sentence long and doesn't even use Lemma 10.3.

Proof 2: Zagier's Proof. The involution on the finite set $S = \{(x, y, z) \in \mathbb{N}^3 : x^2 + 4yz = p\}$ defined by

$$(x, y, z) \mapsto \begin{cases} (x + 2z, z, y - x - z) & \text{if } x < y - z \\ (2y - x, y, x - y + z) & \text{if } y - z < x < 2y \\ (x - 2y, x - y + z, y) & \text{if } x > 2y \end{cases}$$

has exactly one fixed point, so $|S|$ is odd and the involution defined by $(x, y, z) \mapsto (x, z, y)$ also has a fixed point. ∎

Okay, so what just happened there? Let's take it apart slowly, defining our terms and analyzing what is going on.

Definition 10.6. Let X be a set and $f : X \to X$ a bijection. Then f is said to be an *involution* if $f \circ f = id_X$.

If f is an involution on a finite set X, then we can pair up the elements of X into subsets of size 1 or 2: $P_x = \{x, f(x)\}$. Note that $P_{f(x)} = \{f(x), f(f(x))\} = \{x, f(x)\} = P_x$. Furthermore, P_x has size 1 if and only if $x = f(x)$. Let us write $\text{Fix}_f(X)$ for the subset of X that is fixed by f: $\text{Fix}_f(X) = \{x \in X : x = f(x)\}$. Then, since $|P_x|$ is even unless $x = f(x)$, we have

$$|X| \equiv |\text{Fix}_f(X)| \pmod 2.$$

An important consequence is that if f and g are *both* involutions on X, then $|\text{Fix}_f(X)| \equiv |\text{Fix}_g(X)| \pmod 2$. In particular, if f has exactly one fixed point, then g must have *at least* one fixed point.

Let us investigate Zagier's weird involution, and call it f. Suppose $x < y - z$, so that $f(x, y, z) = (x + 2z, z, y - x - z)$. Since $x > 0$ for all

$(x, y, z) \in S$, $x + 2z > 2z$, so $f(x, y, z)$ is in the third case. Thus, we have

$$f(f(x, y, z)) = f(x + 2z, z, y - x - z)$$
$$= (x + 2z - 2z, x + 2z - z + y - x - z, z)$$
$$= (x, y, z),$$

so that—at least in this range—f is an involution. Similar checks show that f is a full involution: it switches the first and third cases, and it sends the second case to itself. Thus, if f has any fixed points, they must be in the second case. Let's investigate.

Suppose $y - z < x < 2y$ and $f(x, y, z) = (x, y, z)$, so that $(x, y, z) = (2y - x, y, x - y + z)$. This implies that $x = y$, and since $x^2 + 4yz = p$, we have $y(y + 4z) = p$. But p is prime, so this can only happen if $x = y = 1$, and then $z = \frac{p-1}{4}$. For these values of x, y, z, we do indeed have $y - z < x < 2y$. So, f does in fact have exactly one fixed point, which means that $|S|$ is odd.

Then, there's the other involution $g(x, y, z) = (x, z, y)$. Since $|S|$ is odd, *any* involution on S has an odd number of fixed points, and in particular at least 1. Thus g has some fixed point, which is of the form (x, y, y), so that $x^2 + 4y^2 = p$. But this is a representation of p as a sum of two squares!

In fact, this proof shows even more than we bargained for: not only can we write $p = x^2 + 4y^2$, where x and y are positive integers, but the number of ways of doing that is *odd*. In fact, it turns out that there is exactly one way of doing so. (See problem 9.) However, neither proof gives us much of a clue about how to *find* x and y so that $x^2 + 4y^2 = p$. Actually finding x and y is more challenging. One way of doing so is using the Tonelli–Shanks algorithm; see [NZM91, §2.9].

10.3   A geometric interpretation of Zagier's involution

The involution f in Zagier's proof might feel unmotivated. However, there is an elegant geometric interpretation of it due to Spivak [Spi07]. An element $(x, y, z) \in S$ can be thought of as an $x \times x$ square, together with four $y \times z$ rectangles, one based at each corner of the square. For instance, the element $(3, 1, 8) \in S$ for $p = 3^2 + 4 \times 1 \times 8 = 41$ can be pictured as in Figure 10.1.

Given such a pictorial representation, we can eliminate the boundary lines of all the rectangles, as shown in Figure 10.2.

But given such a diagram, there are two ways of resolving it into a square with four rectangles emanating from it, with the other one being shown in

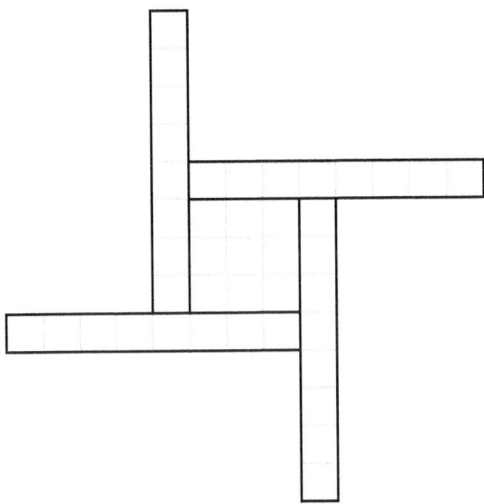

Figure 10.1. A pictorial representation of the element $(3, 1, 8) \in S$.

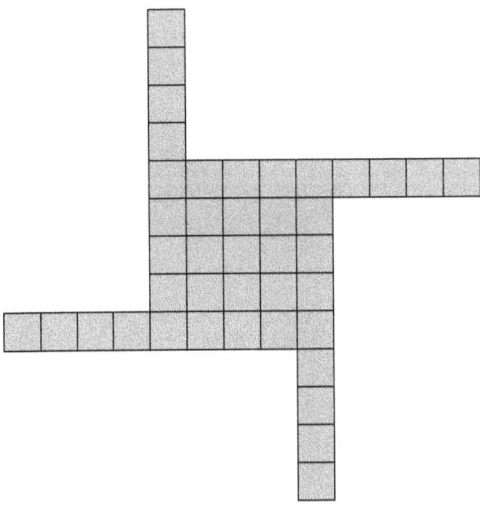

Figure 10.2. The representation without the boundary lines.

Figure 10.3. There is one exception, when there is only one resolution, and this is the triple $(1, 1, \frac{p-1}{4})$.

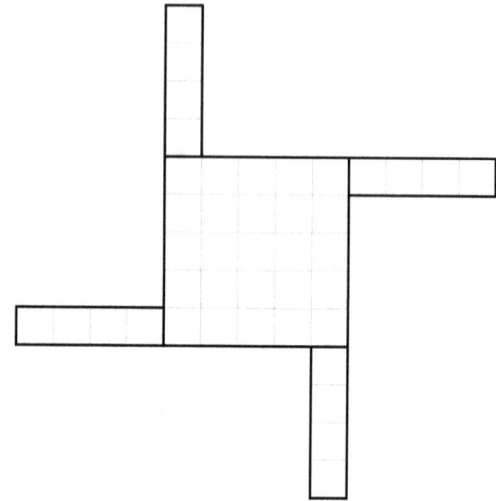

Figure 10.3. The other way of decomposing the figure.

What Zagier's involution does is to swap the two representations. See Figure 10.4 for the involution when $p = 41$. In this case, we have

$$S = \{(1, 1, 10), (1, 2, 5), (1, 5, 2), (1, 10, 1), (3, 1, 8), (3, 2, 4),$$
$$(3, 4, 2), (3, 8, 1), (5, 1, 4), (5, 2, 2), (5, 4, 1)\}.$$

10.4　Composite numbers

Now we are ready to tackle the original question: which positive integers can be represented as the sum of two squares? And we can answer it completely:

Theorem 10.7. *A positive integer n can be written as a sum of two squares if and only if every prime $p \equiv 3 \pmod{4}$ occurring in the prime factorization of n occurs with even exponent.*

This means the following: suppose the prime factorization of n is $p_1^{e_1} p_2^{e_2} \cdots p_k^{e_k}$, where the p_i's are distinct primes. Then n can be written as the sum of two squares if and only if, for each i with $p_i \equiv 3 \pmod{4}$, e_i is even.

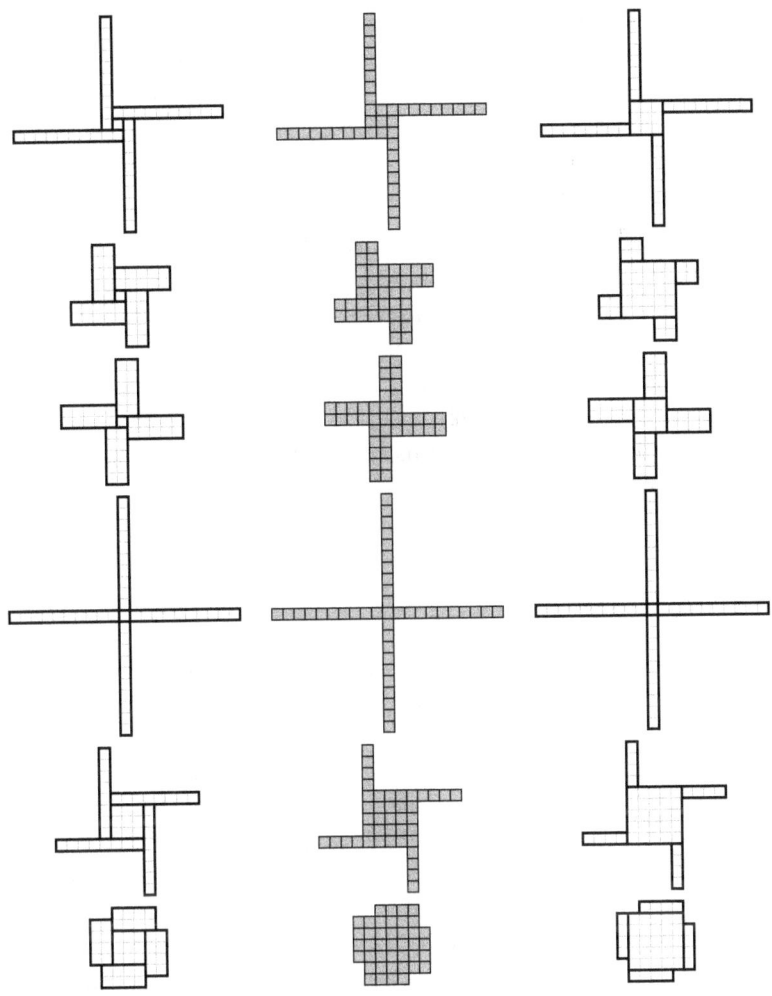

Figure 10.4. Zagier's involution when $p = 41$.

Proof. Let us show first that if every prime $p \equiv 3 \pmod 4$ occurring in the prime factorization of n occurs with even exponent, then n is a sum of two squares. This is the easy direction, because of everything we have already done. If $p \equiv 1 \pmod 4$ is a factor of n, then we know that p can be written as a sum of two squares. If $p \equiv 3 \pmod 4$ is a factor of n, then certainly p^2 can be written as a sum of two squares, namely as $p^2 + 0^2$. Finally, 2 can be written as a sum of two squares, as $2 = 1^2 + 1^2$. Thus n is a product of numbers that can each be written as sums of two squares, and hence by Theorem 10.2 n can be too.

Now let's do the other direction. Suppose that n can be written as the sum of two squares, as $n = x^2 + y^2$. It suffices to show that if $p \equiv 3$ (mod 4) divides n, then p must divide both x and y. This is because, if that happens, then $\frac{n}{p^2}$ is also a sum of two squares, namely $\frac{n}{p^2} = \left(\frac{x}{p}\right)^2 + \left(\frac{y}{p}\right)^2$, and the result follows by induction on n. In order to see why p must divide both x and y, let us suppose otherwise: $p \nmid x$. Then we can find some $\bar{x}$ such that $x\bar{x} \equiv 1$ (mod p). Let us multiply the equation $n = x^2 + y^2$ by $\bar{x}^2$ to obtain $n\bar{x}^2 = (x\bar{x})^2 + (y\bar{x})^2$. Now, reducing modulo p, we have $0 \equiv 1 + (y\bar{x})^2$ (mod p). But this means that $(y\bar{x})^2 \equiv -1$ (mod p), and we saw earlier in Lemma 10.3 that this cannot happen when $p \equiv 3$ (mod 4). And we're done!   ∎

10.5   Further reading

We have seen how to determine the numbers that can be written in the form $x^2 + y^2$, where x and y are integers. We can ask infinitely many related questions. For instance, given an integer d, which integers can be written in the form $x^2 + dy^2$? Indeed, it turns out that there is an entire (excellent) book [Cox13] on the topic.

Can Zagier's proof be generalized to these other cases? Elsholtz in [Els10] used an argument similar to Zagier's to determine the primes that can be written as $x^2 + 2y^2$ and $x^2 + 3y^2$.

Another interesting approach is Conway's topograph method. This is a pictorial method, in which one draws a graph to determine the numbers represented by $x^2 + y^2$, or some similar expression. See Conway's book [CF97] to learn about it.

10.6   Problems

(1) Determine the primes $p < 100$ for which p can be written as $a^2 + ab + b^2$ for some integers a and b. Make a reasonable conjecture (that you expect to be true) about which primes can be written as $a^2 + ab + b^2$ based on your data.

(2) The number 3024961 factors as $3024961 = 29 \times 104309$, where both of those factors are prime. You are given that $3024961 = 519^2 + 1660^2$. Use this to find a and b such that $a^2 + b^2 = 104309$.

(3) Use the fact that a prime $p \equiv 1 \pmod 4$ can be written as a sum of two squares and the fact that the product of two integers that can be written as sums of two squares can also be written as a sum of two squares to conclude that if $p \equiv 1 \pmod 4$, then there is a primitive Pythagorean triple (a, b, p), i.e. relatively prime integers a and b with $a^2 + b^2 = p^2$. (For example, we have the familiar triples $(3, 4, 5)$, $(5, 12, 13)$, $(8, 15, 17)$, $(20, 21, 29)$.)

(4) We showed that if m and n are sums of two squares, then so is mn. Show this is not true for sums of three squares by finding m and n that are sums of three squares, but mn is not a sum of three squares.

(5) Make a list of numbers up to 100 and determine which ones can be written as sums of three squares. Make a reasonable conjecture (that you expect to be true) based on your data about which numbers can be written as a sum of three squares.

(6) Show that if n can be represented as a sum of squares of two *rational* numbers, then it can also be represented as a sum of squares of two integers.

(7) If $p \equiv 1 \pmod 4$ is prime, then use Wilson's Theorem to find an integer n such that $n^2 \equiv -1 \pmod p$.

(8) Prove that a prime $p \equiv 1 \pmod 8$ can be written in the form $x^2 + 2y^2$. (Hint: break up the numbers $\{1, 2, \ldots, p-1\}$ into sets of the form $\{x, \bar{x}, -x, -\bar{x}, ix, -ix, i\bar{x}, -i\bar{x}\}$, where $i^2 \equiv -1 \pmod p$. Then analyze how many elements these sets have.) It turns out that primes $p \equiv 3 \pmod 8$ can also be written in the form $x^2 + 2y^2$, but I don't know of such a simple proof of this fact.

(9) We showed that a prime $p \equiv 1 \pmod 4$ can be written as a sum of two squares. Show that this representation is essentially unique, by explaining how to factor n if we know two representations $n = a^2 + b^2 = c^2 + d^2$ as a sum of two squares. This is known as Euler's factorization method. (Hint: if $n = a^2 + b^2 = c^2 + d^2$, then factor both sides using $\sqrt{-1}$, and then take gcds of appropriate factors.) Use this to factor $1000009 = 1000^2 + 3^2 = 972^2 + 235^2$.

Chapter 11

An introduction to p-adic numbers

11.1   Lifting congruences

In general, reduction modulo m is a very common way to guarantee that some equation has no integer solutions. This doesn't always work, as there are examples of equations that have solutions modulo m for every m, yet still have no integer solutions. But it's a good start. What is less clear is whether we can use reduction to find solutions when they do exist.

Let us consider the polynomial $f(x) = x^2 - 31x - 102$. We can easily find the roots of this polynomial using the quadratic formula, but in the interest of doing something new, let us attempt to find its roots differently. Let us pick some number m and try to find the roots of this sample polynomial modulo m. In general, this sort of thing works best when m is a prime, so then it's more natural to call it p. Let's (arbitrarily) try $p = 5$.

Does the equation $x^2 - 31x - 102$ have any solutions modulo 5? It's easier to tell when we reduce all the coefficients modulo 5, so we get $x^2 - x - 2$. The roots of this modulo 5 are 2 and $-1 \equiv 4$. We have two possibilities here, so let's choose 4. (This is not completely at random, and we'll see why soon.) Our next aim is to try to find a root of f congruent to 4 (mod 5), i.e. some a such that $f(a) = 0$ and $a \equiv 4$ (mod 5). To do this, we try to determine what this root should be modulo higher and higher powers of 5. What happens modulo 25? Can we find a "lift" of the root 4 (mod 5) to a root modulo 25?

Any lift of 4 (mod 5) to something modulo 25 can be written as $5a_1 + 4$, where $a_1 \in \{0, 1, 2, 3, 4\}$. So, let's try plugging that into f: we want to find an a_1 such that

$$(5a_1 + 4)^2 - 31(5a_1 + 4) - 102 \equiv 0 \quad (\text{mod } 25).$$

Reducing the coefficients modulo 25, we get

$$15a_1 + 16 - 5a_1 + 1 - 2 \equiv 0 \pmod{25},$$

or

$$10a_1 + 15 \equiv 0 \pmod{25}.$$

Dividing by 5, we get

$$2a_1 + 3 \equiv 0 \pmod{5},$$

which has the unique solution $a_1 = 1$, so $5a_1 + 4 = 9$. Thus, we have now lifted our solution 4 (mod 5) to 9 (mod 25).

Onward! Let us now lift to a solution modulo $5^3 = 125$. This time, we look for a number of the form $25a_2 + 9$ such that

$$(25a_2 + 9)^2 - 31(25a_2 + 9) - 102 \equiv 0 \pmod{125},$$

where $a_2 \in \{0, 1, 2, 3, 4\}$. We find that $a_2 = 1$, so the solution is 34 (mod 125). We can keep on going, modulo 625, 3125, and so forth, except that 34 is actually a root of the original polynomial, so we're done.

But remember, there's another root, the one that's 2 (mod 5). If we try the same thing there, we sequentially learn that the solution is 2 (mod 5), 22 (mod 25), 122 (mod 125), 622 (mod 625), and so forth. In fact, the solution is -3.

11.2   The p-adic numbers and their arithmetic

When we use the above method for solving an equation $f(x) = 0$, we are really trying to find integers $a_0, a_1, a_2, \ldots \in \{0, 1, \ldots, p-1\}$ such that

$$f\left(\sum_{i=0}^{n-1} a_i p^i\right) \equiv 0 \pmod{p^n},$$

for all n. In base p, these partial solutions are the numbers $a_{n-1}a_{n-2} \cdots a_2 a_1 a_0$. It makes sense, then, to put together all the a_n's into one big base-p number: $\cdots a_n a_{n-1} a_{n-2} \cdots a_2 a_1 a_0$. If $a_n = 0$ for all sufficiently large n, then we have a nonnegative integer in its usual base-p expansion. If $a_n = p-1$ for all sufficiently large n, then we have a *negative* integer in an unusual form.

An especially closed-minded person would elect to restrict the sorts of numbers we may consider to those two types: the integers that we already know and love. But in this formulation, that seems a bit artificial: there's no reason why we have to place restrictions on what all but finitely many of the

a_n's can be, when the universe doesn't force us to do that. We call a number of the form $\cdots a_n a_{n-1} a_{n-2} \cdots a_2 a_1 a_0$, where each $a_i \in \{0, 1, \ldots, p-1\}$, a *p-adic number* or a *p-adic integer*. We denote the set of *p*-adic integers by $\mathbb{Z}_p$.

Given a new system of numbers, we would like to make sure that we can do arithmetic with it. Fortunately, we can. Addition, subtraction, and multiplication work pretty much the same way as they do for decimal numbers, except that we have to remember that everything is now in base *p*. Let's look at examples of addition, subtraction, and multiplication in $\mathbb{Z}_5$.

Example (Addition). We have

	$\cdots$	4	1	3	2	0	1
+	$\cdots$	2	0	3	3	4	4
	$\cdots$	1	2	2	1	0	0

As usual, we "carry" whenever a sum in a column exceeds $p-1$, and all addition is modulo p.[1]

Example (Subtraction). We have

	$\cdots$	3	1	2	2	0	3
−	$\cdots$	4	1	0	4	2	4
	$\cdots$	4	0	1	2	2	4

As usual, we "borrow" when computing $a - b$ when $a < b$.

Note also that $-1 = \cdots 44444$:

	$\cdots$	0	0	0	0	0	0
−	$\cdots$	0	0	0	0	0	1
	$\cdots$	4	4	4	4	4	4

---

[1] What is carrying? It turns out that it is most naturally interpreted as a cocycle, i.e. something that represents a class in group cohomology. See Daniel Isaksen's beautiful paper [Isa02] for an explanation.

Example (Multiplication). We have

$$
\begin{array}{rccccccc}
       & \cdots & 3 & 4 & 2 & 1 & 0 & 2 \\
\times & \cdots & 2 & 4 & 4 & 1 & 0 & 3 \\
\hline
       & \cdots & 1 & 3 & 1 & 3 & 1 & 1 \\
       & \cdots & 0 & 0 & 0 & 0 & 0 &   \\
       & \cdots & 2 & 1 & 0 & 2 &   &   \\
       & \cdots & 4 & 1 & 3 &   &   &   \\
       & \cdots & 1 & 3 &   &   &   &   \\
       & \cdots & 4 &   &   &   &   &   \\
\hline
       & \cdots & 3 & 4 & 0 & 0 & 1 & 1 \\
\end{array}
$$

Division is a little bit different, because of the unfortunate algorithm most of us learned when we were very young.[2] For addition, subtraction, and multiplication, our usual algorithms start with the least significant (rightmost) digits and work their way up to the most significant (leftmost) digit. But our algorithm for division works differently, starting with the most significant digit and working down. Since numbers in $\mathbb{Z}_p$ do not have most significant digits—they go on to the left forever—this approach does not work.

However, we can also perform division starting from the least significant digit, when our numbers are written in a *prime* base. I don't know how to make a neat table for it like the way we originally get taught how to do long division, so we'll have to do it without the picture.

Let us compute $\frac{1}{3}$ in $\mathbb{Z}_5$. This means solving the equation $3x = 1$ in $\mathbb{Z}_5$. First, we solve it modulo 5, and we find that the (unique) solution is $x \equiv 2$ (mod 5). So we write $x = 2 + 5a_1 + \cdots$. To find a_1, we solve $3(2 + 5a_1) \equiv 1$ (mod 25), or $6 + 15a_1 \equiv 1$ (mod 25) or $1 + 3a_1 \equiv 0$ (mod 5), so $a_1 = 3$. Next, we find a_2 such that $2 + 3 \times 5 + 25a_2$ is an inverse of 3 modulo 125, so $3(2 + 3 \times 5 + 25a_2) \equiv 1$ (mod 125), or $51 + 75a_2 \equiv 1$ (mod 125), or $2 + 3a_2 \equiv 0$ (mod 5), so $a_2 = 1$. Continuing on, we find that, in base 5,

$$
\frac{1}{3} = \cdots 3131313132.
$$

Let's check that this makes sense: if the last digit were a 1 instead of a 2, it would just be 313131 repeating forever. We should then add 1 to get

---

[2]When you are actually doing arithmetic, I recommend avoiding these algorithms whenever possible, as they are slow and error-prone. Instead, I recommend first trying to find tricks specific to the numbers at hand to perform the computations more quickly and accurately.

$\frac{1}{3}$. In other words,

$$\frac{1}{3} = \sum_{n=0}^{\infty} (3 \times 5 + 1) \times 25^n + 1,$$

which is good, because we know how to sum geometric series:

$$\sum_{n=0}^{\infty} (3 \times 5 + 1) \times 25^n + 1 = \frac{16}{1 - 25} + 1 = -\frac{16}{24} + 1 = \frac{1}{3},$$

as desired.[3]

11.3 $\mathbb{Q}_p$

The division algorithm we discussed works whenever the denominator is not a multiple of p. More precisely, if $\frac{m}{n}$ is a fraction in lowest terms, then it is in $\mathbb{Z}_p$ as long as $p \nmid n$. In fact, this works even if $m, n \in \mathbb{Z}_p$, rather than just $\mathbb{Z}$, although "lowest terms" loses at least a bit of its meaning in that case. But what if $p \mid n$? For instance, what is $\frac{1}{p}$ in $\mathbb{Z}_p$?

Well, nothing, because our familiar process can't get started. To find the first digit, we have to solve $a_0 p \equiv 1 \pmod{p}$, which has no solutions. But we shouldn't have expected that *everything* would be in $\mathbb{Z}_p$, should we? Inspired by what we do with ordinary real numbers, we introduce a ... p-mal point? (It would be silly to call it a decimal point when we're writing our numbers in base p, wouldn't it?) As a base-p number, $\frac{1}{p}$ is $\cdots 000.1$.

More generally, a p-adic rational number is a base-p number of the form

$$\cdots a_3 a_2 a_1 a_0 . a_{-1} a_{-2} \cdots a_{-k},$$

where there are only *finitely* many digits to the *right* of the p-mal point. In terms of a sum, a p-adic rational number can be written as

$$\sum_{n=-k}^{\infty} a_n p^n.$$

We write $\mathbb{Q}_p$ for the set of all p-adic rational numbers; the relationship between $\mathbb{Q}_p$ and $\mathbb{Z}_p$ is the same as that between $\mathbb{Q}$ and $\mathbb{Z}$, namely that every element of $\mathbb{Q}_p$ can be expressed as $\frac{a}{b}$ where $a, b \in \mathbb{Z}_p$ and $b \neq 0$, and that every such fraction is in $\mathbb{Q}_p$.

---

[3]Is this okay? Should we be concerned about convergence? Yes, it is okay, and yes, we should be concerned about convergence! Convergence in $\mathbb{Z}_p$ means the same "sort" of thing as it does in $\mathbb{R}$, but our notion of "closeness" is different. However, we will not develop this point further in this book.

If $a, b \in \mathbb{Z}_p$ and $b \neq 0$, then $\frac{a}{b} \in \mathbb{Q}_p$. To see this, write $b = p^\beta b'$, where b' is not divisible by p. Then

$$\frac{a}{b} = \frac{a}{p^\beta b'} = \frac{1}{p^\beta}\frac{a}{b'},$$

so that $\frac{a}{b'} \in \mathbb{Z}_p$ and multiplication by $\frac{1}{p^\beta}$ just shifts the p-mal point by β positions.

11.4 Some p-adic irrational numbers

Now, not all p-adic numbers are rational. This is because (see problem 7) the rational numbers have eventually periodic digits, but we can certainly find sequences of digits which are not eventually periodic. Some of them are irrational and algebraic: roots of polynomial equations. Let's have a look and try to find a root of the quadratic polynomial $f(x) = x^2 - x + 3$ in $\mathbb{Z}_5$. Note that the roots in $\mathbb{C}$ are $\frac{1 \pm \sqrt{-11}}{2}$.

We solve this equation just as we did in the introductory example: by looking for a solution modulo 5, then trying to lift it modulo higher and higher powers of 5. To do the first step, we just check all five possibilities and see if any of them work:

$$0^2 - 0 + 3 = 3 \not\equiv 0 \pmod 5$$
$$1^2 - 1 + 3 = 3 \not\equiv 0 \pmod 5$$
$$2^2 - 2 + 3 = 5 \equiv 0 \pmod 5$$
$$3^2 - 3 + 3 = 9 \not\equiv 0 \pmod 5$$
$$4^2 - 4 + 3 = 15 \equiv 0 \pmod 5.$$

Thus the two solutions modulo 5 are 2 and 4. Let's work with the one that is 2 (mod 5).

To lift it, we attempt to find an a_1 such that $f(5a_1 + 2) \equiv 0 \pmod{25}$. Plugging this in, we need that

$$(5a_1 + 2)^2 - (5a_1 + 2) + 3 \equiv 0 \pmod{25},$$

or, after expanding and simplifying,

$$20a_1 + 4 - 5a_1 - 2 + 3 = 15a_1 + 5 \equiv 0 \pmod{25}.$$

Dividing by 5, we need that $3a_1 + 1 \equiv 0 \pmod 5$, so $a_1 = 3$. Thus the solution is $3 \times 5 + 2 \pmod{25}$, or 17 (mod 25). Note that we didn't get stuck here. We will investigate why very shortly, after we go through just one more step of the procedure.

Continuing along, we need to find a_2 such that $f(25a_2 + 17) \equiv 0$ (mod 125). Plugging this into the polynomial, we get

$$(25a_2 + 17)^2 - (25a_2 + 17) + 3 \equiv 0 \quad (\text{mod } 125),$$

or

$$100a_2 + 39 - 25a_2 - 17 + 3 \equiv 75a_2 + 25 \equiv 0 \quad (\text{mod } 125),$$

or $3a_2 + 1 \equiv 0$ (mod 5). So $a_2 = 3$. Doing this procedure infinitely many times, we eventually find a 5-adic root of $x^2 - x + 3$. Thus we have found a quadratic irrational number in $\mathbb{Z}_5$.

Now, let us stop for a moment to think about why we always have a unique way to lift a solution modulo 5^n to a solution modulo 5^{n+1}. Note that, at each step, we only have to solve a *linear* equation modulo 5, i.e. an equation of the form $ca + d \equiv 0$ (mod 5), where c and d are fixed. This equation *always* has a unique solution, *unless* $c = 0$. So, why isn't $c = 0$? This amounts to finding the coefficient of a_{n-1} modulo 5^n in $f(a_{n-1}5^{n-1} + \cdots + a_1 \times 5 + a_0)$. We can lump all the rest of the terms together, so that we have $f(5^{n-1}a + b) \equiv 0$ (mod 5^n). The coefficient of a is then $2 \times 5^{n-1}b - 5^{n-1}$; this is nonzero modulo 5^n *unless* $2b - 1 \equiv 0$ (mod 5), or $b \equiv 3$ (mod 5). But that is not the case: $b \equiv 2$ (mod 5). Thus we can never get stuck, and in fact we will always have a *unique* lift. As soon as we have a solution modulo 5, we can *guarantee* that we will have a *unique* solution all the way up to $\mathbb{Z}_5$.

11.5   Hensel's Lemma

Let us now think more generally about when we will get stuck. Let us start with a polynomial $f(x)$, and we are looking for a p-adic root of $f(x) = 0$. Let us assume that we have found a root a_0 modulo p, and we wish to lift it all the way to a root in $\mathbb{Z}_p$. Can we do so?

Let us suppose that $f(x) = c_d x^d + c_{d-1}c^{d-1} + \cdots + c_1 x + c_0$, and let us suppose that we have a root b modulo p^n above a_0 that we wish to lift to a root modulo p^{n+1} by finding some a_n such that $f(a_n p^n + b) \equiv 0$ (mod p^{n+1}). When we expand out, we won't be concerned about the coefficient of a_n^2 or higher powers of a_n, because those will all be divisible by p^{n+1}; all that matters is the coefficient of a_n. So, we compute it: by the Binomial Theorem (see Chapter 17), it is

$$p^n \left(dc_d b^{d-1} + (d-1)c_{d-1}b^{d-2} + \cdots + 2c_2 b + c_1 \right).$$

We need to make sure that this is nonzero modulo p^{n+1}, i.e. that

$$\left(dc_d b^{d-1} + (d-1)c_{d-1}b^{d-2} + \cdots + 2c_2 b + c_1\right) \not\equiv 0 \pmod{p}.$$

If you know calculus, you will recognize this as saying that $f'(b) \not\equiv 0$ (mod p). We have now proven *Hensel's Lemma*:

Theorem 11.1 (Hensel's Lemma). *Let $f(x)$ be a polynomial with coefficients in $\mathbb{Z}_p$, and suppose a_0 is such that $f(a_0) \equiv 0$ (mod p). If $f'(a_0) \not\equiv 0$ (mod p), then there is a unique $a \in \mathbb{Z}_p$ such that $a \equiv a_0$ (mod p) and $f(a) = 0$.*

Note that some of the hypotheses here can fail. One thing that can go wrong is that we might not have a root modulo p. For example, if $f(x) = x^2 + 2$ and $p = 5$, then there is no root modulo 5, so there is certainly no root of f in $\mathbb{Z}_5$. Another thing that can go wrong is that f has a root a_0 modulo p, but the derivative $f'(a_0)$ is 0 (mod p). For example, $f(x) = x^2 + 2$ has no root in $\mathbb{Z}_2$, even though $f(0) \equiv 0$ (mod 2).

11.6 Further reading

The p-adic numbers are very important in modern number theory, as important as the real and complex numbers. Thus there is a lot to be said about them. An in-depth study of them will involve a blend of number theory, analysis, and topology, and in some ways they behave quite differently from the real and complex numbers. For instance, in the p-adic numbers, all triangles are isosceles! A good introductory book on the p-adic numbers is [Gou20], while a magical more advanced book is [Kob12].

11.7 Problems

(1) Compute $\frac{1}{7}$ in $\mathbb{Z}_3$.
(2) Is 3 a multiple of 5 in $\mathbb{Z}_{11}$? Prove it.
(3) Prove that every nonzero element of $\mathbb{Q}_p$ can be written uniquely in the form $p^n r u$, where $n \in \mathbb{Z}$, $r \in \{1, 2, \ldots, p-1\}$, and $u \in 1 + p\mathbb{Z}_p = \{u \in \mathbb{Z}_p : u \equiv 1 \pmod{p}\}$. What change do you have to make to represent only the nonzero elements of $\mathbb{Z}_p$?
(4) Describe the elements of $\mathbb{Q} \cap \mathbb{Z}_p$. Prove that your answer is correct.
(5) For which values of $a \in \mathbb{Z}$ does the polynomial $x^2 + x + a$ have a root in $\mathbb{Q}_2$? Prove that your answer is correct.
(6) How many roots does $x^4 + 3x^2 + 3$ have in $\mathbb{Z}_7$?

(7) Show that an element of $\mathbb{Q}_p$ (or $\mathbb{Z}_p$) is in $\mathbb{Q}$ if and only if its digits are eventually periodic, i.e. if they eventually repeat like we saw for $\frac{1}{3}$ in $\mathbb{Z}_5$.

(8) Find a polynomial $f(x)$ with integer coefficients such that f has a root $\alpha \in \mathbb{Z}/p\mathbb{Z}$ that does not lift to a root in $\mathbb{Z}_p$. Find a polynomial $g(x)$ with integer coefficients such that g has a root $\beta \in \mathbb{Z}/p\mathbb{Z}$ that lifts to two or more distinct roots in $\mathbb{Z}_p$.

(9) Let p be a prime. Show that there are exactly p solutions to $x^p = x$ in $\mathbb{Z}_p$. Let T denote the set of solutions. Show that T is closed under multiplication: if $x, y \in T$, then $xy \in T$. The elements of T are called *Teichmüller representatives*.

(10) Let $a \in \mathbb{Z}_p$, and let t_a be the Teichmüller representative congruent to a modulo p. Show that, for each positive integer n, $t_a \equiv a^{p^n}$ (mod p^{n+1}). Thus you can approximate the Teichmüller representative in some congruence class by taking an arbitrary element of that congruence class and raising it to some large power of p.

(11) Show that if p is an odd prime and $a \equiv 1$ (mod p), then a has a square root in $\mathbb{Z}_p$. On the other hand, show that 3 and 5 are not squares in $\mathbb{Z}_2$.

PART 2
Combinatorics

Chapter 12

Additive and multiplicative problems

12.1   The art of counting

Combinatorics is all about counting: determining the number of ways to perform some task, order a list, and so forth. There are many ways of counting. The simplest, of course, is to make a list of all the possibilities, and then, well, count: 1, 2, 3, and so on. For very simple problems, when the number is quite small, this is a completely satisfactory method of counting.

However, there are several issues that prevent this simple approach from being a good general problem-solving strategy. One problem is that the number we're trying to count to may be quite large. A typical example of this is Archimedes's stomachion problem (Οστομάχιον in Greek), as shown in Figure 12.1. The challenge here is to determine the number of ways the displayed polygons (triangles, quadrilaterals, and pentagons) can be arranged into a square, with no holes and no overlaps. Archimedes, the great mathematician from ancient Greece, asked this question. Only recently, Cutler determined that the number of such arrangements is exactly 17152 if we count all solutions, or 536 if we count them up to symmetry. We don't want to have to count that high!

Another issue with this approach is the challenge of actually making a list. When we list out all the possibilities, we want to do so in such a way that we can be completely sure that every possibility is on the list, exactly once. This requires good organization; without a method for constructing our list in an organized manner, we are likely to miss some possibilities, or else list some possibilities twice. Try to imagine for a moment how we would construct a list for the stomachion problem so as to be absolutely certain that we have accounted for all possibilities, and there are no duplicates. It's not an easy problem at all!

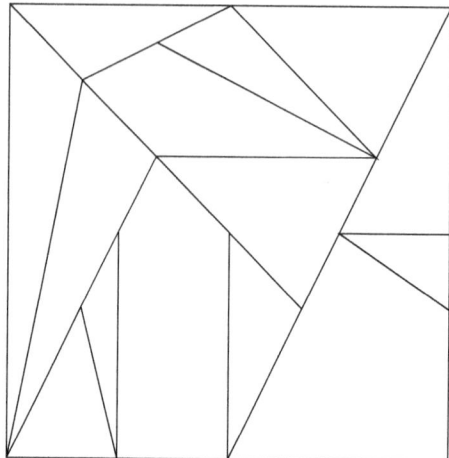

Figure 12.1. Archimedes's stomachion problem: how many ways are there to arrange the pieces to form a square?

There is one more issue with the enumerated list approach. What happens if, instead of being asked for the number of solutions to a *single* problem, we are asked for a *function* that describes the number of solutions to a *family* of problems? A typical simple example, which we'll examine in detail very shortly, is to determine the number of subsets of the set $[n] = \{1, 2, \ldots, n\}$. We can make lists to work out small cases:

$n = 1$: $\varnothing$, $\{1\}$
$n = 2$: $\varnothing$, $\{1\}$, $\{2\}$, $\{1, 2\}$
$n = 3$: $\varnothing$, $\{1\}$, $\{2\}$, $\{3\}$, $\{1, 2\}$, $\{1, 3\}$, $\{2, 3\}$, $\{1, 2, 3\}$.

We find that the number of subsets of $[n]$ when $n = 1, 2, 3$ is $f(1) = 2$, $f(2) = 4$, and $f(3) = 8$, simply by counting the lists we created. But we can't make a list for *each* n, because there are infinitely many. Instead, we need a better strategy.

The most prized enumeration method in combinatorics is the *bijective* method. We have some set S, possibly described in a complicated manner, and we want to know how many elements it has. To do this, we find some other set T, such that we *already* know how many elements T has, and then we find a function $f : S \to T$ that is *bijective*. Let's recall exactly what that means.

Definition 12.1. Let S and T be two sets, and let $f : S \to T$ be a function.

- We say that f is *surjective* or *onto* if, for every $t \in T$, there is some $s \in S$ such that $f(s) = t$.
- We say that f is *injective* or *one-to-one* if, whenever $s \neq s'$, we have $f(s) \neq f(s')$.
- We say that f is *bijective* if it is both injective and surjective.

See Figure 12.2 for pictures of surjective and injective functions.

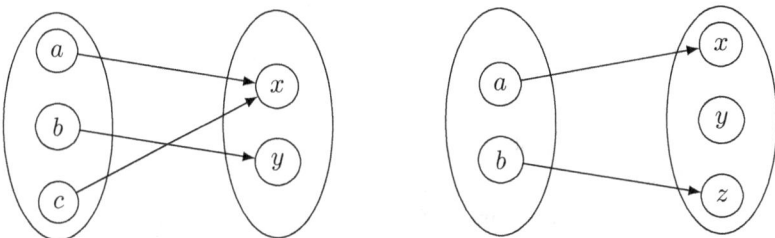

Figure 12.2. Left: a surjective function. Right: an injective function.

The key point here is that *if there is a bijection between two sets S and T, then they have the same number of elements.* The challenge, of course, is coming up with a suitable bijection, i.e. an easily countable T and a bijective function $f : S \to T$. Sometimes, this can be an extremely difficult problem, depending on the nature of S.

12.2 Additive problems

Sometimes, when solving counting problems, we can break the problem down into several sub-problems, and then add the results. Let's start with a typical (easy) example.

Question 12.2. *You have three cats and four dogs. How many cats and dogs total do you have?*

The answer is 7, of course, but let's delve a bit deeper and understand exactly what's going on, and how to interpret this in terms of bijections. The non-bijective way of doing this would be to take the three cats and four dogs, line them up, and count them. That's a pretty silly idea, because they probably won't sit still for long enough to count them. As soon as you have done even something as simple as giving the animals names (or representing them as pictures), you've already created a bijection.

So, let's give them names. Instead of giving them normal cat and dog names, let's give them names that will be easier to work with from a mathematical point of view. We'll call the cats C_1, C_2, C_3 and the dogs D_1, D_2, D_3, D_4. Thus we have created a bijection from our set of animals to the set $A = \{C_1, C_2, C_3, D_1, D_2, D_3, D_4\}$. Thus the two sets have the same number of elements. If we feel that the set A is easy enough to count, then we can stop there. If not, we can make things even clearer. Consider the set $[7] = \{1, 2, 3, 4, 5, 6, 7\}$. We create a bijection between A and $[7]$ as follows: define a function $f : A \to [7]$ by setting $f(C_i) = i$ for $1 \le i \le 3$ and $f(D_i) = i + 3$ for $1 \le i \le 4$.

To check that f is a bijection, we show that it is both surjective and injective. Let's start by showing that f is surjective. This means that for every $j \in [7]$, we must show that there is some $a \in A$ such that $f(a) = j$. If $1 \le j \le 3$, then $j = f(C_j)$, and if $4 \le j \le 7$, then $j = f(D_{j-3})$. That takes care of all cases, so f is surjective. To show that f is injective, we need to show that if $a \ne b$, then $f(a) \ne f(b)$. Up to symmetry, we have three cases:

Case 1: a and b are both cats. Say $a = C_i$ and $b = C_j$. Then $f(a) = i$ and $f(b) = j$, so if $f(a) = f(b)$, then $i = j$, and so they are the same animal.

Case 2: a is a cat and b is a dog. Then $f(a) \le 3$ and $f(b) \ge 4$, so $f(a) \ne f(b)$.

Case 3: a and b are both dogs. Say $a = D_i$ and $b = D_j$. Then $f(a) = i+3$ and $f(b) = j + 3$. If $f(a) = f(b)$, then $i + 3 = j + 3$, or $i = j$, so they are the same animal.

This completes the proof of injectivity. Since we have shown that f is both injective and surjective, it is bijective.

More generally, if we had c cats and d dogs, then we would have a total of $c + d$ animals.

Needless to say, we won't usually be *quite* that detailed with our proofs or quite so pedantic about our bijections.

Remark 12.3. We will often use the notation $[n]$ for the integers from 1 to n. From now on, we will simply use that notation whenever needed, rather than specifying what it means on each occasion.

12.3   Multiplicative problems

Sometimes when solving counting problems, we break the problem down into several sub-problems, then *multiply* the results.

Question 12.4. *You have three cats and four dogs. How many ways are there to pick one cat and one dog?*

Clearly, there are three ways to pick a cat, and four ways to pick a dog, but then what? How do we create a bijection between cat–dog pairs and something else?

Similarly to what we did before, we have a bijection between the cats and [3], and we have a bijection between the dogs and [4]. We therefore have a bijection between the cat–dog pairs and the set $[3] \times [4]$, which consists of all ordered pairs (i, j) where $1 \leq i \leq 3$ and $1 \leq j \leq 4$.

However, it might still not be entirely obvious how many elements $[3] \times [4]$ has, so we'd like to describe a bijection between $[3] \times [4]$ and $[n]$ for some integer n. It turns out that the right integer is $n = 12$, so let's write down a bijection between $[3] \times [4]$ and $[12]$. One possible bijection is as follows:

$$f(1, 1) = 1 \qquad f(2, 1) = 5 \qquad f(3, 1) = 9$$
$$f(1, 2) = 2 \qquad f(2, 2) = 6 \qquad f(3, 2) = 10$$
$$f(1, 3) = 3 \qquad f(2, 3) = 7 \qquad f(3, 3) = 11$$
$$f(1, 4) = 4 \qquad f(2, 4) = 8 \qquad f(3, 4) = 12.$$

Or, if you want a "formula" for f,

$$f(i, j) = 4(i - 1) + j.$$

Evidently, f is a bijection, so there are 12 cat–dog pairs.

More generally, of course, if we had c cats and d dogs, then we would have cd cat–dog pairs.

How could we have expected in advance that this problem would be a multiplicative problem, whereas the previous one would be an additive problem? This comes from experience, but a pretty good heuristic[1] is that "or" problems (pick one of these things, or one of those things, but not both) are additive, whereas "and" problems (pick one of these things, and one of those things) are multiplicative.

---

[1] A heuristic is a method for making educated guesses about how to solve problems. Heuristics aren't proofs and aren't acceptable substitutes for proofs, but generally when working on a math problem, you'll want to start with a heuristic and then turn it into a proof later.

Multiplicative problems are more common than additive problems, so let's look at some more. Recall that in §12.1 we considered the problem of determining the number of subsets of $[n]$. A subset T of a set S is a set such that all the elements of T are elements of S. In other words, it consists of some elements from S and nothing from outside of S. A subset can be empty, i.e. it is allowed not to have any elements.

Theorem 12.5. *If S contains n elements, then S has 2^n subsets.*

Proof. Let us assume that $S = [n]$, since the number of subsets of S depends only on the number of elements of S. We construct a bijection from the set[2] $\mathscr{P}(S)$ of subsets of S to $[2] \times [2] \times \cdots \times [2]$, where there are n copies of $[2]$, also denoted $[2]^n$. To do this, we construct a function $f : \mathscr{P}(S) \to [2]^n$, defined by

$$f(T) = (a_1, a_2, \ldots, a_n),$$

where $a_i = 1$ if $i \notin T$ and $a_i = 2$ if $i \in T$.

Let us show that f is a bijection. We begin by checking that it is surjective. Let $(a_1, a_2, \ldots, a_n)$ be an arbitrary element of $[2]^n$, where each $a_i \in [2]$. Then define T to be $T = \{i \in [n] : a_i = 2\}$. Then $f(T) = (a_1, a_2, \ldots, a_n)$. Since $(a_1, a_2, \ldots, a_n) \in [2]^n$ was arbitrary, this shows that f is surjective.

Now let's show that f is injective. Suppose that $f(T) = f(U)$, for some $T, U \subseteq S$. Let us say that $f(T)$ and $f(U)$ are equal to $(a_1, a_2, \ldots, a_n)$. Then $T = \{i \in [n] : a_i = 2\}$, and U is also equal to the same thing. Thus $T = U$. This implies that f is injective.

Thus we have shown that the function f is a bijection from $\mathscr{P}(S)$ to $[2]^n$, so $\mathscr{P}(S)$ and $[2]^n$ have the same number of elements. Since $[2]^n$ has 2^n elements, so does $\mathscr{P}(S)$. ∎

Example. Let us suppose that $S = [5]$ and $T = \{1, 2, 5\}$. Then $f(T) = (2, 2, 1, 1, 2)$. Notice that the 2's are in positions 1, 2, and 5, exactly the elements of T.

Remark 12.6. It's more common to biject $\mathscr{P}(S)$ with $\{0,1\}^n$, i.e. n-tuples of elements that are 0 and 1, rather than 1 and 2. However, our way works just as well.

Let's look at another similar multiplicative problem. We would like to determine the number of ways of ordering the elements of $[n]$. For instance,

---

[2]The power set of S.

there are 6 ways of ordering the elements of 3, namely

$$123 \quad 132 \quad 213 \quad 231 \quad 312 \quad 321.$$

Theorem 12.7. *The number of ways of ordering the elements of $[n]$ is $n! = n \times (n-1) \times (n-2) \times \cdots \times 3 \times 2 \times 1$.*

One can prove Theorem 12.7 using a bijection with $[n] \times [n-1] \times \cdots \times [3] \times [2] \times [1]$, but it is slightly annoying to describe. So we'll do it a bit differently.

Proof. We have n choices for which number goes first. After we have determined that, we have $n-1$ choices for which number goes second, since it can be anything except for the number that went first. Similarly, we have $n-2$ choices for which number goes third, and so forth. Putting all this together, there are $n \times (n-1) \times (n-2) \times \cdots \times 3 \times 2 \times 1 = n!$ ways of arranging the numbers in $[n]$. ∎

Next up, what happens if we want to repeat some elements in the list? For instance, perhaps we'd like to know the number of ways of arranging the letters in the word COMBINATORICS. That's a 13-letter word, but with a bunch of repeated letters, so the answer isn't just 13!.

Theorem 12.8. *Let n and k be positive integers, and suppose that $a_1, \ldots, a_k$ are nonnegative integers with $a_1 + \cdots + a_k = n$. The number of ways of ordering a_i copies of i for each i with $1 \leq i \leq k$ is*

$$\frac{n!}{a_1! a_2! \cdots a_k!}.$$

Proof. Let us suppose the number of such orderings is R. Now, consider the a_i copies of i to be different, labeled as $i_1, i_2, \ldots, i_{a_i}$, for each i. The number of ways of ordering these n objects is $(a_1 + a_2 + \cdots + a_k)! = n!$. For each of the R orderings of the indistinguishable 1's, 2's, and so forth, we can arrange the 1's in $a_1!$ ways, the 2's in $a_2!$ ways, and so forth. Thus we have

$$a_1! a_2! \cdots a_k! R = n!,$$

or

$$R = \frac{n!}{a_1! a_2! \cdots a_k!},$$

as claimed. ∎

Another common type of problem is about ordering only *some* of the elements of a set.

Theorem 12.9. *Let n and k be positive integers, with $k \leq n$. The number of ways of ordering k of the numbers in $[n]$ is $\frac{n!}{(n-k)!}$.*

Proof. There are n ways to choose the first item in the list, then $n-1$ ways to choose the second item in the list, and so forth. In general, there are $n - i + 1$ ways to choose the ith item in the list. Thus the number of ways of ordering k of the numbers from $[n]$ is

$$n \times (n-1) \times (n-2) \times \cdots \times (n-k+1)$$
$$= \frac{n!}{(n-k) \times (n-k-1) \times \cdots 3 \times 2 \times 1} = \frac{n!}{(n-k)!},$$

as claimed. ∎

12.4 Complementary counting

Sometimes, in order to count the number $|A|$ of elements in a set A, it is easier to count the elements that are not in A instead. The formal setup is that we have two finite sets A and B, with $A \subseteq B$, and we wish to count $|A|$. Now, every element of B is in exactly one of A or $B \setminus A = \{b \in B : b \notin A\}$. Thus we have

$$|A| = |B| - |B \setminus A|.$$

Thus if we can determine $|B|$ and $|B \setminus A|$, then we know what $|A|$ is. Let's see an example.

Question 12.10. *How many subsets of $[2n]$ are there that contain at least one even number?*

With the notation as above, we let B be the set of subsets of $[2n]$ and A the set of subsets of $[2n]$ containing at least one even number. Then $B \setminus A$ is the set of subsets of $[2n]$ not containing at least one even number, i.e. those that only contain odd numbers. As we have already seen, we have $|B| = 2^{2n}$. There are n odd numbers in $[2n]$, so the number of subsets of $[2n]$ containing only odd numbers is 2^n, i.e. $|B \setminus A| = 2^n$. Thus we have $|A| = 2^{2n} - 2^n$.

A good rule of thumb for when to use complementary counting is that it is likely to be useful if you think that $|B \setminus A| < |A|$. It tends to be easier to count the sizes of smaller sets than the sizes of larger ones. Of course, there are many instances where this rule of thumb fails, but this is at least something to keep in mind when solving combinatorial problems.

12.5 Bijections and inverses

So far, in order to check that a function is a bijection, we have had to prove that it is both injective and surjective. However, there is another way to check that a function is a bijection.

Theorem 12.11. *Let A and B be sets, and let $f : A \to B$ be a function. Then f is a bijection if and only if there is a function $g : B \to A$ such that $g \circ f(a) = a$ for all $a \in A$, and $f \circ g(b) = b$ for all $b \in B$.*

This function g "undoes" whatever f does. For instance, in the bijection we gave to answer Question 12.4, we have a function f from the set of cat–dog pairs to $[3] \times [4]$, and the function g goes the other way: from $[3] \times [4]$ to the set of cat–dog pairs. We call g the *inverse* function of f.

Proof. We must show two things: if f is a bijection, then it has an inverse; and if f has an inverse, then it is a bijection.

Let's start with the first one. Suppose f is a bijection. We must show that f has an inverse function g. To do this, we construct g. Pick any $b \in B$. Since f is surjective, there is some $a \in A$ such that $f(a) = b$. Since f is injective, if $f(a) = f(a') = b$, then $a = a'$. It follows that there is exactly one element $a \in A$ such that $f(a) = b$. We define $g(b) = a$.

Let us now check that g is an inverse function to f. Take any element $a \in A$. We must show that $g(f(a)) = a$. If $f(a) = b$, then $g(b) = a$, so $g(f(a)) = a$. Similarly, if b is any element of B and $g(b) = a$, then $f(a) = b$, so $f(g(b)) = f(a) = b$. Thus g is indeed an inverse function to f.

Now, suppose that f has an inverse function g. We must show that f is a bijection, i.e. it is both surjective and injective. Let us show that f is surjective. Take any $b \in B$; we must show that there is some $a \in A$ such that $f(a) = b$. If $a = g(b)$, then we have $f(a) = f(g(b)) = b$, so a satisfies the desired property. Since b was arbitrary, this shows that f is surjective. Finally, we must show that f is injective. Suppose that $f(a) = f(a')$; we must show that $a = a'$. Since $f(a) = f(a')$, we have $g(f(a)) = g(f(a'))$. Since g is an inverse of f, $g(f(a)) = a$ and $g(f(a')) = a'$. Thus $a = a'$. This completes the proof. ∎

Often it is easier to write down an inverse function than it is to check both injectivity and surjectivity directly, so this gives us another way to check that functions are bijective; one that is often more convenient.

12.6   Problems

In all of these problems, write out a complete solution. Writing down an answer without giving a full solution does not qualify as having solved the problem.

(1) You have 20 cubes, of which 7 of them are blue, 8 of them are green, 2 are red, and 3 are yellow. How many ways are there of arranging the cubes into a $2 \times 2 \times 5$ block? Two arrangements are said to be the same if they have the same colored cubes in every place.

(2) (a) How many subsets of $[10]$ contain 1?
    (b) How many subsets of $[10]$ contain 1 or 2?[3]
    (c) How many subsets of $[10]$ contain at least one odd number?

(3) Which of the following functions are injective, surjective, and bijective?

   (a) $f : \mathbb{Z} \to \mathbb{Z}$, $f(n) = n$.
   (b) $f : \mathbb{Z} \to \mathbb{Z}$, $f(n) = n^2$.
   (c) $f : \mathbb{N} \to \mathbb{N}$, $f(n) = n^2$.
   (d) $f : \mathbb{Z} \to \mathbb{Z}$, $f(n) = \lfloor \frac{n}{3} \rfloor$.

(4) Suppose we have finite sets $S_1, \ldots, S_k$, which have $n_1, \ldots, n_k$ elements in them, respectively. How many ways are there to pick at most one element from each S_i, if you have to pick an element from at least one of the S_i's? Give the simplest bijective proof that you can.

(5) In how many ways can we place $k \le \min(m, n)$ identical rooks on an $m \times n$ chessboard in such a way that no two attack each other? (That is, no two are allowed to be in the same row (rank) or column (file).) Your answer should be expressed in terms of k, m, and n.

(6) (a) How many functions are there from $[m]$ to $[n]$?
    (b) How many injective functions are there from $[m]$ to $[n]$?

(7) A *composition* of a number n is a way of writing $n = a_1 + a_2 + \cdots + a_k$ for some k, where each a_i is a positive integer, called a *part* of the composition. For instance, there are four compositions of 3: 3, $2 + 1$,

---

[3] In mathematics, "or" is always inclusive: A or B includes the possibility of both A and B.

1+2, and 1+1+1. Determine, with proof, the number of compositions of n, as a function of n. Give a bijection with some other set you already know how to count.

(8) Determine, with proof, the number of compositions of n such that an even number of the parts are even. For instance, when $n = 3$, there are 2, namely 3 and $1 + 1 + 1$.

(9) How many compositions of n are there with k parts?

(10) Let k and n be positive integers. Determine the number of k-tuples $(S_1, S_2, S_3, \ldots, S_k)$ of subsets of $[n]$ such that $S_1 \subseteq S_2 \subseteq \cdots \subseteq S_k$, as a function of k and n.

(11) Let m and n be positive integers. How many ways are there to fill in an $m \times n$ grid with 0's and 1's such that the number of 1's in each row and each column is even?

(12) Consider an equilateral triangle of sidelength n, broken down into equilateral triangles of sidelength 1, as shown in Figure 12.3 when $n = 5$. How many paths are there from the top triangle to the middle triangle (or the left of middle in the case that n is even) in the bottom row, such that adjacent triangles in the path share a common edge and never go up or revisit a triangle? One such path is shown in Figure 12.3.

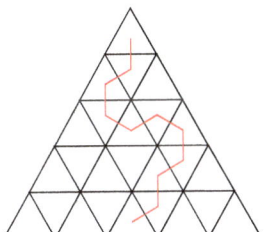

Figure 12.3. An equilateral triangle of sidelength 5 broken down into equilateral triangles of sidelength 1.

Chapter 13

The pigeonhole principle and matching problems

13.1 The pigeonhole principle

Let's review the pigeonhole principle, that we initially looked at in Chapter 4, with the aim of using it to help us with combinatorial problems. We start with two versions of the pigeonhole principle. Both of them are very simple, and the second version is a generalization of the first.

Theorem 13.1 (Pigeonhole principle, Version 1). *Let n and k be positive integers. If $n > k$, and we have n objects placed into k boxes, then there must be at least two objects in some box.*

Proof. We prove this by contradiction. Assume that there is no box with at least two objects, i.e. each box has at most one object in it. Let us suppose that m of the boxes have one object in them, and the other $k - m$ boxes have no objects in them. Thus the total number of objects is $m \leq k < n$, which is a contradiction because the total number of objects is n. ∎

And here is the more general version:

Theorem 13.2 (Pigeonhole principle, Version 2). *Let n, k, and r be positive integers, with $n > kr$. If we have n objects placed into k boxes, then there must be at least $r + 1$ objects in some box.*

The proof is very similar to before.

Proof. Assume that there is no box with at least $r + 1$ objects. Then every box has at most r objects. Thus the maximum number of objects there could be is $kr \leq n$, the total number of objects. This is a contradiction. ∎

13.2   The pigeonhole principle in graph theory

One place in combinatorics where the pigeonhole principle often comes up is in graph theory. Let's first explain what a graph is, in the combinatorial sense.

Definition 13.3. A *graph* G consists of a set V of vertices, together with a set E of edges, where each edge is a set containing exactly two distinct vertices.

Frequently, we represent a graph pictorially, drawing dots for each of the vertices, and lines (or curves) for each of the edges. See Figure 13.1 for a drawing of a famous graph, known as the *Petersen graph*.

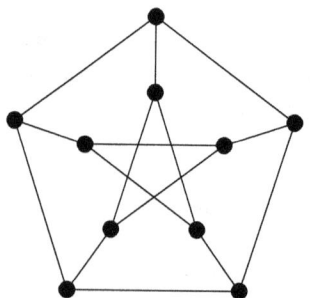

Figure 13.1. The Petersen graph.

Definition 13.4. If v and w are two vertices of V, we say that v and w are *adjacent* if $\{v, w\} \in E$.

Definition 13.5. The *degree* $d(v)$ of a vertex v is the number of edges containing v.

Typically we will be interested in graphs that are *simple* and *finite*. A graph is said to be *simple* if, whenever e_1 and e_2 are distinct edges, the vertices of e_1 are not both the same as the vertices of e_2. That is, we do not allow multiple edges between the same pair of vertices. A graph is said to be *finite* if it has a finite number of vertices. When not specified, we will always assume that our graphs are both simple and finite. In the case that G is a simple graph, $d(v)$ is the number of vertices adjacent to v.

Theorem 13.6. *Let G be a graph with at least two vertices. Then there are two vertices with the same degree.*

This is an interesting example, because at first it doesn't appear that the pigeonhole principle is applicable: we're trying to fit n objects (the vertices) into n boxes (the degrees). However, it turns out that there are two boxes that can't be filled simultaneously, and this is what allows us to use the pigeonhole principle.

Proof. Suppose that G has n vertices. Then the possible degrees are $0, 1, 2, \ldots, n-1$. There are n possible degrees. However, if one vertex has degree 0, then it is not adjacent to any other vertex. In particular, this prevents any vertex from having degree $n-1$. Thus we cannot have both a vertex of degree 0, and a vertex of degree $n-1$. So the possible degrees are either $0, 1, \ldots, n-2$, or $1, 2, \ldots, n-1$. Either way, we have $n-1$ possible degrees and n vertices, so by the pigeonhole principle, there must be two vertices with the same degree. ∎

13.3 Matching problems

A classic pigeonhole problem is the *mutilated chessboard problem*.

Question 13.7. *Take an 8×8 chessboard, and remove two opposite corners, as shown in Figure 13.2. Is it possible to cover the remaining 62 squares with 31 dominoes, such that every square is covered exactly once? (Each domino covers two squares that share an edge.)*

Figure 13.2. A mutilated chessboard.

The answer is no: we cannot tile the resulting board with dominoes. The reason is that if we alternately color the squares white and black, as shown in Figure 13.2, then there are 30 black squares and 32 white squares. Since a domino covers one white square and one black square, we cannot cover the board with 31 dominoes.

This argument is reminiscent of the pigeonhole principle. We consider the white squares to be the pigeons and the dominoes to be the pigeonholes. Since there are 32 white squares that must be covered by 31 dominoes, one domino must cover two white squares. But that's impossible!

Now, we might wonder more generally when a given figure can be tiled with dominoes. For example, can the figure in Figure 13.3 be tiled with dominoes? Here, the black square is deleted.

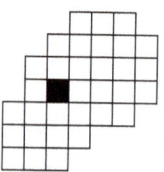

Figure 13.3. The black square is removed. Can the remaining region be tiled by dominoes?

If we were to tile the figure like a chessboard, we would have 16 black squares and 16 white squares. (Note that the marked square isn't a black square; it's just a hole in the board.) That doesn't imply that we can necessarily cover the figure with dominoes; it only implies that we can't conclude that it is impossible to do it *using the same argument as for the mutilated chessboard*. But maybe there's some other reason why we can't cover it.

Indeed, it is impossible to cover this figure using dominoes. To see this, let's color some of the squares red and blue, as shown in Figure 13.4. We have colored 7 squares red and 6 squares blue. However, the red squares can only share dominoes with the blue squares. Thus the red and blue region alone cannot be covered with dominoes, which certainly implies that the full figure cannot be so covered.

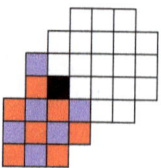

Figure 13.4. Each red square must share a domino with a blue square.

As you can see, it requires some cleverness to come up with these arguments. It would be nice if there were a streamlined way of solving these problems. And indeed, there is a beautiful theorem that tells us exactly which figures can be tiled with dominoes—and much more. This is the *Hall Marriage Theorem*, which we will state and prove in a moment.

But before we do that, we generalize from the problem of domino tilings to the problem of graph matchings. Given a board, we produce a graph, as follows. Put a vertex of the graph in the center of every square of the board. Then connect two centers if the squares share an edge, i.e. if it is possible to put a domino on those two squares. See Figure 13.5 for an example of this graph; the board is drawn in black, and the graph is drawn in blue. Once we have drawn the graph, we are free to remove the underlying board, since the graph captures all the relevant information.

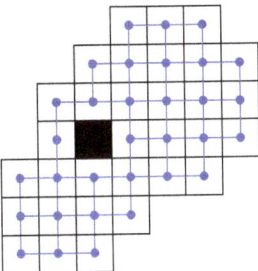

Figure 13.5. A graph from a board.

Definition 13.8. Given a graph $G = (V, E)$, a *perfect matching* of G is a subset $M \subseteq E$ of edges, such that every vertex $v \in V$ is contained in exactly one edge of M.

In the context of domino tilings, the edges in M are the edges containing dominoes. (See Figure 13.6 for an example of how this works.) Perfect matchings are more general than domino tilings, because there are graphs that do not come from subsets of rectangular boards. So if we can determine which graphs have a perfect matching, we can solve the problem of determining whether it is possible to tile any board—and much more.

In fact, we don't want to work with *all* graphs, but only ones that behave a bit more like chessboards. Recall that in a domino tiling, each domino has to cover one black square and one white square. So we want to consider graphs that have an analogue of black squares and white squares. These are called *bipartite* graphs.

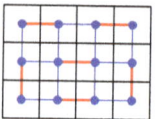

Figure 13.6. A domino tiling as a graph. The red edges correspond to the dominoes.

Definition 13.9. A graph $G = (V, E)$ is said to be *bipartite* if we can partition the set V of vertices into two subsets A and B, with $A \cap B = \varnothing$ and $A \cup B = V$, such that every edge in E has one endpoint in A and the other in B.

See Figure 13.7 for an example of a bipartite graph.

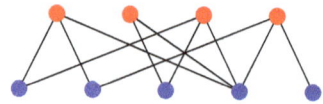

Figure 13.7. A bipartite graph. The red vertices are A, and the blue vertices are B.

This is exactly what we have in the case of chessboards: we let A be the set of black squares and B the set of white squares. We only have edges between black squares and white squares, never between two black squares or two white squares.

13.4   The Hall Marriage Theorem

The Hall Marriage Theorem gives us a definitive answer to when a bipartite graph has a perfect matching. The theorem doesn't tell us how many perfect matchings a graph has, only if the number is at least one or not.

Theorem 13.10 (Hall Marriage Theorem). *Let $G = (V, E)$ be a bipartite graph, with parts A and B. For a subset $W \subseteq A$ of vertices in A, let $N(W)$ denote the set of vertices in B that share an edge with some $w \in W$. Similarly, we can define $N(W)$ for a subset of B. Then G has a perfect matching if and only if for every $W \subseteq A$ and for every $W \subseteq B$, we have $|N(W)| \geq |W|$.*

Here $|W|$ and $|N(W)|$ denote the number of vertices in W and $N(W)$, respectively. Observe that this is just where we got into trouble in Figure 13.3: we had a set W (the red squares in Figure 13.4) of size 7, with only six neighbors in $N(W)$ (the blue squares). The Hall Marriage Theorem tells us that that's the only thing that can possibly go wrong to prevent a perfect matching.

The proof of the Hall Marriage Theorem relies on a common proof strategy in mathematics: When we're trying to show that something exists, we create a partial version of it, and then show that there is some process we can do to create a larger partial version. In this case, what we'll do is to assume that we can produce a partial matching, matching k of the vertices in A with k of the vertices in B, then extend to a larger partial matching, matching $k+1$ of the vertices in A with $k+1$ of the vertices in B. By doing this process repeatedly, we eventually end up matching all the vertices in A with all the vertices in B.

Proof of Theorem 13.10. We have two things to prove. First, that if there is a perfect matching, then $|N(W)| \geq |W|$ for all $W \subseteq A$ and $W \subseteq B$. Second, we must prove that if $|N(W)| \geq |W|$ for all $W \subseteq A$ and $W \subseteq B$, then there is a perfect matching.

We start by assuming that G has a perfect matching, say with matching $M \subseteq E$. This means that M consists of some of the edges of E, and every vertex in G is in exactly one of the edges of M. Pick any subset $W \subseteq A$ or $W \subseteq B$; without loss of generality, we'll assume that $W \subseteq A$. Let $X \subseteq B$ be the set of vertices in B that are matched with vertices in W, i.e.

$$X = \{v \in B : \{v, w\} \in M \text{ for some } w \in W\}.$$

Since every $v \in X$ is only matched with a single element of W, we have $|X| = |W|$. But $X \subseteq N(W)$, so $|X| \leq |N(W)|$. Thus $|W| = |X| \leq |N(W)|$, as desired.

Now we suppose that $|N(W)| \geq |W|$ for all $W \subseteq A$ and $W \subseteq B$. We will prove that G has a perfect matching. Note that the condition $|N(W)| \geq |W|$ for all $W \subseteq A$ and $W \subseteq B$ implies that $|A| = |B|$, for $|A| \leq |N(A)| \leq |B|$, and similarly $|B| \leq |A|$.

Suppose that for some integer k with $0 \leq k < |A|$, we have a partial matching with k edges, i.e. we have a set $M_k \subseteq E$ with $|M_k| = k$, such that M_k matches k of the vertices in A with k of the vertices in B. Since $k < |A|$, there is some unmatched vertex $x_0 \in A$. Since $|N(\{x_0\})| \geq |\{x_0\}| = 1$, x_0 must be adjacent to some vertex, say y_1, in B. If y_1 is unmatched in M_k, then we connect x_0 to y_1, and we have extended the matching M_k to M_{k+1}.

On the other hand, suppose that y_1 is already matched in M_k, say to x_1. Since

$$|N(\{x_0, x_1\})| \geq |\{x_0, x_1\}| = 2,$$

we know that either x_0 or x_1 is adjacent to some vertex y_2 of B other than y_1. Let's write $x_{r(2)}$ for this vertex, so that $r(2) \in \{0, 1\}$. If y_2 is unmatched in M_k, then we modify M_k by connecting $x_{r(2)}$ to y_2 and replacing any other edges as needed. On the other hand, if y_2 is matched in M_k, say to x_2, then we repeat the process and find some $y_3 \notin \{y_1, y_2\}$ adjacent to $x_{r(3)}$, where $r(3) \in \{0, 1, 2\}$. In general, if we have already found vertices $y_1, \ldots, y_i$ adjacent to $x_1, \ldots, x_i$, respectively, then $|N(\{x_0, x_1, \ldots, x_i\})| \geq |\{x_0, \ldots, x_i\}| = i + 1$, so $N(\{x_0, \ldots, x_i\})$ contains more than just $y_1, \ldots, y_i$. Eventually this process must terminate, when we find an unmatched y_{i+1}, because there are only finitely many (in fact, k) matched vertices in B, and we find a new one at each step. Once we find an unmatched vertex y_{i+1}, we can produce D_{k+1}, as follows. We follow the path $y_{i+1}, x_{r(i+1)}, y_{r(i+1)}, x_{r(r(i+1))}, y_{r(r(i+1))}, \ldots$, which terminates with x_0. Then we connect y_{i+1} with $x_{r(i+1)}$, and $y_{r(i+1)}$ with $x_{r(r(i+1))}$, and so forth—deleting any edges in M_k containing any of those vertices. When this is done, we have produced M_{k+1}.

It follows that $M_{|A|}$ exists, and this is a perfect matching. ∎

The process is illustrated in Figures 13.8 and 13.9.

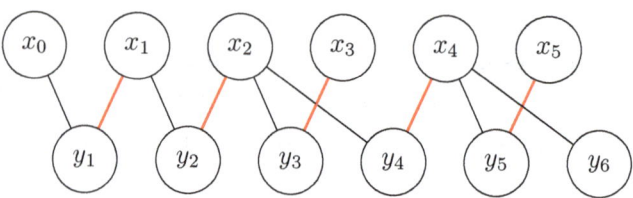

Figure 13.8. The red edges are in M_k, and we're trying to construct M_{k+1}.

13.5   Problems

(1) Give an example of a bipartite graph with 12 vertices, 6 in each part, with a perfect matching. Show the matching. Give one where there is no perfect matching, even though each vertex has at least two neighbors.

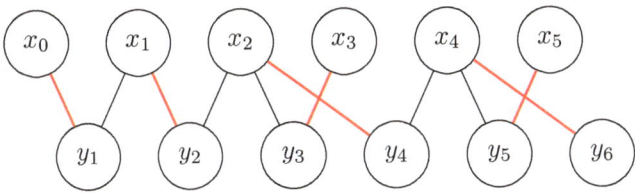

Figure 13.9. To construct M_{k+1} from M_k, we must first match y_6 to x_4. This breaks the match between y_4 and x_4, so we have to match y_4 to x_2. Continuing on this way, breaking matches and rematching vertices as needed, we eventually get to M_{k+1}.

(2) Find all triples (a, b, c) of positive integers with $a \le b \le c$ such that $\frac{1}{a} + \frac{1}{b} + \frac{1}{c} = 1$.

(3) Let S be a subset of $[2n]$, consisting of $n + 1$ elements. Show that there are two distinct elements $a, b \in S$ such that a is a multiple of b.

(4) A *cycle* in a graph G is a sequence $v_0, v_1, \ldots, v_n$ of vertices in G such that $v_0 = v_n$, and for every i with $0 \le i \le n - 1$, there is an edge between v_i and v_{i+1}. We call n the *length* of the cycle. Prove that G is bipartite if and only if it has no cycles of odd length.

(5) Consider an $n \times n$ grid of points. Connect every pair of points with an edge, and color the edge red if the points are in the same row, and blue otherwise. Show that among any $n + 1$ points, there is at least one red edge between two of the points.

(6) Suppose we have $mn + 1$ numbers $a_1, a_2, \ldots, a_{mn+1}$. Show that we can either find integers $1 \le i_1 < i_2 < \cdots < i_{m+1} \le mn + 1$ such that $a_{i_1} \le a_{i_2} \le \cdots \le a_{i_{m+1}}$, or we can find integers $1 \le j_1 < j_2 < \cdots < j_{n+1} \le mn + 1$ such that $a_{j_1} \ge a_{j_2} \ge \cdots \ge a_{j_{n+1}}$. That is, we can either find an increasing subsequence of length $m + 1$ or a decreasing subsequence of length $n + 1$. (For example, suppose $m = n = 3$. The $mn + 1 = 10$ numbers $3, 9, 1, 4, 2, 10, 8, 5, 7, 6$ contain an increasing subsequence of length 4, namely $1, 2, 5, 7$. There are also other increasing and decreasing subsequences of length 4.)

(7) Suppose we have an art gallery, in the shape of a not-necessarily-convex polygon with n vertices. A guard placed at a point p in the gallery can guard any point q in the gallery as long as the line segment connecting p and q is contained entirely within the gallery. Show that it is possible to place $\lfloor \frac{n}{3} \rfloor$ guards, at some of the corners of the gallery such that—among all of them—they can guard the entire gallery.

(8) Suppose we have a $2n \times 2n$ chessboard, and we place n rooks on each row and each column. Prove that we can find $2n$ rooks such that no two lie in the same row or column.

(9) Prove that if all the vertices of a bipartite graph have the same nonzero degree, then it has a perfect matching.

(10) We are given two square sheets of paper with integer area n. Suppose we divide each of these papers into n regions, each of area 1. (The divisions for the two piece of papers may be distinct.) Then we place the two sheets of paper directly on top of each other. Show that we can place n pins on the pieces of paper such that the interiors of all $2n$ regions have been pierced.

Chapter 14

Double counting

14.1 Introduction to double counting

One very important way of proving theorems in combinatorics is to use double counting. What this means is that we count the elements of a set S in two ways. If, counting in the first way, we find that S contains m elements, and counting in the second way we find that S contains n elements, then it must be the case that $m = n$. Typically, S is not exactly the set we are interested in, but rather a modification of another set T that is easier to work with. Let's see a simple example of this technique in action.

Theorem 14.1. *If G is a graph with a finite number of vertices and edges, then the number of vertices with odd degree is even.*

Proof. We calculate the sum of the degrees of all the vertices in G in two ways. First, we calculate the sum of degrees based on the vertices: it is equal to

$$\sum_{v \in V} d(v).$$

On the other hand, we can calculate the sum of the degrees based on the edges. Each edge contributes a count of 1 to each vertex it connects, for a total of 2. Thus we have

$$\sum_{v \in V} d(v) = \sum_{e \in E} 2.$$

The sum on the right is obviously even, so the sum on the left must be as well.

Furthermore, we have

$$\sum_{v \in V} d(v) = \sum_{\substack{v \in V \\ d(v) \text{ odd}}} d(v) + \sum_{\substack{v \in V \\ d(v) \text{ even}}} d(v).$$

The second sum on the right is even, so the first sum on the right must be also. Since each summand in that sum is odd, there must be an even number of summands, i.e. the number of vertices with an odd degree is even. ∎

14.2   Incidences

Formally, what we did above was to count the pairs (v, e) where v is one of the vertices of e, in two different ways. One way was to count by vertex: this gives us $\sum_{v \in V} d(v)$. The other way is to count by edge, which gives us $\sum_{e \in E} 2$. More generally, what we typically do when double counting is to take two sets A and B, where we're interested in certain ordered pairs (a, b) called *incident pairs*, where $a \in A$ and $b \in B$. We can count the incident pairs by first counting A, which gives us $\sum_{a \in A} n(a)$, where $n(a)$ is the number of $b \in B$ such that (a, b) is an incident pair. Or, we can count by B, which gives us $\sum_{b \in B} n(b)$. These may have a very different feel, as we saw in Theorem 14.1. One of them is just a sum of a bunch of 2's, whereas the other is some unknown quantity we're trying to understand.

Definition 14.2. Let A and B be two sets. Then we call a subset $S \subseteq A \times B$ an *incidence*. We call the pairs $(a, b) \in S$ *incident pairs*.

One common way of representing an incidence S is as a bipartite graph. We make a graph whose vertex set is $A \cup B$, when we connect $a \in A$ and $b \in B$ with an edge if and only if $(a, b) \in S$.

Example. Suppose $A = \{a, b, c, d\}$ and $B = \{x, y, z\}$, and
$$S = \{(a, x), (a, y), (b, y), (c, x), (c, z), (d, x), (d, y), (d, z)\}.$$
The corresponding bipartite graph is shown in Figure 14.1.

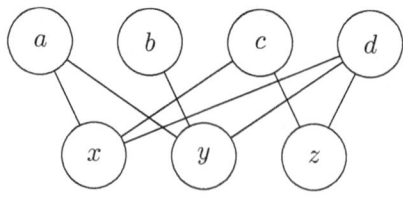

Figure 14.1. A graph representing an incidence.

Sometimes it is fruitful to represent an incidence as a matrix, known as the *incidence matrix*. Given an incidence $S \subseteq A \times B$, we build a matrix whose rows correspond to the elements of A, and whose columns correspond to the elements of B. Then we fill the matrix with 0's and 1's. We put a 1 in the (a, b) entry if $(a, b) \in S$, and we put a 0 in the (a, b) entry if $(a, b) \notin S$. For instance, the matrix representation of the incidence in the example and Figure 14.1 is

$$\begin{pmatrix} 1 & 1 & 0 \\ 0 & 1 & 0 \\ 1 & 0 & 1 \\ 1 & 1 & 1 \end{pmatrix},$$

where the rows are a, b, c, d from top to bottom, and the columns are x, y, z from left to right.

14.3   Sperner's Lemma

Another famous proof involving double counting is that of Sperner's Lemma. Sperner's Lemma is a result about cutting up polygons into triangles and coloring the vertices of these triangles. It was first used to prove the Brouwer Fixed-Point Theorem, one of the most famous theorems in algebraic topology, just using combinatorics. Some interesting applications of Sperner's Lemma can be found in [AZ18, Chapter 28] and [Su99].

Theorem 14.3 (Sperner's Lemma). *Suppose we have a polygon in the plane, subdivided into triangles in such a way that two triangles either don't intersect, intersect at a vertex of each, or intersect at an edge of each. Color each of the vertices of the small triangles red, blue, or green. Call an edge of a small triangle a* purple edge *if one of its vertices is red and the other is blue, and call a triangle a* rainbow triangle *if it has a red vertex, a blue vertex, and a green vertex. Then the number of purple edges on the boundary of the polygon is congruent to the number of rainbow triangles modulo 2.*

See Figure 14.2 for an example of Sperner's Lemma.

Proof. Put a dot on each side of each purple edge, whether it is on the boundary or in the interior of the polygon. Now, let us count the number of dots in the *interior* of the polygon, by looking at each edge. Each interior purple edge contributes 2 to the count (one on each side), and each other interior edge contributes 0. On the other hand, each boundary purple edge contributes 1 to the count. (See Figure 14.3.) Thus, modulo 2, the number of dots is congruent to the number of purple edges on the boundary.

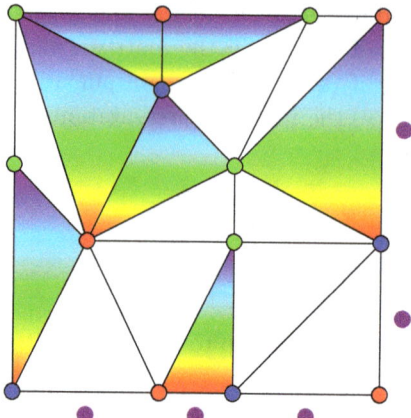

Figure 14.2. A square dissection with five purple edges on the boundary and seven rainbow triangles. (Note that $5 \equiv 7 \pmod 2$.)

Now let us count the number of dots by looking at triangles. Each rainbow triangle contains exactly one dot, and each non-rainbow triangle contains either 0 or 2. Thus, modulo 2, the number of dots is congruent to the number of rainbow triangles. But the number of dots is the same as what we got counting by purple edges. Thus we find that the number of boundary purple edges is congruent to the number of rainbow triangles modulo 2, as desired.   ∎

Let's express this proof in terms of an incidence. The thing we double counted was the number of interior dots. The interior dots correspond to pairs (e, t), where e is a purple edge and t is a triangle containing the edge e. For a triangle t, let $n(t)$ denote the number of purple edges in t; and for a purple edge e, let $n(e)$ denote the number of triangles containing that edge. Counting the dots in the interior of the polygon in terms of triangles, we get

$$D = \sum_t n(t) = \sum_r 1 + \sum_{!r} (0 \text{ or } 2),$$

where $\sum_r$ denotes the sum over rainbow triangles and $\sum_{!r}$ denotes the sum over nonrainbow triangles. On the other hand, summing over the purple edges, we have

$$D = \sum_{pe} n(e) = \sum_{bpe} 1 + \sum_{ipe} 2,$$

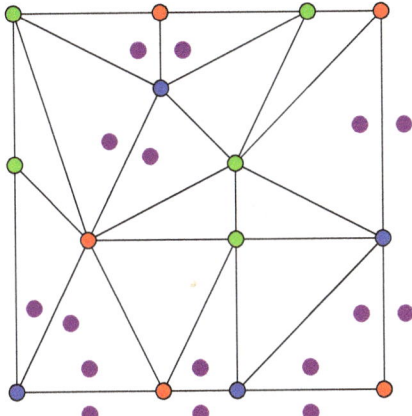

Figure 14.3. Put a dot on each side of a purple edge.

where $\sum_{bpe}$ denotes the sum over boundary purple edges and $\sum_{ipe}$ denotes the sum over interior purple edges.

If we draw the graph version of this incidence, our vertex set consists of all the triangles and all the purple edges. We put an edge between the vertex corresponding to a purple edge and a vertex corresponding to a triangle, if and only if that purple edge is one of the edges of that triangle.

14.4 Average number of divisors

Definition 14.4. Let n be a positive integer. We define the *divisors function* $d(n)$ to be the number of positive integers that divide n, including both 1 and n.

For example, $d(8) = 4$, since the divisors of 8 are 1, 2, 4, and 8.

There are some large numbers n for which $d(n)$ is very small. For instance, if n is prime, then the only divisors of n are 1 and n, so $d(n) = 2$. On the other hand, $d(n)$ can get as large as desired. But how big is it *on average*?

In order to make that a more precise question, let's fix some large number x and ask how many divisors numbers up to x have, on average. That is, we are interested in computing

$$\frac{1}{x} \sum_{n=1}^{x} d(n). \tag{14.1}$$

We won't be able to get a nice exact answer for this, but instead we'll be able to get a pretty good approximation. It isn't the best we can do. But, as we'll see, it's not bad. We'll get the right answer to within a range of 1.

The first thing we need to know is how to express $d(n)$ better. Each divisor of n contributes 1 to $d(n)$, so we have

$$d(n) = \sum_{m \mid n} 1.$$

This is the first key ingredient we need to estimate (14.1). Thus we have

$$\frac{1}{x} \sum_{n=1}^{x} d(n) = \frac{1}{x} \sum_{n=1}^{x} \sum_{m \mid n} 1.$$

The next thing we need to do here—and almost everywhere else in mathematics when we encounter a double sum—is to switch the order of summation. When the sums are infinite, this isn't always legal, but with finite sums we are always welcome to switch the order of summation.

In this case, when we switch the order of summation, we want the outer sum to be over m and the inner sum to be over n. But now we have to figure out what their ranges are. Initially, m can be any divisor of any integer from 1 to x, so m can be any integer from 1 to x. Furthermore, n has to be a multiple of m between 1 and x. This means that $n = mk$, where $n \leq x$, or $k \leq \frac{x}{m}$—and that instead of summing over only the n's that are multiples of m, we sum over k. (See the Interlude following this chapter for a more thorough discussion of switching the order of summation.) Thus we have

$$\frac{1}{x} \sum_{n=1}^{x} d(n) = \frac{1}{x} \sum_{m=1}^{x} \sum_{k=1}^{\frac{x}{m}} 1.$$

The inner sum

$$\sum_{k=1}^{\frac{x}{m}} 1$$

is easy to evaluate: it consists of $\lfloor \frac{x}{m} \rfloor$ terms, each of which is 1, so it's just $\lfloor \frac{x}{m} \rfloor$. Thus we have

$$\frac{1}{x} \sum_{n=1}^{x} d(n) = \frac{1}{x} \sum_{m=1}^{x} \sum_{k=1}^{\frac{x}{m}} 1 = \frac{1}{x} \sum_{m=1}^{x} \left\lfloor \frac{x}{m} \right\rfloor.$$

Floor functions are often difficult to work with exactly, but at least they are easy to bound: for any y, we have $y - 1 < \lfloor y \rfloor \leq y$. Thus we have

$$\frac{1}{x} \sum_{n=1}^{x} d(n) = \frac{1}{x} \sum_{m=1}^{x} \left\lfloor \frac{x}{m} \right\rfloor \leq \frac{1}{x} \sum_{m=1}^{x} \frac{x}{m} = \sum_{m=1}^{x} \frac{1}{m},$$

and similarly

$$\frac{1}{x} \sum_{n=1}^{x} d(n) = \frac{1}{x} \sum_{m=1}^{x} \left\lfloor \frac{x}{m} \right\rfloor > \frac{1}{x} \sum_{m=1}^{x} \left(\frac{x}{m} - 1 \right) = \sum_{m=1}^{x} \frac{1}{m} - 1.$$

That's pretty good. We have bounded the answer to within just 1:

$$\sum_{m=1}^{x} \frac{1}{m} - 1 < \frac{1}{x} \sum_{n=1}^{x} d(n) \leq \sum_{m=1}^{x} \frac{1}{m}.$$

If you aren't satisfied with the number $\sum_{m=1}^{x} \frac{1}{m}$ appearing in an answer, you might be happier knowing that there is some number $\gamma \approx 0.577$, called the *Euler–Mascheroni constant*, such that

$$\sum_{m=1}^{x} \frac{1}{m} \approx \log(x) + \gamma,$$

where log is the natural logarithm, to base e.

Again, let's explain how this proof can be expressed in terms of an incidence. Our two sets are both the integers from 1 to x, and our incidence consists of pairs (k, n) where k is a factor of n. If we sum first over n, we get

$$\sum_{n=1}^{x} d(n).$$

On the other hand, if we sum first over k, we get

$$\sum_{k=1}^{x} \left\lfloor \frac{x}{k} \right\rfloor,$$

where $\lfloor \frac{x}{k} \rfloor$ is the number of n's such that (k, n) is in the incidence.

The study of averages of arithmetic functions like $d(n)$ is important in analytic number theory. See [Apo76] for much more on this topic.

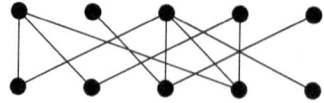

Figure 14.4. A graph.

14.5   Problems

(1) Write down the incidence matrix of the graph in Figure 14.4.
(2) Let k be a positive integer. A graph is said to be k-*regular* if every vertex has degree k. Given a matrix, how can you tell whether it is the incidence matrix of a bipartite k-regular graph? Prove your answer is correct.
(3) Suppose that P is a polyhedron, all of whose faces are triangles. Suppose that P has F faces and E edges. Prove that $2E = 3F$.
(4) Let $G = (V, E)$ be a graph. Show that

$$\sum_{v \in V} d(v)^2 = \sum_{e = \{v, w\} \in E} (d(v) + d(w)).$$

(5) In a certain committee, each member belongs to exactly three subcommittees, and each subcommittee has exactly three members. Prove that the number of members equals the number of subcommittees.
(6) Suppose $G = (V, E)$ is a bipartite graph, where X and Y are the parts of V, and that every vertex has degree ≥ 1. Suppose that for every edge $e = \{x, y\}$, where $x \in X$ and $y \in Y$, we have $d(x) \geq d(y)$. Show that $|X| \leq |Y|$.
(7) A *permutation* of a set S is a bijection $f : S \to S$. A *fixed point* of a permutation f is some $s \in S$ such that $f(s) = s$. Let $p_k(n)$ be the number of permutations of $[n]$ with exactly k fixed points. Prove that

$$\sum_{k=0}^{n} k p_k(n) = n!.$$

(8) There are n points in the plane such that no three of them are collinear. Prove that the number of triangles whose vertices are chosen from these n points and whose area is 1, is at most $\frac{2}{3}(n^2 - n)$. (Hint: given two points, how many possible third points can there be such that the triangle they form has area 1?)

Interlude: Swapping double sums

Often in mathematics, we encounter double sums, such as the double sum

$$\sum_{n=1}^{x} \sum_{m|n} 1$$

that we encountered in §14.4. For illustrative purposes, it will actually be more convenient to look at a slightly more general sum, namely

$$\sum_{n=1}^{x} \sum_{m|n} f(m,n),$$

where f is some function of two variables. If $x = 10$, then this sum is

$$
\begin{aligned}
f(1,1) &+ f(1,2) + f(2,2) + f(1,3) + f(3,3) + f(1,4) + f(2,4) + f(4,4) \\
&+ f(1,5) + f(5,5) + f(1,6) + f(2,6) + f(3,6) + f(6,6) + f(1,7) + f(7,7) \\
&+ f(1,8) + f(2,8) + f(4,8) + f(8,8) + f(1,9) + f(3,9) + f(9,9) \\
&+ f(1,10) + f(2,10) + f(5,10) + f(10,10).
\end{aligned}
$$

Note the order that we do the sum in: we take $n = 1$ and sum over all $m \mid 1$ (i.e. just $m = 1$), then we move on to $n = 2$ and sum over all $m \mid 2$, then $n = 3$, and so forth. But we can change the order of the terms without affecting the final answer. Almost all the time when we encounter double sums in nature, they start off being in the wrong order, and we have to switch them. Before we look at how to switch the order of summation in this example, let's look at some easier cases.

The first case to consider is a double sum of the form

$$\sum_{n=1}^{y} \sum_{m=1}^{x} f(m,n).$$

This means that we first take $n = 1$ and sum over all values of m from 1 to x. Then we move on to $n = 2$ and sum over all values of m from 1 to x, and so forth. We can also sum in the other order. This means that we first take $m = 1$ and sum over all values of n from 1 to y. Then we take $m = 2$ and sum over all values of n from 1 to y, and so forth. We write this as a double sum as

$$\sum_{m=1}^{x} \sum_{n=1}^{y} f(m, n).$$

The answer is the same, but sometimes one way of doing this double sum will be easier than the other way.

One way to think about how to switch the sum is by considering Figure 14.5. The index m denotes the x-coordinate, whereas the index n denotes the y-coordinate. Originally, the outer sum is over n, meaning that we look at the y-coordinate first. For each value of n (or y-coordinate), we sum over the range of m marked by dots, which is all of them from 1 to y. When we switch the sums, we first look at each value of m (or x-coordinate), and sum over the values of n indicated by the dots, which is $\sum_{m=1}^{x} \sum_{n=1}^{y} f(m, n)$.

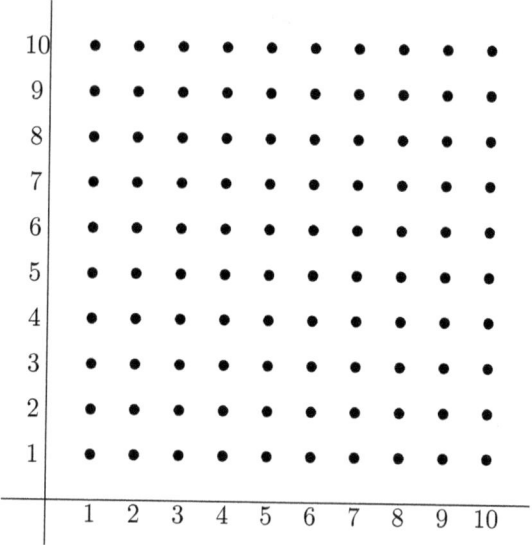

Figure 14.5. The region of summation for $\sum_{n=1}^{10} \sum_{m=1}^{10} f(m, n) = \sum_{m=1}^{10} \sum_{n=1}^{10} f(m, n)$.

Exchanging the double sum becomes trickier when the bound of the inner sum depends on the index variable of the outer sum. For example, we might consider the sum

$$\sum_{n=1}^{x} \sum_{m=1}^{n} f(m,n).$$

Note that the upper bound of the inner sum is n rather than x, so it varies. For instance, if $x = 3$, then we get the sum

$$f(1,1) + f(1,2) + f(2,2) + f(1,3) + f(2,3) + f(3,3).$$

Observe that we're summing over all pairs (m,n) where $1 \leq m \leq n \leq x$. If we want the outer sum to be over m, then that means we need the possible values of m to range from 1 to x, and then n must be $\geq m$, while still remaining $\leq x$. Thus we have

$$\sum_{n=1}^{x} \sum_{m=1}^{n} f(m,n) = \sum_{m=1}^{x} \sum_{n=m}^{x} f(m,n).$$

See Figure 14.6.

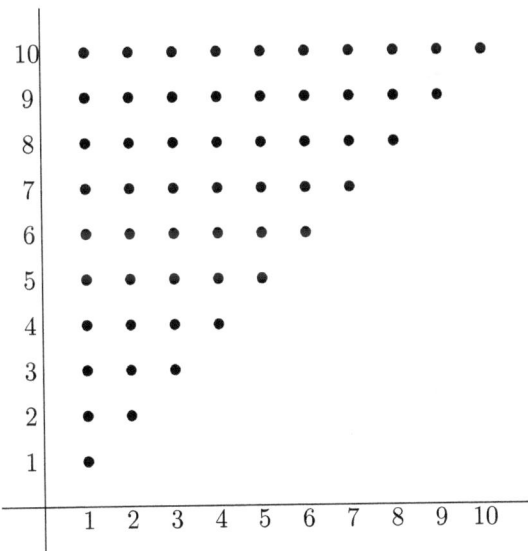

Figure 14.6. The region of summation for $\sum_{n=1}^{10} \sum_{m=1}^{n} f(m,n) = \sum_{m=1}^{10} \sum_{n=m}^{10} f(m,n)$.

Example. Let us evaluate the sum

$$S = \sum_{k=1}^{n} k 2^k.$$

This isn't a double sum, but we can turn it into one by writing $k 2^k$ as $\sum_{m=1}^{k} 2^k$. Thus we have

$$S = \sum_{k=1}^{n} \sum_{m=1}^{k} 2^k.$$

If we switch the sums, we get

$$S = \sum_{m=1}^{n} \sum_{k=m}^{n} 2^k.$$

The inner sum can be evaluated directly: we have

$$\sum_{k=m}^{n} 2^k = 2^{n+1} - 2^m.$$

Thus we have

$$S = \sum_{m=1}^{n} (2^{n+1} - 2^m) = \sum_{m=1}^{n} 2^{n+1} - \sum_{m=1}^{n} 2^m.$$

Both of the sums on the right can be evaluated directly as well: we have

$$\sum_{m=1}^{n} 2^{n+1} = n 2^{n+1},$$

whereas

$$\sum_{m=1}^{n} 2^m = 2^{n+1} - 2.$$

Thus we have

$$S = n 2^{n+1} - 2^{n+1} + 2 = (n-1) 2^{n+1} + 2.$$

Finally, we have the case that we saw in §14.4, where we have

$$\sum_{n=1}^{x} \sum_{m|n} f(m,n).$$

Here n ranges from 1 to x, while m ranges over the divisors of n. If we switch the sums, then m can be any number from 1 to x, so the outer sum has to be $\sum_{m=1}^{x}$. Now, before we switch the sums, m ranges over divisors of n. Saying that m is a divisor of n is the same as saying that

n is a multiple of m, so after switching the sum, n must range over the multiples of m. However, we have an additional constraint on n, which is $n \leq x$. Thus n only ranges over those multiples of m that are $\leq x$. It is not particularly convenient to write the condition that n is a multiple of m directly as a subscript of a summation, but fortunately, we have another approach: saying that n is a multiple of m is the same as saying that there exists an integer k such that $n = km$. Thus, summing over values of n that are multiples of m is the same as summing over k, except that now k must range from 1 to $\frac{x}{m}$. Thus we have

$$\sum_{n=1}^{x} \sum_{m|n} f(m,n) = \sum_{m=1}^{x} \sum_{k=1}^{\frac{x}{m}} f(m,n).$$

However, this last expression is not entirely satisfactory, because the expression that we're summing, $f(m,n)$, has an n in it, whereas n is nowhere to be found in the indices of summation. Fortunately, we can recover it, since $n = km$, so we can rewrite this last sum as

$$\sum_{n=1}^{x} \sum_{m|n} f(m,n) = \sum_{m=1}^{x} \sum_{k=1}^{\frac{x}{m}} f(m,km).$$

See Figure 14.7.

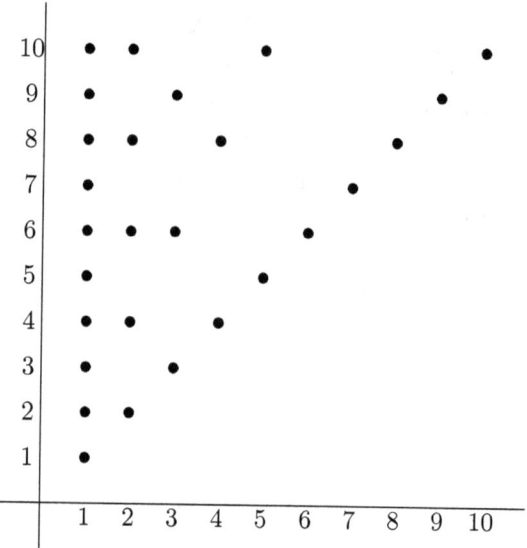

Figure 14.7. The region of summation for $\sum_{n=1}^{10} \sum_{m|n} f(m,n) = \sum_{m=1}^{10} \sum_{k=1}^{\frac{10}{m}} f(m, km)$.

Chapter 15

Introduction to binomial coefficients

15.1 Forming committees

The binomial coefficients are the most important numbers in combinatorics. They show up in a huge number of problems, in a wide range of contexts. Let us see how they are defined.

Definition 15.1. Let n and k be nonnegative integers. The *binomial coefficient* $\binom{n}{k}$, pronounced "n choose k," is the number of ways of selecting an unordered k-member committee from among n people.

Equivalently, $\binom{n}{k}$ is the number of k-element subsets of a set of size n. The first thing to do is to find a formula for $\binom{n}{k}$.

Theorem 15.2. *If $n \geq k \geq 0$ are nonnegative integers, then*

$$\binom{n}{k} = \frac{n!}{k!(n-k)!}.$$

Proof. Instead of counting committees, we'll count the top k finishers in a race with n people. If we have n people running in a race, there are n people who could be the first-place winner, then $n-1$ people who could be the second-place winner, then $n-2$ people who could be the third-place winner, and so forth: in general there are $(n-i+1)$ people who could be the ith-place winner. Thus the number of ways of distributing awards to the top k finishers is

$$n(n-1)(n-2)\cdots(n-k+1) = \frac{n!}{(n-k)!}.$$

On the other hand, the number of ways of choosing the top k finishers without ranking them is $\binom{n}{k}$, and then the number of ways of ranking them is $k!$. Thus we have

$$\frac{n!}{(n-k)!} = \binom{n}{k} k!,$$

so

$$\binom{n}{k} = \frac{n!}{k!(n-k)!},$$

as claimed.   ∎

Remark 15.3. Sometimes, it will also be convenient to define $\binom{n}{k}$ when $n < k$. To do this, we simply set $\binom{n}{k} = 0$ when $n < k$. This is logical enough by the combinatorial definition as the number of k-member committees that we can form out of n people: there are no ways to do this.

We already have an interesting consequence:

Corollary 15.4. *The number* $\frac{n!}{k!(n-k)!}$ *is an integer.*

Proof. It counts something, so it must be an integer.   ∎

At this point, it is only fair to point out that the fact that $\binom{n}{k}$ has a nice formula in terms of familiar functions is more than we really deserve. Most of the time in combinatorics, we end up with functions that cannot be expressed in terms of more familiar functions, at least not in a convenient way. We will see a typical example of this in Chapter 20, when we discuss the Stirling numbers.

15.2   Fundamental binomial identities

Proposition 15.5. *For each nonnegative integer n, we have*

$$\sum_{k=0}^{n} \binom{n}{k} = 2^n.$$

Proof. Consider n people. The right side counts the number of committees we can form from those n people: each person can either be in the committee or not be in the committee, and we allow the committee to be empty. The $\binom{n}{k}$ term on the left side counts the number of committees with k people. If we sum over k, then we're counting the number of committees with any number of people, which is the same as what the right side counts.   ∎

Proposition 15.6. $\sum_{k=0}^{n}(-1)^k \binom{n}{k} = 0$ *if $n \geq 1$ is an integer.*

Proof. Move all the negative terms to the other side, so that we must show that

$$\sum_{k=0}^{\lfloor \frac{n}{2} \rfloor} \binom{n}{2k} = \sum_{k=0}^{\lfloor \frac{n-1}{2} \rfloor} \binom{n}{2k+1}.$$

The left side counts the number of committees with an even number of members, and the right side counts the number of committees with an odd number of members. To see that these are equal, we give a bijection between the set of committees with an even number of members, and the set of committees with an odd number of members. To do this, let us pick one of the n people, whom we shall call person 1. Given a committee with an even number of members, we produce a committee with an odd number of members by toggling person 1's membership: if person 1 was already in the committee, we form a new committee without person 1; and if person 1 wasn't in the committee, we form a new committee by adding in person 1. Either way, we end up with a committee of odd size. This operation is invertible: we can toggle person 1's membership again starting with an odd-sized committee to form an even-sized committee. ∎

Example. Suppose $n = 7$. The committee consisting of people 1, 4, and 6 has an odd number of members. If we remove person 1, then we end up with the committee with people 4 and 6, which has an even number of members.

Theorem 15.7 (Pascal's Identity). $\binom{n}{k} = \binom{n-1}{k-1} + \binom{n-1}{k}$.

Proof. The left side counts the number of committees consisting of k people, from among n people total. Let us again distinguish one person, named person 1. If a committee includes person 1, then there are $k - 1$ spots left, from among $n - 1$ people, for a total of $\binom{n-1}{k-1}$ committees. On the other hand, if the committee doesn't include person 1, then there are k spots left from among $n - 1$ people, for a total of $\binom{n-1}{k}$ committees. ∎

Pascal's identity highlights the connection between the binomial coefficients and *Pascal's triangle*. To form Pascal's triangle, we start with a 1, and then we calculate further numbers in the triangle by adding the two numbers directly above it. See Figure 15.1 for the first 8 rows of Pascal's triangle beginning with $n = 0$.

Because of Pascal's identity, the entries in Pascal's triangle are simply the binomial coefficients: the kth entry in the nth row is $\binom{n}{k}$. This is because

$n = 0$:						1					
$n = 1$:					1		1				
$n = 2$:				1		2		1			
$n = 3$:			1		3		3		1		
$n = 4$:		1		4		6		4		1	
$n = 5$:	1		5		10		10		5		1
$n = 6$:	1	6		15		20		15		6	1
$n = 7$:	1	7	21		35		35		21	7	1

Figure 15.1. Pascal's triangle.

the entries in Pascal's triangle and the binomial coefficients are constructed in the same way from previous entries: they are both sums of the same two earlier numbers, so they must be equal. This is actually a common way of proving combinatorial theorems: show that two things are equal because they satisfy the same recurrence (and the same initial conditions).

Here's a more complicated binomial identity, although still one with a simple combinatorial interpretation.

Theorem 15.8 (Vandermonde's Identity).

$$\sum_{k=0}^{r} \binom{m}{k}\binom{n}{r-k} = \binom{m+n}{r}.$$

Proof. Suppose we have m cats and n dogs. The right side counts the number of ways we can choose r animals total from the $m + n$. The kth term on the left counts the number of ways we can choose k cats and $r - k$ dogs, which is also r total. If we sum over all k, then we obtain all ways of choosing r animals, regardless of how many are cats and how many are dogs. Thus the two sides are counting the same thing. ∎

Sometimes Vandermonde's identity is used in conjunction with the obvious identity $\binom{n}{k} = \binom{n}{n-k}$:

Corollary 15.9.

$$\sum_{k=0}^{n} \binom{n}{k}^2 = \binom{2n}{n}.$$

Proof. Replace $\binom{n}{k}^2$ with $\binom{n}{k}\binom{n}{n-k}$ and use Vandermonde's identity with $m = n = r$. ∎

15.3 Stars and bars

The following problem might not look like a binomial coefficient problem, but it secretly is, thanks to a clever bijection.

Question 15.10. *How many ways are there to represent n as a sum of k positive integers, where order matters?*

For instance, there are 3 ways of writing 4 as a sum of 2 positive integers: $3 + 1$, $2 + 2$, and $1 + 3$.

To work this out in general, we set up n stars, and then we put $k - 1$ bars in some of the gaps between the stars. Once we have done that, we read off the representation of $n = a_1 + a_2 + \cdots + a_k$, as follows: a_1 is the number of stars to the left of the first bar, then a_2 is the number of stars between the first and second bars, and so forth, until a_k is the number of stars to the right of the $(k - 1)^{\text{st}}$ bar. For example, if $n = 15$ and $k = 6$, then one such stars-and-bars representation is

$$* * \mid * \mid * * * \mid * * * * * * \mid * * \mid *,$$

which corresponds to $15 = 2 + 1 + 3 + 6 + 2 + 1$.

In order to count the number of such representations, just note that we select $k - 1$ of the $n - 1$ gaps to put bars in. Hence there are $\binom{n-1}{k-1}$ such representations, known as *compositions*.

Perhaps we would like to allow the parts to be 0. There are two ways of handling this small change, and both are good to know. It doesn't seem so easy to modify the argument above, because we'd have to allow multiple bars to go in the gaps (and also allow bars to go at the beginning and end). But we can still do it, with just a bit of cleverness.

The first way is simply to increase n. We claim that the number of nonnegative compositions of n into k parts is equal to the number of positive compositions of $n + k$ into k parts. Why? Given a nonnegative composition of n into k parts, we can convert it into a positive composition of $n + k$ into k parts, simply by adding 1 to each part. Similarly, given a positive composition of $n + k$ into k parts, we can convert it into a nonnegative composition of n into k parts by subtracting 1 from each part. These two operations are clearly inverses: if we start with a nonnegative composition of n into k parts, convert it into a positive composition of $n + k$ into k parts, and then convert it back, we get back to the same composition we started with, and vice versa. Thus we have just proven that our map is a bijection, so the two sets have the same size. Thus there are $\binom{n+k-1}{k-1} = \binom{n+k-1}{n}$ nonnegative compositions of n into k parts.

The other method, using stars and bars directly, is to start with only bars and convert some of them into stars. Start with a row of $n+k-1$ bars. Now, turn n of them into stars. Reading off the number of stars between consecutive bars gives us a nonnegative composition of n into k parts. For instance,

$$* \mid\mid * * * * \mid\mid * \mid ** \mid$$

corresponds to the composition $1 + 0 + 4 + 0 + 1 + 2 + 0$ of 8 into 7 parts.

One way to interpret the count of nonnegative compositions is to note that they are in bijection with the number of ways of choosing $k-1$ elements (the bars) from a set of size $n+1$ (the gaps), except that we are allowed to choose gaps as many times as we wish.

Definition 15.11. Let n and k be nonnegative integers. The *multichoose number* $\left(\!\binom{n}{k}\!\right)$ is the number of ways of selecting k objects from a set of size n, where we are allowed to choose objects as many times as desired.

Definition 15.12. A collection of unordered elements with repetition allowed is called a *multiset*.

We have already essentially proven the following theorem, but let's make it more precise:

Theorem 15.13. *For any nonnegative integers n and k, we have*

$$\left(\!\binom{n}{k}\!\right) = \binom{n+k-1}{k}.$$

Proof. We give a bijection between sets counted by the left and right sides. The left side counts the number of multisets of size k from $[n]$, whereas the right side counts the number of sets—with no repetition allowed—of size k from $[n+k-1]$. Let us write $\left(\!\binom{[n]}{k}\!\right)$ and $\binom{[n+k-1]}{k}$ for these two sets.

Let us start with a multiset $S \in \left(\!\binom{[n]}{k}\!\right)$ and produce a set $f(S) \in \binom{[n+k-1]}{k}$ out of it. Write S in nondecreasing order as $a_1 \leq a_2 \leq \cdots \leq a_k$. Now define $b_i = a_i + i - 1$. Since $b_{i+1} - b_i = (a_{i+1} + i) - (a_i + i - 1) = a_{i+1} - a_i + 1 \geq 1$, we have $b_{i+1} > b_i$, so the b_i's form a strictly increasing sequence, i.e. the b_i's are distinct. Furthermore, $b_i \leq n+i-1$, so the largest any b_i can be is $b_k \leq n+k-1$. Thus the b_i's form a set in $\binom{[n+k-1]}{k}$, which we call $f(S)$.

Now we have to produce an inverse map to f, which we call g: out of a set $T \in \binom{[n+k-1]}{k}$, we must produce a multiset $g(T) \in \left(\!\!\binom{[n]}{k}\!\!\right)$. Write T in increasing order as $b_1 < b_2 < \cdots < b_k$. Then define $a_i = b_i - i + 1$. The multiset $g(T)$ is simply the set of a_i's. Since the b_i's are strictly increasing, the a_i's are nondecreasing, and they lie in $[n]$. Thus $g(T) \in \left(\!\!\binom{[n]}{k}\!\!\right)$, as desired.

Finally, we must show that f and g are inverses. Given $S \in \left(\!\!\binom{[n]}{k}\!\!\right)$, we must show that $g(f(S)) = S$, and similarly, if $T \in \binom{[n+k-1]}{k}$, then we must show that $f(g(T)) = T$. Let us take some $S \in \left(\!\!\binom{[n]}{k}\!\!\right)$, say with $S = (a_1 \leq a_2 \leq \cdots \leq a_k)$. Then $f(S) = (a_1 < a_2+1 < a_3+2 < \cdots < a_k+k-1)$, and $g(f(S)) = (a_1 \leq (a_2+1)-1 \leq (a_3+2)-2 \leq \cdots \leq (a_k+k-1)-(k-1)) = S$, as desired. Similarly, $f(g(T)) = T$ for all $T \in \binom{[n+k-1]}{k}$. This completes the proof. ∎

Example. Let S be the multiset $\{1,1,4,4,6,7,7,7\}$ in $\left(\!\!\binom{[8]}{8}\!\!\right)$. The corresponding subset in $\binom{[15]}{8}$ is $\{1,2,6,7,10,12,13,14\}$.

15.4 Further reading

Since the binomial coefficients are the most important numbers in combinatorics, there is no shortage of identities involving them. A good place to look for far more of them is [GKP94, Chapter 5].

15.5 Problems

(1) Prove the following identities combinatorially, i.e. by explaining why they count the same thing:

(a)
$$\sum_{k=0}^{n} k\binom{n}{k} = n \cdot 2^{n-1}.$$

(b)
$$\sum_{n=r}^{m} \binom{n}{r} = \binom{m+1}{r+1}.$$

(This is known as the *hockey-stick identity*.)

(c)
$$\binom{n}{k} = \frac{n}{k}\binom{n-1}{k-1},$$
assuming $k \neq 0$.

(d)
$$\binom{n}{k}\binom{k}{m} = \binom{n}{m}\binom{n-m}{k-m}.$$

(e)
$$\sum_{k=0}^{\lfloor n/2 \rfloor} \binom{n-k}{k} = F_{n+1},$$

where F_r is the rth Fibonacci number, defined by $F_0 = 0$, $F_1 = 1$, and $F_r = F_{r-1} + F_{r-2}$ for $r \geq 2$.

(2) Given nonnegative integers k and n, how many solutions are there to
$$a_1 + a_2 + \cdots + a_k \leq n,$$
where each a_i must be a nonnegative integer? Give your answer in closed form, without a $\sum$ or $\cdots$.

(3) (a) There are n chairs laid out in a row. In how many ways can k people sit in the chairs, such that no two people sit in the same chair or in adjacent chairs?

(b) Same question, but now the chairs are arranged in a circle instead of in a row.

(4) Prove the following generalization of Vandermonde's Identity:
$$\sum_{k_1 + \cdots + k_d = r} \binom{n_1}{k_1}\binom{n_2}{k_2}\cdots\binom{n_d}{k_d} = \binom{n_1 + \cdots + n_d}{r}.$$

(5) Evaluate the sum
$$\sum_{0 \leq i \leq j \leq n} \binom{n}{i}\binom{n}{j}.$$

(6) Prove that, for every positive integer k and every nonnegative integer n, $\binom{kn}{n}$ is divisible by $(k-1)n + 1$. If you can, find a combinatorial interpretation of this fact by explaining what $\frac{1}{(k-1)n+1}\binom{kn}{n}$ counts.

(7) Given nonnegative integers k, n and r, how many solutions are there to the equation
$$a_1 + a_2 + \cdots + a_k = n,$$
where each a_i is a nonnegative integer greater than or equal to r?

(8) In terms of n and k, determine when $\binom{n}{k}$ is even, and when it is odd. (Fun aside for mathematical culture: make a very large Pascal's triangle, i.e. one with many rows, and shade the even entries in the triangle to get an interesting pattern!)

(9) Give a *combinatorial* proof to show that $\binom{n}{k}^2 \geq \binom{n}{k-1}\binom{n}{k+1}$ for all n and k. This means showing that both sides count something, and explaining why there are at least as many elements counted by the left side as by the right side. A sequence $a_0, a_1, \ldots, a_n$ such that $a_k^2 \geq a_{k-1}a_{k+1}$ is said to be *log-concave*, so this shows that the binomial coefficients $\binom{n}{\cdot}$ are log-concave.

(10) How many ways are there to write a positive integer n as the sum of k positive integers, if an integer m can be colored with m different colors? (Order matters, so for instance $8 = 3 + 2 + 3$ is different from $8 = 2 + 3 + 3$.)

Chapter 16

Lattice paths

16.1 Lattice paths

One place where binomial coefficients show up is in counting lattice paths.

Question 16.1. *How many paths are there from* $(0,0)$ *to* (m,n), *where each step goes one unit to the right or one unit up?*

See Figure 16.1 for an example of such a path from $(0,0)$ to $(5,4)$.

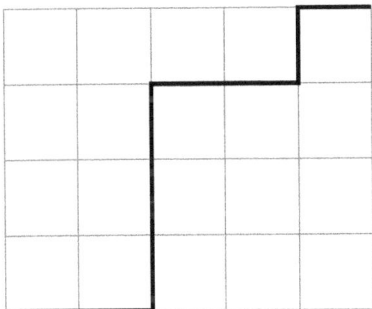

Figure 16.1. A lattice path from $(0,0)$ to $(5,4)$.

It is easy to count such paths in terms of binomial coefficients. To do this, let us write down a path as a string of letters: we write R whenever we take a right step and U whenever we take an up step. Thus the path shown in Figure 16.1 is RRUUURRUR.

Observe that getting from $(0,0)$ to $(5,4)$ takes 9 steps, and of those 5 of them are right steps and 4 are up steps. More generally, in order to get from $(0,0)$ to (m,n), we must take $m+n$ steps. Of those, m must be right steps and n must be up steps. Thus the number of paths is the same as the number of ways of choosing m letters from $m+n$ to be R's, forcing the rest to be U's. This number is simply $\binom{m+n}{m}$.

16.2   Dyck paths and Catalan numbers

That was too easy, so let's up the ante a bit. Let's place a restriction on the paths.

Question 16.2. *How many paths are there from $(0,0)$ to (n,n), taking steps one unit to the right or one unit up, that never cross above the diagonal line $y = x$?*

We call such paths *Dyck paths*. An example is shown in Figure 16.2.

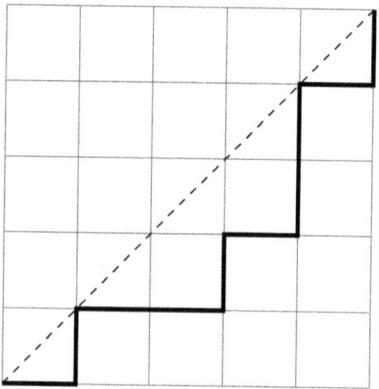

Figure 16.2. A Dyck path.

These are more difficult to count. However, it turns out that there is still an elegant formula for them.

Theorem 16.3. *The number of paths from $(0,0)$ to (n,n), taking steps to the right or up and never going above the line $y = x$, is equal to $\frac{1}{n+1}\binom{2n}{n}$.*

There are many proofs of Theorem 16.3. We'll see two of them.

Proof 1. Our approach is based on the fact that $\frac{1}{n+1}\binom{2n}{n} = \binom{2n}{n} - \binom{2n}{n-1}$, which can be checked by straightforward computation.

Let us consider an arbitrary path from $(0,0)$ to (n,n) using right and up steps, regardless of whether it goes above the line $y = x$. If it never goes above the line, then it's one of the paths we want. If it ever does go above the line, then it does so for the first time at some point, moving from (k,k) to $(k,k+1)$. Up to that point, the path consists of k right steps and $k+1$ up steps. After that, there are $n-k$ right steps and $n-k-1$ up steps. Let us now modify the path, in the following way: the part of the path from $(0,0)$ to $(k,k+1)$ will be left unchanged, but after that, every up step becomes a right step, and every right step becomes an up step. (See Figure 16.3 for an example.) After this transformation, we have a total of $k+(n-k-1) = n-1$ right steps and $(k+1)+(n-k) = n+1$ up steps, so the path ends at $(n-1, n+1)$. All paths ending at $(n-1, n+1)$ are obtainable in this way, and every bad path leads to a different path to $(n-1, n+1)$, so there are $\binom{2n}{n-1}$ such bad paths. Thus there are $\binom{2n}{n} - \binom{2n}{n-1}$ good paths, as desired. ∎

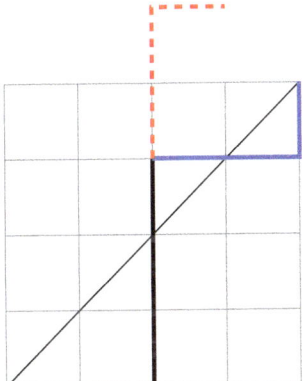

Figure 16.3. The dashed red path is the flip of the blue part of the original path.

The numbers $\frac{1}{n+1}\binom{2n}{n}$ appear all over the place in combinatorics. They are known as the *Catalan numbers* and denoted C_n.

Proof 2. Observe that $C_n = \frac{1}{n+1}\binom{2n}{n} = \frac{1}{2n+1}\binom{2n+1}{n}$, by straightforward calculation. We will give a combinatorial interpretation for $\frac{1}{2n+1}\binom{2n+1}{n}$. To do this, let us consider paths from $(0,0)$ to $(n+1,n)$ that never go above the line connecting $(0,0)$ and $(n+1,n)$; see Figure 16.4. The first step must be a right step, but after that we obtain a Dyck path from $(1,0)$ to $(n+1, n)$, never passing above the line connecting $(1,0)$ and $(n+1, n)$. Furthermore, any Dyck path from $(0,0)$ to (n, n) can be converted into one of these new types of paths by starting with a right step and then following the Dyck path—taking a right step whenever the Dyck path does and an up step whenever the Dyck path does; see Figure 16.5. Thus the number of such paths is also counted by the Catalan numbers C_n.

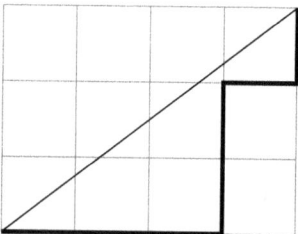

Figure 16.4. A good path from $(0,0)$ to $(4,3)$, never going above the line.

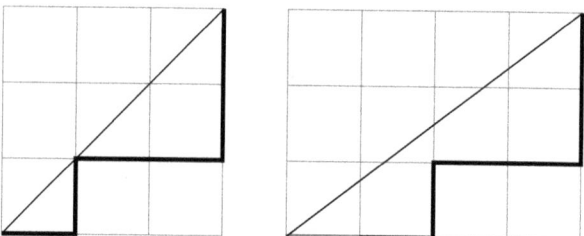

Figure 16.5. Converting a Dyck path from $(0,0)$ to $(3,3)$ to a good path from $(0,0)$ to $(4,3)$ by adding a right step at the beginning.

Now, let us suppose we have any path of up and right steps from $(0,0)$ to $(n+1, n)$, whether or not it stays below the line. Let us write it as a sequence $a_1, a_2, \ldots, a_{2n+1}$, where each a_i is either "U" or "R" depending on whether it is an up step or a right step. We now consider $2n+1$ paths obtainable from this path, all going from $(0,0)$ to $(n+1, n)$, by performing cyclic shifts of the path. The first path is the original one. The second path is $a_2, a_3, a_4, \ldots, a_{2n+1}, a_1$. The third path is $a_3, a_4, \ldots, a_{2n+1}, a_1, a_2$, and so forth. Note that all these paths are necessarily different.

We claim that *exactly* one of these paths stays below the line, which then proves the theorem. To see this, let us take two copies of the original path and concatenate them, yielding a path from $(0,0)$ to $(2n+2, 2n)$. At the start of each of the $2n+1$ original steps, draw a line parallel to the line from $(0,0)$ to $(n+1, n)$ starting at the current location. All of them end at another point on the extended path. All of these lines are different, so one of them is above all the rest of them. The path starting there is a good path. See Figure 16.6 for an example. ∎

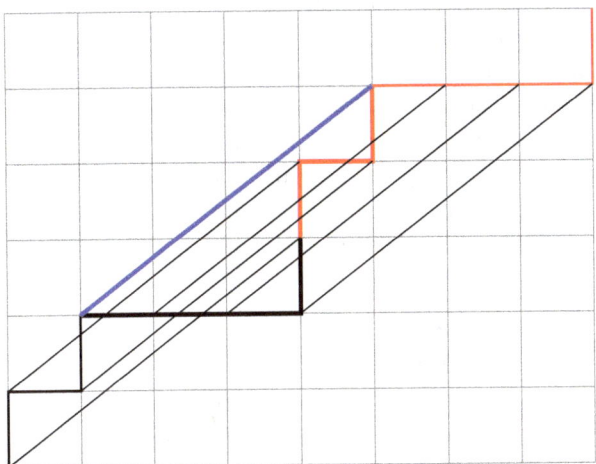

Figure 16.6. Starting with the path URURRRU, we obtain the good path RRRUURU, the unique good cyclic shift of the original path.

16.3 The ubiquitous Catalan numbers

The Catalan numbers count much more than just Dyck paths. Richard Stanley, in his famous 2-volume book [Sta12, Sta99] on enumerative combinatorics, has an exercise (Exercise 6.19 from Volume 2) that lists 66 combinatorial interpretations of the Catalan numbers; the exercise is to take every pair of them and give an explicit bijection between them. Thus this "exercise" is actually $\binom{66}{2}$ exercises: quite a serious undertaking! If that weren't enough, he has expanded his list to 214 entries and has written an entire book [Sta15] on the Catalan numbers and the many, many things they count. We won't look at all of them, but let's see just a handful of them.

One of these other interpretations of the Catalan numbers is that they are the number of parenthesizations of an expression involving $n + 1$ symbols into pairs. For example, if $n = 3$, we can look at various ways of parenthesizing the product $abcd$; they are

$$(((ab)c)d), \quad ((ab)(cd)), \quad ((a(bc))d), \quad (a((bc)d)), \quad (a(b(cd))).$$

We require that each corresponding pair of parentheses contains two symbols or symbol groups, where a symbol group is either a single symbol or a parenthesized expression. For example, in $(a(bc))$, the outermost pair of parentheses parenthesizes two groups, the first of which is the symbol a, and the second of which is the parenthesized group (bc).

We claim that there are C_n ways of parenthesizing $n + 1$ symbols, such that expressions are parenthesized in pairs, as above. To see this, we give a bijection with Dyck paths from $(0,0)$ to (n, n). To form the bijection, we start with a parenthesized product, and then we delete all the right parentheses as well as the rightmost variable, which in this case is d. We obtain the following expressions, which may be painful to read:

$$(((abc, \quad ((ab(c, \quad ((a(bc, \quad (a((bc, \quad (a(b(c.$$

Observe that these are all different. We can always recover the fully parenthesized expression by putting the last variable back at the end, and then, working from left to right, reinserting a right parenthesis whenever we have two consecutive symbols or groups of symbols.

Example. Let's give an example in the case of a longer string. Consider the string

$$(((a(bc((de(fg.$$

How do we insert the remaining symbols? First, we put the last character h back in, to get

$$(((a(bc((de(fgh.$$

Now, we work from the left. The first time we have two adjacent symbols is bc, so we close off the parenthesis before bc and make (bc) into a group. This gives

$$(((a(bc)((de(fgh.$$

Now that (bc) counts as a group, we have two adjacent groups: a and (bc), so that pair needs to be correctly parenthesized. When we do so, we get

$$(((a(bc))((de(fgh.$$

Next up is the pair (de), so we cap that off to get

$$(((a(bc))((de)(fgh.$$

Next is fg, yielding

$$(((a(bc))((de)(fg)h.$$

Now (de) and (fg) are adjacent blocks, so we pair them to get

$$(((a(bc))((de)(fg))h.$$

The next set of adjacent blocks are the blocks $(a(bc))$ and $((de)(fg))$, so we group them to get

$$(((a(bc))((de)(fg)))h.$$

Finally, the h gets grouped with everything else, so that the final parenthesization is

$$(((a(bc))((de)(fg)))h).$$

If we go backward, deleting the h and all the right parentheses, we do indeed get

$$(((a(bc((de(fg,$$

which is what we started with.

Now it's easy to get a bijection with Dyck paths. Starting with a Dyck path, replace every right step with a left parenthesis—the (symbol—and replace every up step with a variable. Then perform the above process to reinsert the right parentheses and last variable. This process is invertible, i.e. we can start with a parenthesized expression and get back a Dyck path, inverse to the other direction, so this is a bijection.

Another thing counted by Catalan numbers is binary trees with $n + 1$ leaves.

Definition 16.4. A *tree* is a connected graph with no cycles. Alternatively, a tree is a connected graph in which $|V| = |E| + 1$. A *rooted tree* is a tree together with a distinguished vertex, called the *root*. Given a rooted tree and a vertex v, each neighbor of v is either closer to the root or further away from the root than v. Ones that are further away are called *children*. A *binary tree* is a rooted tree in which every vertex has either 0 or 2 children. In the latter case, one of the children is called the *left child*, and the other is called the *right child*. A *leaf* in a binary tree is a vertex with no child nodes emanating from it.

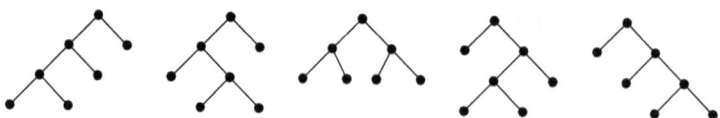

Figure 16.7. The five binary trees with four leaves.

It's easier to understand what binary trees are pictorially. The five binary trees with 4 leaves are shown in Figure 16.7.

Why are these counted by the Catalan numbers? This is because there is a bijection with parenthesized expressions. We label the leaves from left to right by variables $x_1, \ldots, x_{n+1}$ (or a, b, c, d when there are four of them), and we use the tree structure to get a parenthesization. This is best shown by example, and we illustrate one in Figure 16.8.

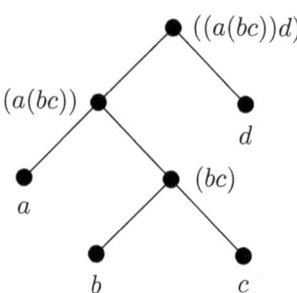

Figure 16.8. Converting a tree to a parenthesization.

16.4   A recurrence for Catalan numbers

The Fibonacci numbers satisfy the familiar recurrence $F_{n+2} = F_n + F_{n+1}$. Similarly, the Catalan numbers satisfy a recurrence, albeit a slightly more complicated one:

Theorem 16.5. *For all integers $n \geq 0$, we have*

$$C_{n+1} = C_0 C_n + C_1 C_{n-1} + \cdots + C_n C_0 = \sum_{k=0}^{n} C_k C_{n-k}.$$

Note that the subscript on the left side is $n + 1$, whereas in each term on the right, the sum of the subscripts is only n, i.e. they are off by one.

Proof. Given a Dyck path from $(0,0)$ to $(n+1, n+1)$, let us look at the first time after $(0,0)$ that it hits the line $y = x$. Suppose it next hits at $(k+1, k+1)$, where $0 \le k \le n$. Consider the subpath of the Dyck path from $(0,0)$ to $(k+1, k+1)$. Clearly, this path starts with a right step and ends with an up step. Between those, we have a path from $(1,0)$ to $(k+1, k)$. Since we assume that $(k+1, k+1)$ is the *first* time the path hits the line $y = x$, the subpath from $(1,0)$ to $(k+1, k)$ cannot hit the line $y = x$, i.e. it stays strictly below it. In other words, it is a path from $(1,0)$ to $(k+1, k)$ never crossing the line $y = x - 1$. But that's just a shifted Dyck path of length k, and so there are C_k such subpaths.

After the point $(k+1, k+1)$, we have a path from $(k+1, k+1)$ to $(n+1, n+1)$ which never crosses above the line $y = x$. This is a shifted version of a path from $(0,0)$ to $(n-k, n-k)$ not crossing above the $y = x$ line, so there are C_{n-k} of those. In total then, there are $C_k C_{n-k}$ Dyck paths from $(0,0)$ to $(n+1, n+1)$ that first hit the line $y = x$ at $(k+1, k+1)$. (This works even when $k = n$, when the path stays entirely below the diagonal before finishing at $(n+1, n+1)$.)

Summing from $k = 0$ to n now gives the desired result. ∎

16.5 Problems

(1) $2n$ (distinguishable) people are waiting to buy tickets to a show that costs \$5. n of them have only a \$5 bill, and the other n of them have only a \$10 bill. The teller has no change to start with. In how many ways can the people line up such that each person can buy a ticket and get the correct change (if necessary) without having to switch their order?

(2) How many lattice paths are there from $(0,0)$ to (n,n) that do not touch the diagonal at (i, i) for any i with $1 \le i \le n - 1$?

(3) How many paths are there in $\mathbb{Z}^3$ from $(0,0,0)$ to (a, b, c), with steps of the form $(1,0,0)$, $(0,1,0)$, and $(0,0,1)$?

(4) Suppose you start at $(0,0)$, and at every second, you make a random step of the form $(1,0)$ or $(0,1)$, each with probability $\frac{1}{2}$. In terms of m and n, what is the probability that you reach (m, n) at some point?

(5) Show that the number of elements in each of the following sets is C_n, the nth Catalan number, by giving a bijection with the set of Dyck paths or some other Catalan object you know.

(a) Sequences of length $2n$ consisting of n a's and n b's, such that for each $1 \leq k \leq 2n$, the number of a's among the first k letters is at least as large as the number of b's among the first k letters.

(b) Triangulations of a regular $(n + 2)$-gon into n triangles. (See Figure 16.9.)

(c) Ways of filling in a $2 \times n$ grid with numbers from 1 to $2n$ (each number used exactly once), such that the rows are increasing from left to right and the columns are increasing from top to bottom. (See Figure 16.10.)

(d) Sequences $1 \leq a_1 \leq a_2 \leq \cdots \leq a_n$ of integers with $a_i \leq i$ for all i.

(e) Sequences $a_1 < a_2 < \cdots < a_{n-1}$ of integers such that $1 \leq a_i \leq 2i$ for all i.

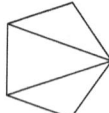 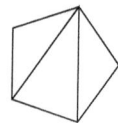  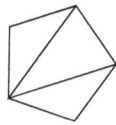 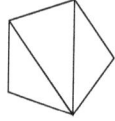

Figure 16.9. The five triangulations of a regular pentagon.

1	2	3
4	5	6

1	2	4
3	5	6

1	2	5
3	4	6

1	3	4
2	5	6

1	3	5
2	4	6

Figure 16.10. The five ways of filling a $2 \times n$ grid with numbers from 1 to 6 with increasing rows and columns.

(6) Suppose that $D_0, D_1, D_2, \ldots$ is a sequence of numbers such that $D_0 = 1$ and, for all $n \geq 0$, we have

$$D_{n+1} = D_0 D_n + D_1 D_{n-1} + \cdots + D_n D_0. \qquad (16.1)$$

Prove that D_n is the nth Catalan number for all n. (Make sure you understand what is being asked here: you are *not* being asked to prove that the Catalan numbers satisfies this recurrence, something we already did in Theorem 16.5.)

(7) Prove directly that the number of binary trees with $n + 1$ leaves satisfies the Catalan recurrence (16.1). This, together with the previous problem, gives another proof that the number of binary trees with $n + 1$ leaves is equal to C_n.

(8) Let a and b be two positive integers with $\gcd(a, b) = 1$. How many paths are there from $(0, 0)$ to (a, b) that never go above the line connecting $(0, 0)$ and (a, b)? When $a = b + 1$, this is a Catalan number, as shown in proof 2. In general, this number is known as a *rational Catalan number*.

(9) Prove combinatorially that if $\gcd(a, b) = 1$, then $\binom{a+b}{a}$ is divisible by $a + b$. Show that this is not necessarily the case if $\gcd(a, b) \neq 1$.

Chapter 17

The Binomial Theorem

17.1   The Binomial Theorem

The Binomial Theorem, one of the most frequently used theorems in mathematics, concerns the expansion of an expression of the form $(x+y)^n$, where n is a nonnegative integer.

Theorem 17.1 (Binomial Theorem). *If n is a nonnegative integer, then*

$$(x + y)^n = \sum_{k=0}^{n} \binom{n}{k} x^k y^{n-k}.$$

We'll give two proofs of the Binomial Theorem.

Proof 1. We will prove it by induction on n. Our base case is $n = 0$, where we have $(x + y)^0 = 1$, which is indeed the case. Now, let's do the inductive step. Suppose that the result is true for n, i.e.

$$(x + y)^n = \sum_{k=0}^{n} \binom{n}{k} x^k y^{n-k}.$$

We must show that it is true for $n + 1$, i.e.

$$(x + y)^{n+1} = \sum_{k=0}^{n+1} \binom{n + 1}{k} x^k y^{n+1-k}.$$

199

To do this, we simply calculate:

$$(x+y)^{n+1} = (x+y)(x+y)^n$$

$$= (x+y) \sum_{k=0}^{n} \binom{n}{k} x^k y^{n-k}$$

$$= \sum_{k=0}^{n} \binom{n}{k} x^{k+1} y^{n-k} + \sum_{k=0}^{n} \binom{n}{k} x^k y^{n+1-k}$$

$$= \sum_{k=1}^{n+1} \binom{n}{k-1} x^k y^{n-k+1} + \sum_{k=0}^{n} \binom{n}{k} x^k y^{n+1-k}$$

$$= \sum_{k=0}^{n+1} \left[\binom{n}{k-1} x^k y^{n+1-k} + \binom{n}{k} x^k y^{n+1-k} \right]$$

$$= \sum_{k=0}^{n+1} \left[\binom{n}{k-1} + \binom{n}{k} \right] x^k y^{n+1-k}$$

$$= \sum_{k=0}^{n+1} \binom{n+1}{k} x^k y^{n+1-k}.$$

In the last step, we used Pascal's identity $\binom{n}{k-1} + \binom{n}{k}$, and in a previous step we extended our sums. They were originally from $k=1$ to $n+1$ and $k=0$ to n. Afterwards, both went from $k=0$ and $n+1$. This does not change anything, because the extra terms are zero. This completes the proof. ∎

And now for the second proof:

Proof 2. Let us compute the coefficient of $x^k y^{n-k}$ in $(x+y)^n$. When we expand out $(x+y)^n$, we must choose either an x or a y in each factor of $x+y$. Among the n factors, we must choose k of the x's and $n-k$ of the y's. The number of ways of doing this is $\binom{n}{k}$, so the coefficient of $x^k y^{n-k}$ in $(x+y)^n$ is $\binom{n}{k}$. Summing over all k, we have

$$(x+y)^n = \sum_{k=0}^{n} \binom{n}{k} x^k y^{n-k},$$

as desired.   ∎

A simple consequence of the Binomial Theorem is the following result, which we already saw in Chapter 15:

Corollary 17.2. *For any nonnegative integer n, we have*

$$\sum_{k=0}^{n} \binom{n}{k} = 2^n.$$

Proof. Plug in $x = y = 1$ into the Binomial Theorem. ∎

This says that the total number of subsets of a set with n elements is 2^n. Another familiar consequence of the Binomial Theorem is that, if $n \geq 1$, there are the same number of subsets of $[n]$ with an even number of elements as with an odd number of elements.

Corollary 17.3. *For each integer $n \geq 1$, we have*

$$\sum_{k=0}^{n} (-1)^k \binom{n}{k} = 0.$$

Proof. Let $x = -1$ and $y = 1$ in the Binomial Theorem. ∎

17.2 Proofs using the Binomial Theorem

One thing we can do with the Binomial Theorem is to use it to prove combinatorial identities. Let's see how to use it to prove a familiar theorem that we already proved in Chapter 15.

Theorem 17.4 (Vandermonde's Identity). *Let m and n be nonnegative integers, and suppose $r \leq m + n$. Then we have*

$$\sum_{k=0}^{r} \binom{m}{k}\binom{n}{r-k} = \binom{m+n}{r}.$$

We previously proved Vandermonde's identity using a direct combinatorial argument, but this time we'll prove it using the Binomial Theorem.

Proof. Since $(x+1)^{m+n} = (x+1)^m (x+1)^n$, we have

$$\sum_{r=0}^{m+n} \binom{m+n}{r} x^r = (x+1)^{m+n}$$

$$= (x+1)^m (x+1)^n$$

$$= \left(\sum_{k=0}^{m} \binom{m}{k} x^k \right) \left(\sum_{\ell=0}^{n} \binom{n}{\ell} x^\ell \right)$$

$$= \sum_{r=0}^{m+n} \left(\sum_{k=0}^{r} \binom{m}{k}\binom{n}{r-k} \right) x^r,$$

where in the final step we grouped terms in the product based on their power of x. Comparing the coefficients of x^r, we get

$$\binom{m+n}{r} = \sum_{k=0}^{r} \binom{m}{k}\binom{n}{r-k},$$

as desired.   ∎

17.3   Binomial Theorem tricks

The Binomial Theorem tells us how to handle sums of the form $\sum_{k=0}^{n} \binom{n}{k}x^k$, and related things. But what happens if we don't want *all* the terms in the sum, but only some of them? For example, what happens if we only want the even terms? That is, we'd like to evaluate the sum

$$\sum_{\substack{0 \le k \le n \\ k \text{ even}}} \binom{n}{k} x^k.$$

To do this, we use a classic Binomial Theorem trick. Consider the sum $(1+x)^n + (1-x)^n$. By the Binomial Theorem, we have

$$(1+x)^n + (1-x)^n = \sum_{k=0}^{n} \binom{n}{k}x^k + \sum_{k=0}^{n} \binom{n}{k}(-x)^k$$

$$= \sum_{k=0}^{n} \binom{n}{k}\left(x^k + (-x)^k\right)$$

$$= 2 \sum_{\substack{0 \le k \le n \\ k \text{ even}}} \binom{n}{k}x^k.$$

Dividing by 2, we get

$$\sum_{\substack{0 \le k \le n \\ k \text{ even}}} \binom{n}{k} x^k = \frac{(1+x)^n + (1-x)^n}{2}.$$

If we want the multiples of 4, we can use a similar trick, known as the *roots-of-unity filter*. In order to evaluate

$$\sum_{\substack{0 \le k \le n \\ 4 \mid k}} \binom{n}{k},$$

we consider the sum $(1+1)^n + (1-1)^n + (1+i)^n + (1-i)^n$, where $i^2 = -1$. By the Binomial Theorem, we have

$$(1+1)^n + (1-1)^n$$

$$+(1+i)^n + (1-i)^n = \sum_{k=0}^{n} \binom{n}{k} + \sum_{k=0}^{n} (-1)^k \binom{n}{k}$$

$$+ \sum_{k=0}^{n} i^k \binom{n}{k} + \sum_{k=0}^{n} (-i)^k \binom{n}{k}$$

$$= \sum_{k=0}^{n} \left(1 + (-1)^k + i^k + (-i)^k\right) \binom{n}{k}.$$

Now, note that depending on $k \pmod 4$, we get the following values of 1^k, $(-1)^k$, i^k, and $(-i)^k$:

	1^k	$(-1)^k$	i^k	$(-i)^k$	Σ
0 (mod 4)	1	1	1	1	4
1 (mod 4)	1	-1	i	$-i$	0
2 (mod 4)	1	1	-1	-1	0
3 (mod 4)	1	-1	$-i$	i	0

Thus

$$1^k + (-1)^k + i^k + (-i)^k = \begin{cases} 4 & k \equiv 0 \pmod 4, \\ 0 & k \not\equiv 0 \pmod 4. \end{cases}$$

Thus we have

$$2^n + (1+i)^n + (1-i)^n = 4 \sum_{\substack{0 \le k \le n \\ 4 \mid k}} \binom{n}{k},$$

or

$$\sum_{\substack{0 \le k \le n \\ 4 \mid k}} \binom{n}{k} = \frac{2^n + (1+i)^n + (1-i)^n}{4}.$$

It would be a little nicer if our answer didn't have i's in it. To deal with this, we'll just assume that n is a multiple of 4 for now, since $(1+i)^4 = (1-i)^4 = -4$. In that case, we get the slightly simpler expression

$$\sum_{\substack{0 \le k \le n \\ 4 \mid k}} \binom{n}{k} = \frac{2^n + (-4)^{n/4} + (-4)^{n/4}}{4} = 2^{n-2} + \frac{(-4)^{n/4}}{2}.$$

We can derive similar expressions if n is not a multiple of 4, but they are just a little bit uglier.

17.4   More Binomial Theorem tricks

Here's a cute, but not very deep, application of the Binomial Theorem.
Consider the powers of 11: we have

$$11^1 = 11, \qquad 11^2 = 121, \qquad 11^3 = 1331, \qquad 11^4 = 14641.$$

Note that the digits are just the binomial coefficients—or if you prefer, the
numbers you get by reading across the rows of Pascal's triangle. If you look
at $11^5 = 161051$, it may seem that the pattern breaks down. But this is
because the digits start to blend together. What is going on?

This is just the Binomial Theorem in action: we have

$$11^n = (10+1)^n = \sum_{k=0}^{n} \binom{n}{k} 10^k.$$

So, when $n = 4$, we get

$$11^4 = 10^4 + 4 \cdot 10^3 + 6 \cdot 10^2 + 4 \cdot 10^1 + 1 \cdot 10^0 = 14641.$$

When $n = 5$, we get

$$11^5 = 10^5 + 5 \cdot 10^4 + 10 \cdot 10^3 + 10 \cdot 10^2 + 5 \cdot 10^1 + 1 \cdot 10^0.$$

The problem is that here the digits start to blend together. When $n = 4$
and we write the sum in vertical format, we have

```
                1
              4 0
            6 0 0
          4 0 0 0
      +   1 0 0 0 0
      _____
          1 4 6 4 1
```

Note that we never have to carry, and each term fits into its designated
column. On the other hand, when $n = 5$, we get

```
                  1
                5 0
            1 0 0 0
          1 0 0 0 0
          5 0 0 0 0
      +   1 0 0 0 0 0
      _____
          1 6 1 0 5 1
```

Some of the terms spill over into neighboring columns (in particular, the 10's), leading to the apparent disruption of the pattern. It takes longer to happen if you look at powers of 101 instead. For example, $101^5 = 10510100501$, and you see the 1, 5, 10, 10, 5, 1 in their rightful places, distinguished by colors.

Let's try another Binomial Theorem trick.

Question 17.5. *What is the last digit before the decimal point in* $(2 + \sqrt{5})^{2019}$ *?*

Since we have an expression of the form $(x + y)^n$, we're tempted to use the Binomial Theorem to expand. When we do this, we get

$$(2 + \sqrt{5})^{2019} = 2^{2019} + 2019 \cdot 2^{2018}\sqrt{5} + \frac{2019 \cdot 2018}{2}2^{2017} \cdot 5 + \cdots$$

$$= \sum_{k=0}^{2019} \binom{2019}{k} 5^{k/2} 2^{2019-k}.$$

It's not quite clear what to do with that at the moment, but the trick is to use $2 - \sqrt{5}$ as well. We have

$$(2 - \sqrt{5})^{2019} = \sum_{k=0}^{2019} (-1)^k \binom{2019}{k} 5^{k/2} 2^{2019-k}.$$

When we add the two together, we get

$$(2 + \sqrt{5})^{2019} + (2 - \sqrt{5})^{2019}$$

$$= \sum_{k=0}^{2019} \left(\binom{2019}{k} 5^{k/2} 2^{2019-k} + (-1)^k \binom{2019}{k} 5^{k/2} 2^{2019-k} \right).$$

When k is odd, the stuff in parentheses is 0, and when k is even, the two terms in parentheses are equal. Thus we have

$$(2 + \sqrt{5})^{2019} + (2 - \sqrt{5})^{2019} = 2 \sum_{\substack{0 \le k \le 2019 \\ k \text{ even}}} \binom{2019}{k} 5^{k/2} 2^{2019-k}.$$

Observe that the right side is an integer (being a sum of a bunch of integers), and therefore so is the left side. Since we only care about the last digit, that means we can ignore any term divisible by both a 2 and a 5. That's almost all of them: the only one that isn't is the $k = 0$ term, which is divisible by 2 but not by 5. Thus we have

$$(2 + \sqrt{5})^{2019} + (2 - \sqrt{5})^{2019} \equiv 2 \times 2^{2019} \equiv 2^{2020} \equiv 6 \pmod{10},$$

where the last step comes from the Chinese Remainder Theorem. By Fermat's Little Theorem, we have $2^{2020} \equiv 1 \pmod 5$, and certainly $2^{2020} \equiv 0 \pmod 2$, so by combining them we end up with $2^{2020} \equiv 6 \pmod{10}$.

Now we've figured out the last digit of $(2+\sqrt 5)^{2019} + (2-\sqrt 5)^{2019}$, but we need to get rid of the $(2-\sqrt 5)^{2019}$ term. To do this, we note that $-1 < 2-\sqrt 5 < 0$, so $-1 < (2-\sqrt 5)^{2019} < 0$. Since $(2+\sqrt 5)^{2019}+(2-\sqrt 5)^{2019}$ is an integer ending with 6 and $(2-\sqrt 5)^{2019}$ is between -1 and 0, it follows that the last digit before the decimal point of $(2+\sqrt 5)^{2019}$ is a 6.

17.5   Further reading

So far, we have only considered the Binomial Theorem when n, the power to which $x+y$ is raised, is a nonnegative integer. However, we can plug in other values as well, as long as we are careful. Recall that in the Binomial Theorem, we get the binomial coefficients $\binom{n}{k}$ appearing as coefficients of $x^k y^{n-k}$. Combinatorially, $\binom{n}{k}$ doesn't make much sense when n isn't a nonnegative integer. For example, what is $\binom{-2}{3}$ supposed to mean? The number of ways of choosing three objects from among a set of -2 elements?

The trick is to stop thinking of binomial coefficients combinatorially, and start thinking of them as polynomials, where we can plug in values. For example, if $k = 3$, then we have

$$\binom{n}{3} = \frac{n(n-1)(n-2)}{6} = \frac{n^3}{6} - \frac{n^2}{2} + \frac{n}{3},$$

so we can easily plug in $n = -2$ to get $\binom{-2}{3} = \frac{-8}{6} - \frac{4}{2} - \frac{2}{3} = -4$.

Definition 17.6. Let k be a nonnegative integer. Set

$$n^{\underline{k}} = n(n-1)(n-2)\cdots(n-k+1) = \prod_{i=0}^{k-1}(n-i).$$

We call $n^{\underline{k}}$ the *falling factorial*. We define the *binomial coefficient* $\binom{n}{k}$ to be

$$\binom{n}{k} = \frac{n^{\underline{k}}}{k!}.$$

Notice that this new definition agrees with the old one $\binom{n}{k} = \frac{n!}{k!(n-k)!}$ whenever n and k are nonnegative integers with $n \geq k$, because $\frac{n!}{(n-k)!} = n^{\underline{k}}$: all the factors in $n!$ that are less than or equal to $n-k$ get canceled with the corresponding term in the denominator.

The extension of binomial coefficients allows us to extend the Binomial Theorem, albeit with a bit of care.

Theorem 17.7 (General Binomial Theorem). *If x and y are real numbers, with $|x| < |y|$, and α is a complex number, then*

$$(x+y)^\alpha = \sum_{k=0}^{\infty} \binom{\alpha}{k} x^k y^{\alpha-k}.$$

The condition $|x| < |y|$ is needed to guarantee that the sum converges. Letting $\alpha = \frac{1}{2}$ in the General Binomial Theorem, we find that

$$\sqrt{1+x} = 1 + \frac{x}{2} - \frac{x^2}{8} + \frac{x^3}{16} - \frac{5x^4}{128} + 7\frac{7x^5}{256} - \cdots .$$

The General Binomial Theorem is of crucial importance in the study of generating functions, among many other places. See for instance [Wil05] or [FS09].

17.6 Problems

(1) What is the coefficient of x^3 in $(5x + \frac{2}{x^2})^{24}$?

(2) Which coefficient of $(x+y)^n$ is the largest? What about $(x+2y)^n$? Prove that your answers are correct.

(3) Find a closed-form expression for

$$\sum_{k=0}^{n} \frac{1}{k+1} \binom{n}{k}.$$

(4) Find a closed-form expression for

$$\sum_{\substack{0 \le k \le n \\ k \text{ odd}}} \binom{n}{k} 3^k.$$

(5) Find a closed-form expression for the sum

$$\binom{n}{0} - \binom{n}{2} + \binom{n}{4} - \binom{n}{6} + \cdots = \sum_{k=0}^{\lfloor \frac{n}{2} \rfloor} (-1)^k \binom{n}{2k}.$$

(6) Prove that for all integers $n \ge 2$, we have

$$\frac{2^{2n-1}}{n} < \binom{2n}{n} < 4^n.$$

(7) Prove that if $n \ge 2$, we can find a collection S of at least $\frac{2^n}{n}$ subsets of $[n]$ such that no set in S contains another one.

(8) The n-dimensional cube consists of all points

$$(x_1, x_2, \ldots, x_n) \in \mathbb{R}^n$$

such that $0 \le x_i \le 1$ for all i with $1 \le i \le n$. A 0-dimensional face of a cube is a vertex, a 1-dimensional face is an edge, a 2-dimensional face is a square, and so on. The entire cube does not count as a face. Determine, as a function of n, the number of faces (of any dimension) that the n-dimensional cube has. (For example, the 3-dimensional cube has 8 0-dimensional faces, 12 1-dimensional faces, and 6 2-dimensional faces, for a total of 26 faces.)

(9) A complex number α with $|\alpha| > 1$ is said to be a *Pisot number* if there is a polynomial $f(x)$ with integer coefficients and leading coefficient equal to 1 such that $f(\alpha) = 0$, and all roots β of $f(x)$ other than α satisfy $|\beta| < 1$.

  (a) Show that $\frac{1+\sqrt{5}}{2}$, $1 + \sqrt{2}$, and $\frac{3+\sqrt{13}}{2}$ are Pisot numbers.

  (b) Suppose that $\alpha = a + b\sqrt{d}$, where a, b, d are integers, and $d > 0$ is not a perfect square. Prove that if α is a Pisot number, then α^n is very close to an integer for large values of n, i.e. for all $\varepsilon > 0$, there is an N such that α^n is within ε of an integer for all $n > N$. (We say that a number x is within ε of an integer if there is some integer n such that $|x - n| < \varepsilon$.)

Chapter 18

Inclusion-Exclusion

18.1 Inclusion-Exclusion

Suppose we have a finite set X with three subsets, A, B, and C. We wish to find the number of elements in $A \cup B \cup C$. In order to determine the size of $A \cup B \cup C$, we can start by estimating it as $|A| + |B| + |C|$. However, this is not quite right, because an element in two of these subsets is counted multiple times. For instance, an element in $A \cap B$ is counted once in A and once in B, yet we only want to count it once. So, we need to correct our count by subtracting off $|A \cap B|$. Similarly, we need to adjust for elements of $A \cap C$ and $B \cap C$. So our adjusted count is $|A| + |B| + |C| - |A \cap B| - |A \cap C| - |B \cap C|$. But this is still not quite right, because if we have an element in $A \cap B \cap C$, then we count it once in $|A|$, once in $|B|$, and once in $|C|$ for a total of three counts, but then we subtract it once in $|A \cap B|$, once in $|A \cap C|$, and once in $|B \cap C|$. Thus such elements are counted no times at all. So, we make one final adjustment, by adding back in $|A \cap B \cap C|$. Thus we have

$$|A \cup B \cup C| = |A| + |B| + |C| - |A \cap B| - |A \cap C| - |B \cap C| + |A \cap B \cap C|.$$

See Figure 18.1.

More generally, if we have n subsets, say $A_1, \dots, A_n$, then we have a very similar formula:

Theorem 18.1 (Principle of Inclusion-Exclusion). *Let X be a finite set, and let $A_1, \dots, A_n$ be subsets of X. Then*

$$\left| \bigcup_{i=1}^{n} A_i \right| = \sum_{i=1}^{n} |A_i| - \sum_{1 \le i < j \le n} |A_i \cap A_j| + \dots + (-1)^{n-1} \left| \bigcap_{i=1}^{n} A_i \right|.$$

Proof. Let's suppose we have an element x that is in exactly k of the n subsets $A_1, \dots, A_n$. If $k = 0$, then we do not count it anywhere, which

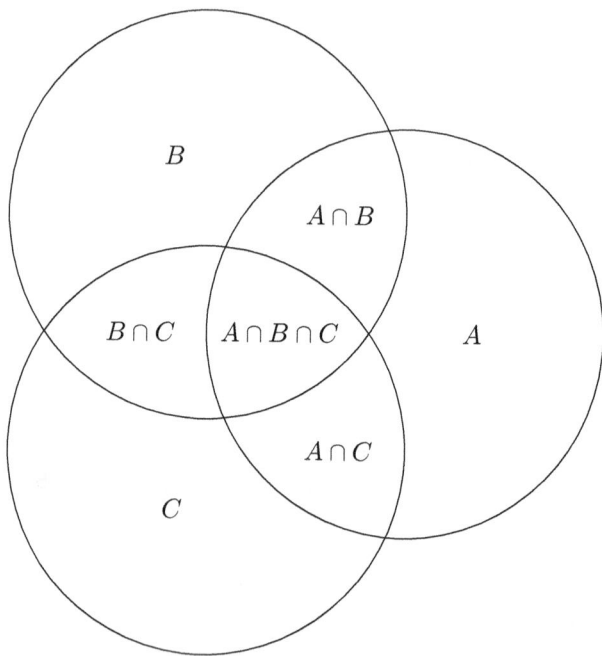

Figure 18.1. The diagram of inclusion-exclusion for three sets.

is what we want, so now let's assume that $k \geq 1$. Since the formula is symmetric in the subsets, we may assume that it is in $A_1, \ldots, A_k$, but not in $A_{k+1}, \ldots, A_n$. Let's see how many times we have counted x in the alternating sum. In the first term, taking one A_i at a time, we count it k times. In the second sum, we count it $\binom{k}{2}$ times, but with a negative sign, and in general in the ith sum, we count it $\binom{k}{i}$ times, with a sign of $(-1)^{i-1}$. Thus the total number of times x is counted is

$$\sum_{i=1}^{k}(-1)^{i-1}\binom{k}{i}.$$

If the sum started from 0 rather from 1, then we would have the full alternating sum of binomial coefficients, which we know to be 0. But we're missing a -1 from it, so the sum is 1, which is just what we want: such an element is counted exactly once. ∎

Example. Let us compute the number of positive integers less than 210 that are relatively prime to 210. We already know how to do this thanks to the formula for the totient function (Theorem 7.11), but we can also do it using the inclusion-exclusion formula, as we will now see. The prime factorization of 210 is $2 \times 3 \times 5 \times 7$, so we can investigate the numbers up to 210 that are divisible by each of 2, 3, 5, and 7. Let A_2, A_3, A_5, A_7 denote the set of positive integers up to 210 that are divisible by 2, 3, 5, and 7, respectively. We wish to count $[210] \setminus (A_2 \cup A_3 \cup A_5 \cup A_7)$. It suffices to count $A_2 \cup A_3 \cup A_5 \cup A_7$, and then subtract the result from 210. Using the inclusion-exclusion formula, we have

$$
\begin{aligned}
|A_2 \cup A_3 \cup A_5 \cup A_7| = & (|A_2| + |A_3| + |A_5| + |A_7|) \\
& - (|A_2 \cap A_3| + |A_2 \cap A_5| + \cdots + |A_5 \cap A_7|) \\
& + (|A_2 \cap A_3 \cap A_5| + \cdots + |A_3 \cap A_5 \cap A_7|) \\
& - |A_2 \cap A_3 \cap A_5 \cap A_7|.
\end{aligned}
$$

Since A_2 is the set of multiples of 2 up to 210, we have $|A_2| = \frac{210}{2} = 105$. Similarly, $|A_3| = \frac{210}{3} = 70$, $|A_5| = \frac{210}{5} = 42$, and $|A_7| = \frac{210}{7} = 30$. The intersections are similar. Thus we have

$$
\begin{aligned}
|A_2 \cup A_3 \cup A_5 \cup A_7| = & (105 + 70 + 42 + 30) \\
& - (35 + 21 + 15 + 14 + 10 + 6) \\
& + (7 + 5 + 3 + 2) \\
& - 1 \\
= & \ 162.
\end{aligned}
$$

So the number of integers up to 210 that are relatively prime to 210 is $210 - 162 = 48$.

One famous problem that is hard to solve without inclusion-exclusion but easy to solve with it is the derangement problem.

Question 18.2 (Derangement Problem). *n guests arrive at a dinner party, each wearing a hat. Each guest, upon arrival, gives eir[1] hat to the host. How many ways are there for the host to return the hats, one to each guest, in such a way that everyone gets a wrong hat?*

---

[1]This is an example of a *Spivak pronoun*, which is a third-person singular gender-neutral pronoun. To form these pronouns, take the third-person plural pronouns, which start with "th," and then remove the "th," producing "ey," "em," "eir," and so forth. The only exception is "themselves," which becomes "emself." One of my missions is to make Spivak pronouns into a commonly used part of the English language. Please do your part to support the cause!

Equivalently, how may permutations[2] of $[n]$ are there, written as functions $f : [n] \to [n]$, such that $f(i) \neq i$ for all i? We call such permutations *derangements*.

To solve the derangement problem using inclusion-exclusion, let A_i denote the set of permutations such that $f(i) = i$. What we want to count is the complement of $A_1 \cup \cdots \cup A_n$. There are $n!$ total permutations, so we'll count $A_1 \cup \cdots \cup A_n$ and then subtract our answer from $n!$.

To use inclusion-exclusion, we must count A_i. Since all the A_i's are the same size, we can count A_n, which consists of those f such that $f(n) = n$, and the rest of f is a permutation of $[n-1]$. Thus $|A_n| = (n-1)!$. Similarly, each $|A_i|$ is $(n-1)!$. Next, we need to count $A_{n-1} \cap A_n$. We have $f \in A_{n-1} \cap A_n$ if and only if $f(n-1) = n-1$ and $f(n) = n$, and the rest of f is a permutation of $[n-2]$, so $|A_{n-1} \cap A_n| = (n-2)!$. In general, if we have k of these A_i's, then the size of their intersection is $(n-k)!$.

Plugging all this into the inclusion-exclusion formula, we find that

$$\left| \bigcup_{i=1}^{n} A_i \right| = n \times (n-1)! - \binom{n}{2} \times (n-2)! + \binom{n}{3} \times (n-3)!$$

$$- \cdots + (-1)^{n-1} \binom{n}{n} \times 0!$$

$$= n! \sum_{i=1}^{n} \frac{(-1)^{i-1}}{i!}.$$

Subtracting from $n!$, we find that the answer to our problem is

$$n! \sum_{i=0}^{n} \frac{(-1)^i}{i!}.$$

If you are familiar with Taylor series, you will recognize the sum as a truncation of the Taylor series expansion of e^x at $x = -1$, so the answer is very close to $\frac{n!}{e}$. In fact, it is always the closest integer to $\frac{n!}{e}$, at least when $n \geq 1$.

Question 18.3. *We know from stars and bars that the number of nonnegative integer solutions to $x_1 + \cdots + x_n = k$ is $\binom{n+k-1}{n-1}$. But how many solutions are there in nonnegative integers $< r$?*

This is another great problem for inclusion-exclusion. Let us write X for the set of all solutions in nonnegative integers, and for a subset $T \subseteq [n]$,

---

[2]A *permutation* of $[n]$ is a bijective function $f : [n] \to [n]$, or equivalently an ordering of the elements of $[n]$.

let us write A_T for the solutions where $x_i \geq r$ for all $i \in T$. Our aim is to count

$$\left| X \setminus \bigcup_{i=1}^{n} A_i \right|,$$

where A_i is shorthand for $A_{\{i\}}$. Note that $|A_T|$ depends only on $|T|$, so we need only work out $|A_T|$ when $T = [i]$ for some i.

In order to compute $A_{[i]}$, we let $y_j = x_j$ for $j > i$ and $y_j = x_j - r$ for $j \leq i$; thus y_j are still nonnegative integers for the elements of $A_{[i]}$. A solution to $x_1 + \cdots + x_n = k$ inside $A_{[i]}$ is equivalent to a solution to

$$y_1 + \cdots + y_n = k - ir.$$

And we know how to count these: there are $\binom{n+k-ir-1}{n-1}$ of them. Now, we simply toss these answers into the inclusion-exclusion formula, and we find that the number is

$$\sum_{i=0}^{n} (-1)^i \binom{n}{i} \binom{n+k-ir-1}{n-1}.$$

There isn't a closed-form expression for that sum, so we've done as well as we can with this problem.[3]

In problems like the last two, we have some homogeneity in the terms in the inclusion-exclusion formula, in that the sizes of the intersections only depend on the number of sets we're intersecting, not the precise choice of sets. When that happens, it makes sense to introduce new notation: for any subset $T \subseteq [n]$ of size i, write $N_{\geq i}$ for $\left| \bigcap_{j \in T} A_j \right|$, the number of elements in all the sets in T (and possibly also others). It is also helpful to write $N_{=i}$ for the number of elements in all the sets in T but in none of the others: $N_{=i} = \left| \bigcap_{j \in T} A_j \cap \bigcap_{j \notin T} A_j^c \right|$. In this new notation, we can reformulate inclusion-exclusion as follows:

Corollary 18.4. *With the notation and assumptions as above, we have*

$$N_{=0} = \sum_{i=0}^{n} (-1)^i \binom{n}{i} N_{\geq i}.$$

Remark 18.5. Note that $N_{\geq i}$ counts the number of elements contained in some *fixed* choice of i of the A_j's. It does *not* count the number of elements in at least i of the A_j's overall.

---

[3] You might wonder how we can be sure that there is no closed-form expression for the sum. A good place to learn about how to *prove* that certain sums have no closed-form expressions is [PWZ96].

Sometimes we may be interested in computing $N_{=j}$ for some $j > 0$. This isn't really any different: we restrict the sets under consideration to those containing $[j]$. Thus we have the following:

Corollary 18.6.

$$N_{=j} = \sum_{i=j}^{n} (-1)^{i-j} \binom{n-j}{i-j} N_{\geq i}.$$

If we want to know how many elements are in exactly two of the sets, but it doesn't matter which two, we can compute that too. We can compute it either by multiplying the right side of Corollary 18.6 with $j = 2$ by $\binom{n}{2}$ in the case that the hypotheses of Corollary 18.6 are satisfied, or else just by thinking, along the lines of our original argument. We start by summing all the $|A_i \cap A_j|$'s, and that's the first approximation. If we have an element in three of the sets, it gets counted three times in the first approximation, but we want to count it 0 times, so we subtract $3 \sum_{1 \leq i < j < k \leq n} |A_i \cap A_j \cap A_k|$. Next, if we have an element in four of the sets, it gets counted $\binom{4}{2} = 6$ times in the first approximation, then $-3 \cdot \binom{4}{3} = -12$ times in the second; so between the first two approximations, we have counted it -6 times, and we want to count it 0 times. So we need to add $6 \sum_{1 \leq i < j < k < \ell \leq n} |A_i \cap A_j \cap A_k \cap A_\ell|$, and so forth.

By thinking in this way, we can solve all sorts of inclusion-exclusion problems, even if the conditions we want to count are weird. For instance, it is not substantially more difficult to count the number of elements that appear in exactly a prime number of the A_i's, or something peculiar like that.

Let's try another problem!

Definition 18.7. A *partition* of n is a way of writing n as a sum of positive integers, where order does not matter. Each of these integers is called a *part*. We let $p(n)$ denote the number of partitions of n.

Because order does not matter, we are free to assume that the parts are arranged in nonincreasing order: $n = a_1 + a_2 + \cdots + a_k$, where $a_1 \geq a_2 \geq \cdots \geq a_k$. For example, there are 7 partitions of 5, namely 5, $4 + 1$, $3 + 2$, $3 + 1 + 1$, $2 + 2 + 1$, $2 + 1 + 1 + 1$, and $1 + 1 + 1 + 1 + 1$.

Theorem 18.8. *The number of partitions of n into only odd parts is equal to the number of partitions of n into distinct parts.*

There are many proofs of Theorem 18.8. We will look at two proofs, the first of which uses inclusion-exclusion, and the second of which is a purely bijective proof.

Proof 1. We start by computing the number $p_o(n)$ of partitions of n into odd parts. The number of partitions containing a 2 is $p(n-2)$, since we can just take any of those and add 2 to it. Similarly, the number of partitions containing a 4 is $p(n-4)$, and so forth. Thus we have

$$
\begin{aligned}
p_o(n) = p(n) & \\
& - p(n-2) - p(n-4) - p(n-6) - \cdots \\
& + p(n-2-4) + p(n-2-6) + p(n-2-8) + \cdots \\
& + p(n-4-6) + p(n-4-8) + \cdots \\
& - p(n-2-4-6) - p(n-2-4-8) - \cdots .
\end{aligned}
$$

On the other hand, if $p_d(n)$ denotes the number of partitions of n into distinct parts, we have

$$
\begin{aligned}
p_d(n) = p(n) & \\
& - p(n-1-1) - p(n-2-2) - p(n-3-3) - \cdots \\
& + p(n-1-1-2-2) + p(n-1-1-3-3) + \cdots \\
& + p(n-2-2-3-3) + p(n-2-2-4-4) + \cdots \\
& - p(n-1-1-2-2-3-3) - \cdots .
\end{aligned}
$$

Since every line of the two equations agree, we have $p_o(n) = p_d(n)$. ∎

Proof 2. Let $P_o(n)$ and $P_d(n)$ denote the set of partitions of n into odd and distinct parts, respectively. We define inverse functions $f : P_o(n) \to P_d(n)$ and $g : P_d(n) \to P_o(n)$, which implies that there is a bijection between $P_o(n)$ and $P_d(n)$, and so there are the same number of each.

Let's start by defining f. Take a partition $A = (a_1 \geq a_2 \geq \cdots \geq a_k)$ of n into odd parts. We produce $f(A)$, a partition into distinct parts, as follows. Suppose we have b_i parts of A equal to i. Suppose that the binary representation of b_i is

$$
b_i = 2^{c_1} + 2^{c_2} + \cdots + 2^{c_r},
$$

where all the c_i's are distinct. Then merge the i's in A into parts of $2^{c_1}i, 2^{c_2}i, \ldots, 2^{c_r}i$ in $f(A)$. Do this for every i to obtain $f(A)$.

Now let's define g. Suppose we have a partition B into distinct parts. We define $g(B)$, a partition of n into odd parts, as follows. Suppose b_i is a part of B. We can write b_i, and indeed any positive integer, uniquely in the form $b_i = 2^e m$, where m is odd. Split the pile of size b_i into 2^e piles, each of size m, and do this for every i to obtain $g(B)$. The function f and g are inverses.   ∎

That was probably a bit tricky to follow, so let's do an example. Let's start with the following partition of 32 into odd parts:

$$7 + 3 + 3 + 3 + 3 + 3 + 1 + 1 + 1 + 1 + 1 + 1 + 1 + 1 + 1 + 1.$$

That is, we have one part equal to 7, five parts equal to 3, and ten parts equal to 1. To deal with the 3's, we write 5 in binary as $4+1$, so we combine 4 of them and 1 of them, to get one part of 12 and one part of 3. Similarly, $10 = 8 + 2$, so we combine the 1's into 8 and 2. The final partition $f(A)$ is $12 + 8 + 7 + 3 + 2$. We can also get that by looking for two parts that are equal and then combining them. Keep doing that until all parts are distinct. This process yields the same result.

To go backward, let's start with $f(A)$, namely $12 + 8 + 7 + 3 + 2$. To convert it into a partition $g(f(A))$ into odd parts, we split each pile in half until it becomes odd. If we split 12 in half, we get $6 + 6$, which isn't odd yet. So we do it again, converting each 6 into $3 + 3$. Similarly, we split 8 into $4 + 4$, then each 4 into $2 + 2$, then each 2 into $1 + 1$. The final result is just A.

18.2   Problems

(1) There are 20 students participating in an after-school program offering classes in yoga, bridge, and painting. Each student must take at least one of these three classes, but may take two or all three. There are 10 students taking yoga, 13 taking bridge, and 9 taking painting. There are 9 students taking at least two classes. How many students are taking all three classes?

(2) How many functions $f : [n] \to [n]$ are there with no fixed points, i.e. such that $f(i) \neq i$ for all i? After you solve this problem using inclusion-exclusion, figure out how to solve it properly!

(3) How many ways are there to choose a hand of 13 cards from a standard deck of 52 in such a way that the hand contains at least one jack, one queen, one king, and one ace?

(4) How many integers are there from 1 to 10000 that are divisible by exactly two of the numbers 4, 5, 6, and 7?

(5) Given a permutation $a = a_1, \ldots, a_n$ of $[n]$, we define its *excedance set*[4] to be $\{i : 1 \le i \le n, a_i > i\}$. How many permutations of $[n]$ are there whose excedance set contains either $n - 1$ or $n - 2$ (or both)?

(6) How many positive integers up to 100000 are not perfect powers? (A number n is a perfect power if there are integers a and b, with $b \ge 2$, such that $n = a^b$.)

(7) Six people of different heights are getting in line to buy donuts. Compute the number of ways they can arrange themselves in line such that no three consecutive people are in increasing order of height, from front to back.

(8) Evaluate the sum $\sum_{i=0}^{n} \binom{n}{i} d(i)$ in closed form, where $d(i)$ denotes the number of derangements of $[i]$.

---

[4]The usual spelling is "exceedance," but the variant "excedance" is standard in combinatorics.

Chapter 19

Recurrences

19.1 Why recurrences?

Suppose we wish to determine the size of some set $S(n)$, where we have one such set for each nonnegative integer n. It might not be easy to count $S(n)$ directly, but perhaps we can express the size of $S(n)$ in terms of the size of $S(n-1)$, and maybe also the sizes of $S(n-2)$, $S(n-3)$, and so forth. Once we have done that, maybe we can use the relation we have written down to solve for the size of $S(n)$ in closed form. Let's see this idea in practice.

Question 19.1. *How many binary sequences are there of length n, such that two 1's never appear consecutively?*

Let us write a_n for the number of such sequences. It is not so straightforward to determine a_n directly, but we can relate different values of a_n. Let us write S_n for the set of sequences enumerated by a_n, i.e. the binary sequences of length n with no two consecutive 1's. We can form a sequence in S_n in one of two ways:

- Take a sequence in S_{n-1}, and put a 0 at the end.
- Take a sequence in S_{n-2}, and put a 01 at the end.

Conversely, starting from a sequence s in S_n, we can produce either a sequence in S_{n-1} or in S_{n-2}:

- If s ends with a 0, remove this final zero to produce a sequence in S_{n-1}.
- If s ends with a 1, then the bit immediately preceding that 1 must be a 0, so we can remove the final 01 to produce a sequence in S_{n-2}.

These two processes, either adding a 0 to a sequence of length $n-1$ or a 01 to a sequence of length $n-2$, and removing a final 0 or 01, are inverses. Thus there is a bijection between S_n and $S_{n-1} \cup S_{n-2}$. The number of elements in S_n is a_n, and the number of elements in $S_{n-1} \cup S_{n-2}$ is $a_{n-1} + a_{n-2}$. Thus the sequence a_n satisfies the recurrence

$$a_n = a_{n-1} + a_{n-2},$$

whenever $n \geq 2$.

This doesn't completely specify the values of a_n though, because there are many sequences of numbers a_n with $a_n = a_{n-1} + a_{n-2}$. For instance, we have the sequence $1, 2, 3, 5, 8, 13, 21, 34, \dots$ and the sequence $3, 4, 7, 11, 18, 29, 47, \dots$ These are different sequences, although they satisfy the same recurrence. In order to specify the recurrence completely, we need to know a few values, typically the first ones. In this case, we can count directly that $a_0 = 1$ and $a_1 = 2$; these are called the *initial conditions* of the recurrence. Thus the actual sequence a_n takes on values $a_0 = 1$, $a_1 = 2$, $a_2 = 3$, $a_3 = 5$, $a_4 = 8$, and so forth. This is almost the Fibonacci sequence, but indexed a little differently. In fact, we have $a_n = F_{n+2}$.

Let's try another example.

Question 19.2. *How many ways are there to place n identical balls into k distinguishable boxes, such that every box gets between 3 and 6 balls?*

This one has two variables, but that's okay. Let $a_{n,k}$ denote the number of such ways, and let $S_{n,k}$ denote the actual distributions, so that $a_{n,k} = |S_{n,k}|$. In order to write down a recurrence, let's look at the kth box. If it contains r balls, then we must have filled the other $k-1$ boxes with a total of $n-r$ balls. Since r can be 3, 4, 5, or 6, we have

$$a_{n,k} = a_{n-3,k-1} + a_{n-4,k-1} + a_{n-5,k-1} + a_{n-6,k-1}.$$

For the initial conditions, we'll let $k = 1$: we have

$$a_{n,1} = \begin{cases} 1 & n \in \{3, 4, 5, 6\}, \\ 0 & n \notin \{3, 4, 5, 6\}. \end{cases}$$

Is this enough information to specify $a_{n,k}$ completely? Let's try to determine $a_{14,3}$ using the recurrence and initial conditions: we have

$$a_{14,3} = a_{11,2} + a_{10,2} + a_{9,2} + a_{8,2}$$

$$= (a_{8,1} + a_{7,1} + a_{6,1} + a_{5,1}) + (a_{7,1} + a_{6,1} + a_{5,1} + a_{4,1})$$

$$\quad + (a_{6,1} + a_{5,1} + a_{4,1} + a_{3,1}) + (a_{5,1} + a_{4,1} + a_{3,1} + a_{2,1})$$

$$= 2 + 3 + 4 + 3$$

$$= 12.$$

So, it worked! All we needed was the recurrence relation and the initial conditions that we already computed. It's clear that, using the same technique, we can compute any $a_{n,k}$, although the computations involved may be rather tedious for large values of n and k.

Sometimes it's helpful to introduce auxiliary sequences, even when they aren't directly part of the problem.

Question 19.3. *How many ternary (i.e. base-3) sequences are there with length n, such that the number of 0's is a multiple of 3?*

Let a_n, b_n, and c_n denote the number of ternary sequences of length n, such that the number of zeros is 0 (mod 3), 1 (mod 3), and 2 (mod 3), respectively. They satisfy the following recurrences:

$$a_{n+1} = 2a_n + c_n, \qquad b_{n+1} = 2b_n + a_n, \qquad c_{n+1} = 2c_n + b_n.$$

The initial conditions are $a_1 = 2$, $b_1 = 1$, and $c_1 = 0$. We only care about a_n, but it would be tricky to write down a recurrence for a_n directly, without the help of b_n and c_n. This system of three recurrences, together with their initial conditions, is enough to describe a_n (and b_n and c_n) entirely. However, we cannot easily compute a_n alone, without computing b_n and c_n simultaneously. (It is possible, using linear algebra, to remove b_n and c_n, but much of the time it's not worth the trouble.)

19.2 Solving linear recurrences

The recurrences we have seen so far have a certain form. They are known as *linear recurrences*.

Definition 19.4. Let $a_0, a_1, a_2, \ldots$ be a sequence, and let k be a positive integer. We say that this sequence satisfies a *linear recurrence relation* of order k if there exist numbers $c_0, c_1, \ldots, c_{k-1}$ such that

$$a_{n+k} = c_{k-1}a_{n+k-1} + c_{k-2}a_{n+k-2} + \cdots + c_1 a_{n+1} + c_0 a_n$$

for all $n \geq 0$.

For example, the Fibonacci sequence satisfies a linear recurrence relation of order 2, by taking $c_0 = c_1 = 1$: we have

$$F_{n+2} = F_{n+1} + F_n.$$

There is a general method for finding closed forms for all sequences satisfying linear recurrence relations.

Definition 19.5. Given a linear recurrence relation

$$a_{n+k} = c_{k-1}a_{n+k-1} + c_{k-2}a_{n+k-2} + \cdots + c_1 a_{n+1} + c_0 a_n,$$

we define its *characteristic polynomial* to be

$$P(x) = x^k - c_{k-1}x^{k-1} - c_{k-2}x^{k-2} - \cdots - c_1 x - c_0.$$

The characteristic polynomial is related to closed form expressions for recurrence relations. It turns out that the situation is simpler when all of the (complex) roots of P are distinct.

Theorem 19.6. *Let $a_0, a_1, \ldots$ be a sequence satisfying a linear recurrence relation*

$$a_{n+k} = c_{k-1}a_{n+k-1} + c_{k-2}a_{n+k-2} + \cdots + c_1 a_{n+1} + c_0 a_n.$$

Let $P(x)$ be the characteristic polynomial, and let $r_1, \ldots, r_k$ be the complex roots of $P(x)$. If $r_1, \ldots, r_k$ are distinct, then there exist complex numbers $s_1, s_2, \ldots, s_k$ such that

$$a_n = s_1 r_1^n + s_2 r_2^n + \cdots + s_k r_k^n$$

for all $n \geq 0$.

Proof. First, we determine the values of $s_1, \ldots, s_k$. Letting $n = 0, 1, \ldots, k-1$, we find $s_1, \ldots, s_k$ so as to satisfy the k simultaneous equations

$$
\begin{array}{ccccccc}
s_1 & + & s_2 & + \cdots + & s_k & = & a_0 \\
s_1 r_1 & + & s_2 r_2 & + \cdots + & s_k r_k & = & a_1 \\
s_1 r_1^2 & + & s_2 r_2^2 & + \cdots + & s_k r_k^2 & = & a_2 \\
& & & \vdots & & & \\
s_1 r_1^{k-1} & + & s_2 r_2^{k-1} & + \cdots + & s_k r_k^{k-1} & = & a_{k-1}.
\end{array}
$$

This is a system of k simultaneous equations in k variables; while not all such systems of equations have solutions, or unique solutions, this one does.[1]

Now, suppose we have found $s_1, \ldots, s_k$ so as to satisfy the above system of equations. That means we know that Theorem 19.6 holds for $n = 0, 1, \ldots, k-1$. We now prove Theorem 19.6 for all n by induction on n. Suppose we already know that Theorem 19.6 is true for all integers

---

[1] Unfortunately, this fact, which relies on linear algebra and in particular the Vandermonde matrix (or determinant), is beyond the scope of this book. See [Mil17, Section 23.2.4] for an explanation of this if you're curious. We shall accept the fact that this system of equations has a unique solution and move on.

less than $n + k$; we will prove it for $n + k$ as our inductive step. Using the recurrence, we have

$$
\begin{aligned}
a_{n+k} &= c_{k-1}a_{n+k-1} + \cdots + c_1 a_{n+1} + c_0 a_n \\
&= c_{k-1}(s_1 r_1^{n+k-1} + \cdots + s_k r_k^{n+k-1}) + \cdots \\
&\quad + c_0(s_1 r_1^n + \cdots + s_k r_k^n) \\
&= s_1 r_1^n (c_{k-1} r_1^{k-1} + \cdots + c_1 r_1 + c_0) + \cdots \\
&\quad + s_k r_k^n (c_{k-1} r_k^{k-1} + \cdots + c_1 r_k + c_0) \\
&= s_1 r_1^{n+k} + \cdots + s_k r_k^{n+k}.
\end{aligned}
$$

In the last step, we used the fact that $r_1, \ldots, r_k$ are roots of $P(x)$, so that $c_{k-1} r_i^{k-1} + \cdots + c_1 r_i + c_0 = r_i^k$ for $1 \le i \le k$. ∎

Example. Consider the Fibonacci sequence $F_{n+2} = F_{n+1} + F_n$ with initial conditions $F_0 = 0$ and $F_1 = 1$. Its characteristic polynomial is $x^2 - x - 1$, whose roots are $\frac{1 \pm \sqrt{5}}{2}$. Thus there exist constants s_1 and s_2 with

$$
F_n = s_1 \left(\frac{1 + \sqrt{5}}{2} \right)^n + s_2 \left(\frac{1 - \sqrt{5}}{2} \right)^n.
$$

In order to find them, we use the initial conditions $F_0 = 0$ and $F_1 = 1$. Thus s_1 and s_2 satisfy the equations

$$
s_1 + s_2 = 0, \qquad s_1 \left(\frac{1 + \sqrt{5}}{2} \right) + s_2 \left(\frac{1 - \sqrt{5}}{2} \right) = 1.
$$

Probably the easiest way of solving these equations is to solve for s_2 in the first equation, since $s_2 = -s_1$, and substitute into the second equation, to get

$$
s_1 \left(\frac{1 + \sqrt{5}}{2} \right) - s_1 \left(\frac{1 - \sqrt{5}}{2} \right) = 1,
$$

or $s_1 \sqrt{5} = 1$, or $s_1 = \frac{1}{\sqrt{5}}$. Thus $s_2 = -\frac{1}{\sqrt{5}}$, so we have

$$
F_n = \frac{1}{\sqrt{5}} \left[\left(\frac{1 + \sqrt{5}}{2} \right)^n - \left(\frac{1 - \sqrt{5}}{2} \right)^n \right],
$$

a formula which is known as *Binet's formula.*

19.3   The Josephus problem

Question 19.7. *n people stand around a circle. Every second person stand-ing, starting with person 2, sits down, until there is only one person left standing. In which position is the last person?*

Let's try it with $n = 7$. In succession, the people who sit down are numbered 2, 4, 6, 1, 5, 3. Thus person 7 remains standing at the end.

Let us write $J(n)$ for the last person left standing. We would like to find a recurrence that J satisfies. It won't be a linear recurrence, so we can't solve it directly using the techniques we have studied so far. But it will be fairly easy to deal with for other reasons.

In order to get a recurrence, note that in the first sweep around the circle, all the people in even-numbered positions have to sit down. When we get to the beginning again, we end up with slightly different situations if n is even versus if n is odd. If n is even, we start again with person 1 remaining standing but person 3 sitting, and so forth. Note that this is essentially just the original problem, but now with $\frac{n}{2}$ people instead of n. If person k is the last one standing in the $\frac{n}{2}$ game, then person $2k - 1$ is the last one standing in the n game. Thus, when n is even, we have

$$J(n) = 2J\left(\frac{n}{2}\right) - 1,$$

or

$$J(2n) = 2J(n) - 1.$$

When n is odd, the situation is slightly different, because person n remains standing and then person 1 has to sit down. After that, we're left with people in positions $3, 5, 7, \ldots, n$, starting with person 3. This game is very similar to the $\frac{n-1}{2}$ game, so we have

$$J(n) = 2J\left(\frac{n-1}{2}\right) + 1,$$

or

$$J(2n + 1) = 2J(n) + 1.$$

Putting these two cases together, we have

$$J(2n) = 2J(n) - 1, \qquad J(2n + 1) = 2J(n) + 1.$$

Let's use this to confirm our answer when $n = 7$. We have $J(7) = 2J(3) + 1$, so we have to compute $J(3)$. We can do this from the recurrence as well: $J(3) = 2J(1) + 1$. Since $J(1) = 1$, we have $J(3) = 3$ and $J(7) = 7$, just what we got computing it by hand.

We can compute values of $J(n)$ for large n quite efficiently using our recurrence. For instance, suppose we wish to compute $J(494)$. Then we have

$$
\begin{aligned}
J(494) &= 2J(247) - 1 \\
&= 2(2J(123) + 1) - 1 = 4J(123) + 1 \\
&= 4(2J(61) + 1) + 1 = 8J(61) + 5 \\
&= 8(2J(30) + 1) + 5 = 16J(30) + 13 \\
&= 16(2J(15) - 1) + 13 = 32J(15) - 3 \\
&= 32(2J(7) + 1) - 3 = 64J(7) + 29 \\
&= 64(2J(3) + 1) + 29 = 128J(3) + 93 \\
&= 128(2J(1) + 1) + 93 = 256J(1) + 221 \\
&= 256 + 221 \\
&= 477.
\end{aligned}
$$

So person number 477 is the last one standing.

While a recurrence of this form doesn't come with a body of theory like that of linear recurrences, in other ways it is preferable. Note that we didn't have to compute $J(n)$ for all $n < 494$ in order to compute $J(494)$: we only had to compute a very small number of values. On the other hand, if we want to compute the nth term of a sequence given by a linear recurrence, we have to compute all the previous terms, rather than just a few of them. (There are certain other benefits to linear recurrences though, such as the fact they have closed-form expressions. Depending on context, these might or might not be easy to work with.)

Usually, recurrences like the one for $J(n)$ do not admit closed-form solutions. This one does, however. Write down the first several values of $J(n)$ and figure out what the pattern is. Then you will prove it in problem 9.

19.4   Further reading

We have just barely scratched the surface on recurrences. A wonderful book on combinatorics with something of a focus on recurrences is [GKP94].

Given a sequence $a_0, a_1, a_2, \ldots$ of numbers, possibly coming from a recurrence, one good way to try to analyze it is to put it into a *generating function.* That is, we construct the series

$$\sum_{n=0}^{\infty} a_n z^n,$$

which is known as the *ordinary power series generating function.* Another possibility is the *exponential generating function*

$$\sum_{n=0}^{\infty} a_n \frac{z^n}{n!}.$$

Depending on the sequence, it may be possible to learn interesting things about the sequence, such as a closed form or growth rate, in terms of properties of the generating function. Two great books focusing on generating functions are [Wil05] and [FS09].

19.5 Problems

(1) Solve the following recurrences:

   (a) $a_{n+2} = 3a_{n+1} - 2a_n$, $a_0 = 1$, $a_1 = 3$.
   (b) $a_{n+2} = 4a_{n+1} - a_n$, $a_0 = 2$, $a_1 = 3$.

(2) Suppose a newly born pair of rabbits, one male and one female, are put in a field. Rabbits are able to mate at the age of one month, so at the end of its second month a female can produce another pair of rabbits. Suppose that our rabbits never die and that the female always produces one new pair (one male, one female) every month from the second month on (so, not when she is one month old, but when she is two months old, or three months old, etc.). How many pairs will there be in one year?

(3) You have 1×1 tiles that come in 5 different colors, and 1×2 tiles that come in 4 different colors. Let a_n be the number of ways of tiling a $1 \times n$ rectangle using the 1×1 and 1×2 tiles. Write down a recurrence for a_n. Then solve it.

(4) Solve the recurrence given by $a_1 = 7$ and for $n > 1$, $a_n = a_{\lfloor \frac{n}{2} \rfloor} + 1$.

(5) Show that the derangement function $d(n)$ satisfies the recurrences

$$d(n) = (n-1)(d(n-1)+d(n-2)) \quad \text{and} \quad d(n) = nd(n-1)+(-1)^n.$$

(6) Write down a recurrence and initial conditions for the number of ternary sequences of length n with no 012 in consecutive positions.

(7) Bob throws a coin repeatedly, scoring one point if it lands heads and two points if it lands tails, adding his score to his previous running total. For each positive integer n, let p_n be the probability that his score is exactly n at some point. Write down a recurrence and initial conditions for p_n, and solve it.

(8) For each positive integer n, we can write $(2 + \sqrt{3})^{2n-1}$ as $a_n + b_n\sqrt{3}$, where a_n and b_n are integers. Prove that $a_n - 1$ is a perfect square for all positive integers n.

(9) Write down and prove a closed formula for $J(n)$, where J is as in the Josephus problem.

(10) Write down a recurrence for the modification of the Josephus function, where every third person has to sit down.

Chapter 20

Stirling numbers

20.1 Stirling numbers of the first kind

Definition 20.1. A *permutation* of $[n]$ is an ordering of the numbers from 1 to n, or equivalently, a bijective function $f : [n] \to [n]$.

There are $n!$ permutations of $[n]$. There are many ways of representing a permutation, and they are valuable for different purposes. One way is to write out $f(1), f(2), \ldots, f(n)$ in a row. For instance, one permutation of $[6]$ is 216453. This means that $f(1) = 2$, $f(2) = 1$, $f(3) = 6$, $f(4) = 4$, $f(5) = 5$, and $f(6) = 3$.

Another way of representing a permutation is as a picture. We write down all the numbers from 1 to n, and then we draw an arrow from k to $f(k)$ for each k. It is generally convenient to draw them such that the arrows do not cross, so as to make it easy to read, as shown in Figure 20.1. This is the same permutation as above.

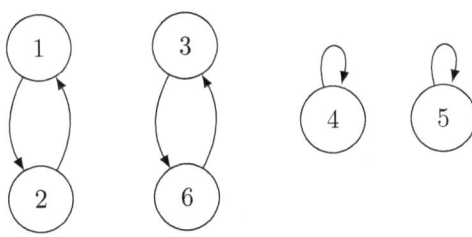

Figure 20.1. Cycle structure for the permutation 216453.

As we can see here, we have some loops, or *cycles*. We have the cycle $[1, 2]$, the cycle $[3, 6]$, the cycle $[4]$, and the cycle $[5]$. Thus there are four cycles.

Note that a cycle is not dependent just on the elements it contains, but also the order they come in. For instance, if we have a cycle $[1, 2, 3]$, meaning that $1 \to 2 \to 3 \to 1$, that's different from $[1, 3, 2]$, where $1 \to 3 \to 2 \to 1$, since these are different functions. However, $[1, 2, 3]$ and $[2, 3, 1]$ and $[3, 1, 2]$ are all the same. Thus we can freely perform cyclic shifts of the elements in the cycle, but otherwise we may not reorder them. Given a cycle of length k, there are k possible cyclic shifts, so it follows that there are $\frac{k!}{k} = (k-1)!$ cycles of length k: order the elements in the cycle in any of $k!$ ways, and there are k cyclic shifts which give the same permutation.

Definition 20.2. Let $n \geq k \geq 0$ be nonnegative integers. The *Stirling number of the first kind* $\left[{n \atop k}\right]$, pronounced "n cycle k," is the number of permutations of $[n]$ with exactly k cycles.

For instance, the permutation shown in Figure 20.1 is one of the things counted by $\left[{6 \atop 4}\right]$.

Unlike the binomial coefficients, the Stirling numbers of the first kind do not have a nice closed-form expression in terms of familiar functions like factorials. (Neither do the Stirling numbers of the second kind, which we will come to shortly.) However, for special values of n and k, we can express $\left[{n \atop k}\right]$ in terms of familiar functions.

Here are some easy ones:

- $\left[{n \atop 0}\right] = 0$ if $n \geq 1$ and $\left[{0 \atop 0}\right] = 1$.
- $\left[{n \atop n}\right] = 1$.
- $\left[{n \atop 1}\right] = (n-1)!$.

Another relatively easy one is $\left[{n \atop n-1}\right] = \binom{n}{2}$. Since there are $n - 1$ cycles, it must be the case that two of the elements are in the same cycle, and all the rest are in different cycles. There are $\binom{n}{2}$ ways of choosing those two elements, and there's only one way of forming a cycle from them. Beyond that, though, it is tedious to express the Stirling number of the first kind in terms of familiar functions, although with sufficient care and patience it can often be done.

More useful is that the Stirling numbers of the first kind satisfy a recurrence, similar to Pascal's identity $\binom{n}{k} = \binom{n-1}{k-1} + \binom{n-1}{k}$ for the binomial coefficients.

To see this, let us imagine that we understand permutations of $[n-1]$ and their cycles. How can we form a cycle of $[n]$ containing n out of them? One possibility is to let n be in a new cycle by itself. If we wish to end up with k cycles, then the permutation of $[n-1]$ should have contained $k - 1$

cycles, so we get a contribution of $\begin{bmatrix} n-1 \\ k-1 \end{bmatrix}$. But it is also possible to append n to an existing cycle. For any of the elements r from 1 to $n-1$, we can modify a permutation f to include n, by creating a new function g such that $g(i) = f(i)$ for $i \neq r$, and then setting $g(r) = n$ and $g(n) = f(r)$. (See Figure 20.2 for a picture.) There are $(n-1)\begin{bmatrix} n-1 \\ k \end{bmatrix}$ ways of doing this. Thus we find the following:

Proposition 20.3.

$$\begin{bmatrix} n \\ k \end{bmatrix} = (n-1)\begin{bmatrix} n-1 \\ k \end{bmatrix} + \begin{bmatrix} n-1 \\ k-1 \end{bmatrix}.$$

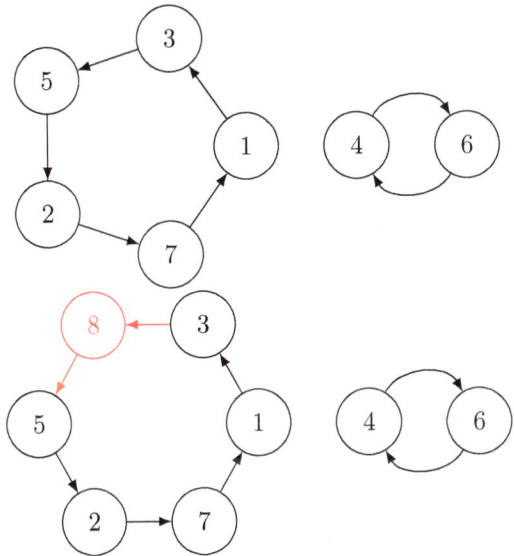

Figure 20.2. Inserting the new number 8 after 3.

Using this recurrence, it is easy to build the Stirling equivalent of Pascal's triangle. See Figure 20.3 for the first few rows.

One easy identity we have is the following:

Proposition 20.4.

$$\sum_{k=0}^{n} \begin{bmatrix} n \\ k \end{bmatrix} = n!.$$

	1	2	3	4	5	6	7	8
n = 0:				1				
n = 1:				0	1			
n = 2:				0	1	1		
n = 3:			0	2	3	1		
n = 4:		0	6	11	6	1		
n = 5:	0	24	50	35	10	1		
n = 6:	0	120	274	225	85	15	1	
n = 7:	0	720	1764	1624	735	175	21	1

Figure 20.3. The triangle of Stirling numbers of the first kind.

Proof. The left side counts the number of permutations with k cycles, and then sums over k, so we are left with the total number of permutations. And that's what the right side counts. ∎

20.2   Stirling numbers of the second kind

Now let's turn to the Stirling numbers of the second kind.

Definition 20.5. Let $n \geq k \geq 0$ be nonnegative integers. The *Stirling number of the second kind* $\left\{ {n \atop k} \right\}$, pronounced "$n$ subset k," is the number of ways of partitioning the elements of $[n]$ into k disjoint subsets. We call a partition of $[n]$ into disjoint subsets a *set partition* of $[n]$.

For instance, $\{1,2,6,8\}, \{3,4,7\}, \{5\}$ is one of the things counted by $\left\{ {8 \atop 3} \right\}$. The order of the subsets does not matter, so $\{5\}, \{1,2,6,8\}, \{3,4,7\}$ is the same as the previous one.

Note that the definitions of the Stirling numbers of the first and second kinds are sort of similar, in that they both involve breaking up $[n]$ into k classes. However, for $\left\{ {n \atop k} \right\}$, that's all there is to it. In the case of $\left[{n \atop k} \right]$, we can further order the elements in each class. It follows that $\left[{n \atop k} \right] \geq \left\{ {n \atop k} \right\}$ for all n and k.

We can explicitly work out a few special cases of the Stirling numbers of the second kind. For instance, $\left\{ {n \atop 0} \right\}$ is 1 if $n = 0$ and 0 if $n > 0$. Similarly, $\left\{ {n \atop 1} \right\} = 1$ for all $n \geq 1$.

Unlike in the case of Stirling numbers of the first kind, $\left\{ {n \atop 2} \right\}$ has a nice simple formula.

Proposition 20.6.

$$\left\{ {n \atop 2} \right\} = 2^{n-1} - 1$$

for all $n \geq 1$.

Proof 1. We need to split $[n]$ up into two nonempty subsets, A and B. For each element $k \in [n]$, we can place k into A or B, so there are two choices for each, or 2^n total. However, both A and B have to be nonempty, so we must subtract off the two where all the elements end up in A and where all the elements end up in B, yielding $2^n - 2$. But since the order of the subsets does not matter, we have overcounted by a factor of 2, giving us our final answer of $2^{n-1} - 1$. ∎

Proof 2. This time, we forcibly distinguish the subsets from the beginning by requiring that 1 gets placed in subset A. The other $n - 1$ elements can go into either subset A or subset B, in 2^{n-1} ways. We must then remove the case where everything goes into subset A, leaving a final answer of $2^{n-1} - 1$. ∎

Furthermore, we have $\left\{ {n \atop n} \right\} = 1$ and $\left\{ {n \atop n-1} \right\} = \binom{n}{2}$, just as was the case for the Stirling numbers of the first kind. This is because if all cycles have length ≤ 2, then there is only one way of arranging them in a cyclic order.

That's about where things end in terms of having useful explicit formulae for $\left\{ {n \atop k} \right\}$, although there is an expression for $\left\{ {n \atop k} \right\}$ as a sum of binomial coefficients. However, like in the case of Stirling numbers of the first kind, there is a recurrence.

Proposition 20.7.

$$\left\{ {n \atop k} \right\} = k \left\{ {n-1 \atop k} \right\} + \left\{ {n-1 \atop k-1} \right\}.$$

Proof. This is much the same as the proof of the recurrence for Stirling numbers of the first kind. We start with some partition of $[n-1]$. If we have a partition of $[n-1]$ into $k-1$ subsets, then we can create a partition of $[n]$ into k subsets by forming a new subset consisting of just n. If we have a partition of $[n-1]$ into k subsets, then we can form a partition of $[n]$ into k subsets in k different ways, by letting n join any of the k existing subsets. The result follows. ∎

This recurrence allows us to build a triangle of Stirling numbers of the second kind, just like we did earlier for Stirling numbers of the first kind. See Figure 20.4.

$n = 0$:   1

$n = 1$:   0   1

$n = 2$:   0   1   1

$n = 3$:   0   1   3   1

$n = 4$:   0   1   7   6   1

$n = 5$:   0   1   15   25   10   1

$n = 6$: 0   1   31   90   65   15   1

$n = 7$: 0   1   63   301   350   140   21   1

Figure 20.4. The triangle of Stirling numbers of the second kind.

20.3   Rising and falling factorials

One unexpected place where Stirling numbers, of both kinds, show up is when converting between ordinary powers and the so-called *rising and falling factorials*.

Definition 20.8. Let n be a nonnegative integer. We define the *falling factorial* $x^{\underline{n}}$ to be

$$x^{\underline{n}} = x(x - 1)(x - 2) \cdots (x - n + 1) = \prod_{i=0}^{n-1} (x - i).$$

Similarly, we define the *rising factorial* $x^{\overline{n}}$ to be

$$x^{\overline{n}} = x(x + 1)(x + 2) \cdots (x + n - 1) = \prod_{i=0}^{n-1} (x + i).$$

When $n = 0$, we have $x^{\underline{0}} = x^{\overline{0}} = 1$, since that's what we always get when we multiply 0 numbers. Let us see how to express the rising and falling powers in terms of ordinary powers.

Theorem 20.9.

$$x^{\overline{n}} = \sum_{k=0}^{n} \left[{n \atop k} \right] x^k,$$

$$x^{\underline{n}} = \sum_{k=0}^{n} (-1)^{n-k} \left[{n \atop k} \right] x^k.$$

Proof. We prove the first identity first, and then we quickly derive the second one from it. We will give an algebraic proof, based on the recurrence for $\begin{bmatrix} n \\ k \end{bmatrix}$ and by induction on n.[1] For the base case, the result is certainly true when $n = 0$, since both sides are equal to 1.[2]

Now, let us suppose that $n \geq 1$ and we know that

$$x^{\overline{n-1}} = \sum_{k=0}^{n-1} \begin{bmatrix} n - 1 \\ k \end{bmatrix} x^k.$$

We have

$$x^{\overline{n}} = x^{\overline{n-1}}(x + n - 1)$$

$$= x \cdot x^{\overline{n-1}} + (n - 1) \cdot x^{\overline{n-1}}$$

$$= x \sum_{k=0}^{n-1} \begin{bmatrix} n - 1 \\ k \end{bmatrix} x^k + \sum_{k=0}^{n-1}(n - 1) \begin{bmatrix} n - 1 \\ k \end{bmatrix} x^k$$

$$= x \sum_{k=1}^{n} \begin{bmatrix} n - 1 \\ k - 1 \end{bmatrix} x^{k-1} + \sum_{k=0}^{n-1}(n - 1) \begin{bmatrix} n - 1 \\ k \end{bmatrix} x^k$$

$$= \sum_{k=1}^{n} \begin{bmatrix} n - 1 \\ k - 1 \end{bmatrix} x^k + \sum_{k=0}^{n-1}(n - 1) \begin{bmatrix} n - 1 \\ k \end{bmatrix} x^k$$

$$= \sum_{k=0}^{n} \left(\begin{bmatrix} n - 1 \\ k - 1 \end{bmatrix} + (n - 1) \begin{bmatrix} n - 1 \\ k \end{bmatrix} \right) x^k$$

$$= \sum_{k=0}^{n} \begin{bmatrix} n \\ k \end{bmatrix} x^k,$$

as desired. Note that we were able to extend the sums from 0 to n (rather than 1 to n in the first case, and 0 to $n - 1$ in the second case) because the extra terms inserted are zero (since $n \geq 1$).

To deduce the second identity, replace x with $-x$ in the first identity, to obtain

$$(-x)^{\overline{n}} = \sum_{k=0}^{n} \begin{bmatrix} n \\ k \end{bmatrix}(-x)^k.$$

Now, observe that $(-x)^{\overline{n}} = (-1)^n x^{\underline{n}}$. Thus we have

$$(-1)^n x^{\underline{n}} = \sum_{k=0}^{n} \begin{bmatrix} n \\ k \end{bmatrix}(-1)^k x^k.$$

Multiplying by $(-1)^n$ and noting that $(-1)^{n+k} = (-1)^{n-k}$ yields the second identity. ∎

---

[1] It is also possible to give a combinatorial proof, but it is more complicated.
[2] Or, if you don't like that, let the base case be $n = 1$, when both sides are equal to x.

So, the Stirling numbers of the first kind tell us how to convert rising factorials and falling factorials into ordinary powers. But what if we want to go the other way around, converting ordinary powers into rising or falling factorials? We can do that too, using Stirling numbers of the *second* kind.

Theorem 20.10.

$$x^n = \sum_{k=0}^{n} \left\{ {n \atop k} \right\} x^{\underline{k}},$$

$$x^n = \sum_{k=0}^{n} \left\{ {n \atop k} \right\} (-1)^{n-k} x^{\overline{k}}.$$

The proof is very similar, so we leave it as an exercise for the reader. But there is a curious application of these two theorems in combination. If we start with ordinary powers, express them in terms of falling factorials, and then express those falling factorials in terms of ordinary powers, we get a new identity. Let's have a look:

$$x^n = \sum_{k=0}^{n} \left\{ {n \atop k} \right\} x^{\underline{k}}$$

$$= \sum_{k=0}^{n} \left\{ {n \atop k} \right\} \sum_{j=0}^{k} (-1)^{k-j} \left[{k \atop j} \right] x^j$$

$$= \sum_{j=0}^{n} \left(\sum_{k=j}^{n} (-1)^{k-j} \left\{ {n \atop k} \right\} \left[{k \atop j} \right] \right) x^j.$$

The left and right sides are both polynomials, and they are equal. This implies that all their coefficients are equal. Thus we have shown the following:

Theorem 20.11.

$$\sum_{k=j}^{n} (-1)^{k-j} \left\{ {n \atop k} \right\} \left[{k \atop j} \right] = \begin{cases} 1 & j = n, \\ 0 & j \neq n. \end{cases}$$

In a certain sense, then, the Stirling numbers of the first and second kinds are inverses. For a follow-up, see problem 7.

20.4   Further reading

After the binomial coefficients, the Stirling numbers are perhaps the next most important numbers in combinatorics. So, there aren't *quite* as many Stirling identities as there are binomial identities, but there are still plenty. You can find quite a few in [GKP94, §6.1].

20.5 Problems

(1) Let N be a squarefree positive integer, i.e. one such that if p is any prime, then $p^2 \nmid N$. Suppose that N has exactly n prime factors. How many ways are there to write N as a product of integers greater than 1, where order doesn't matter? (That is, 5×6 is the same as 6×5, but $2 \times 3 \times 5$ is different.)

(2) Prove that if $n \geq 2$, then

$$n! < \left\{ {2n \atop n} \right\} < (2n)!.$$

(3) How many set partitions of $[n]$ are there in which 1 is not in a block of size 1?

(4) How many set partitions of $[n]$ are there into exactly two parts, such that the parts have different sizes?

(5) How many permutations of $[n]$ have 1 and 2 in the same cycle? How many have 1, 2, and 3 in the same cycle? How many have 1 and 2 in the same cycle, and 3 and 4 in the same cycle (possibly but not necessarily the same one)?

(6) Express the rising factorials in terms of the falling factorials; i.e. identify the numbers $f(n, k)$ defined by

$$x^{\overline{n}} = \sum_{k=0}^{n} f(n, k) x^{\underline{k}}.$$

You should be able to express these numbers in closed form, without using any summations involving a variable number of terms.

(7) Let f be an arbitrary function defined on the nonnegative integers, and define g by the formula

$$g(n) = \sum_{k=0}^{n} \left[{n \atop k} \right] f(k).$$

Show that we can recover f from g via the formula

$$f(n) = \sum_{k=0}^{n} (-1)^{n-k} \left\{ {n \atop k} \right\} g(k).$$

If you are familiar with Möbius inversion in number theory, this is a closely related phenomenon.

(8) Give a combinatorial proof of the identity

$$\left\{ {n+1 \atop m+1} \right\} = \sum_{k=0}^{n} \binom{n}{k} \left\{ {k \atop m} \right\}.$$

Chapter 21

The twelvefold way

21.1 Balls in boxes

How many ways are there to put n balls into k boxes? This type of question, about placing balls into boxes, is one of the most classic combinatorial problems. It often comes with extra rules about how many balls are allowed in each box. Then, the problem breaks up into a bunch more subproblems, because it is not initially clear whether the balls and the boxes are distinguishable or not. That is, if we put ball 1 in box 1 and ball 2 in box 2, is that the same as putting ball 2 in box 1 and ball 1 in box 2? All possible interpretations of this question are interesting, and we will now answer them all.

Let us write N for the set of balls and K for the set of boxes. A way of placing balls in boxes is a function $f : N \to K$. It is often useful to restrict the types of functions we are interested in. Recall the following definitions of some particularly important types of functions we might want to consider.

Definition 21.1. Let $f : A \to B$ be a function.

- We say that f is *injective* (or one-to-one) if, whenever $a_1, a_2 \in A$ with $a_1 \neq a_2$, then $f(a_1) \neq f(a_2)$.
- We say that f is *surjective* (or onto) if, for every $b \in B$, there is at least one $a \in A$ such that $f(a) = b$.
- We say that f is *bijective* if it is both injective and surjective.

See Figure 21.1 for pictures of injective and surjective functions.

In the case of balls in boxes, we can require that the function be injective or surjective. It is injective if each box contains *at most* one ball, whereas it is surjective if each box contains *at least* one ball.

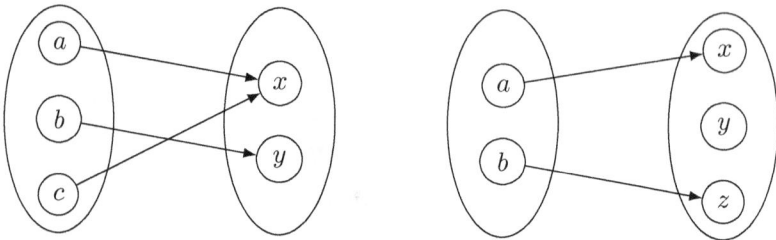

Figure 21.1. Left: a surjective function. Right: an injective function.

Balls	Boxes	f arbitrary	f injective	f surjective
dist	dist	1. $\quad k^n$	2. $\quad k^{\underline{n}}$	3. $\quad k!\left\{{n \atop k}\right\}$
indist	dist	4. $\quad \left(\!\!\left({k \atop n}\right)\!\!\right)$	5. $\quad \left({k \atop n}\right)$	6. $\quad \left(\!\!\left({k \atop n-k}\right)\!\!\right)$
dist	indist	7. $\quad \sum\limits_{i=0}^{k}\left\{{n \atop i}\right\}$	8. $\quad \begin{cases} 1 & n \le k \\ 0 & n > k \end{cases}$	9. $\quad \left\{{n \atop k}\right\}$
indist	indist	10. $\quad \sum\limits_{i=0}^{k}p_i(n)$	11. $\quad \begin{cases} 1 & n \le k \\ 0 & n > k \end{cases}$	12. $\quad p_k(n)$

Table 21.2. The twelvefold way.

We now have everything we need to specify all the standard balls-in-boxes problems. There are twelve such problems: we can specify that the balls are distinguishable or not (2 ways), that the boxes are distinguishable or not (2 ways), and that the function is arbitrary, injective, or surjective (3 ways). Thus we have a total of $12 = 2 \times 2 \times 3$ problems, and we often call this classification of problems the *twelvefold way*. The idea of this classification is due to Gian-Carlo Rota, and the name is due to Joel Spencer.

The answers to all these problems are given in Table 21.2. There is a lot of notation in this table. Most of it we have already seen, but some of it is new.

In cell 1, we find k^n. The reason that the answer is k^n is that, for each of the n balls, there are k places (boxes) to put it in. All of these arrangements are legal.

In cell 2, for injective f, we find the falling factorial $k^{\underline{n}}$, which as we recall is equal to $k(k-1)(k-2)\cdots(k-n+1)$. Let's analyze why the falling factorial $k^{\underline{n}}$ is the right answer for the second cell. Remember that we are restricting ourselves to *injective* functions, which means that there can only be at most one ball per box. The first ball can go into any of the k boxes. Now, since f is injective, the second ball only has $k-1$ boxes it can go to, since it can't end up in the same box as the first ball. Similarly, the third ball can go to $k-2$ boxes, and so forth. Note that this number is 0 if $k < n$.

In cell 3, we have two factors: $k!$ and $\left\{{n\atop k}\right\}$. Why is the answer in the third cell correct? We are restricting ourselves to *surjective* functions, which means that we must have at least one ball per box. This means that we need to break up the balls into k nonempty subsets, which by definition can be done in $\left\{{n\atop k}\right\}$ ways. Then we have to decide how to order the subsets in such a way that the first subset goes into box 1, the second subset goes into box 2, and so forth. That part can be done in $k!$ ways. Thus the final answer is the product of these two terms.

Cells 4 and 6 contain the notation $\left(\!\!\left({k\atop n}\right)\!\!\right)$ and $\left(\!\!\left({k\atop n-k}\right)\!\!\right)$. Recall that $\left(\!\!\left({k\atop n}\right)\!\!\right)$ denotes the number of ways of selecting n elements from a set of size k, but we are allowed to choose the same element more than once if we wish. For instance, one of the things that $\left(\!\!\left({5\atop 4}\right)\!\!\right)$ is counting, assuming the 5 on top refers to [5], is the multiset 1,1,3,4. Since we have a formula for $\left(\!\!\left({k\atop n}\right)\!\!\right)$ in terms of the more familiar binomial coefficients, we can also write the answer to the fourth cell as $\binom{n+k-1}{k-1}$ or $\binom{n+k-1}{n}$.

In cell 6, we need to ensure that each box gets at least one ball. To do that, just place one ball in each box; then the rest of the arrangement, using the remaining $n-k$ balls, follows what we do for cell 4. Thus the answer is indeed $\left(\!\!\left({k\atop n-k}\right)\!\!\right)$. Alternatively, using the formula for multichoose numbers in terms of binomial coefficients, we can write this answer as $\binom{n-1}{k-1}$.

In cell 5, we find our old friend the binomial coefficient $\binom{k}{n}$. Note that this seems slightly nonstandard in that the k is on top and the n is on the bottom. Why is this the right answer for cell 5? Well, the balls are indistinguishable, and the function is injective, meaning that some boxes get 1 ball, and other boxes get 0. The only noticeable feature, then, is which n of the k boxes get a ball, which is the combinatorial interpretation of $\binom{k}{n}$.

Now let's move on to the third row. Let's start with cell 9, with f surjective, since we are already familiar with the Stirling numbers of the second kind. This answer is pretty clear: we just divide up the balls into k sets and place them into boxes. Unlike in the case of cell 3, we don't have to worry about the assignment of subsets to boxes, since the boxes are indistinguishable.

For cell 7, the situation is similar, but now we are allowed to divide them up into fewer than k sets. Note that the $\left\{{n \atop 0}\right\}$ term doesn't seem to contribute much to the discussion, but it's 1 when $n = 0$, which is needed to get the correct answer.

Cells 8 and 11 have very simple answers. We can put all the balls into different boxes only when there are at least as many boxes as balls. There is no way of distinguishing any of the arrangements, so we can only have at most one distinct arrangement.

In cells 10 and 12, we have a new piece of notation: $p_k(n)$. This denotes the number of partitions of n into exactly k parts, i.e. the number of ways of writing n as a sum of exactly k positive integers, where the order of the summation does not matter. That's exactly what we're doing in cell 12: we have to divide the balls up among the k boxes, and the number of balls in the ith box is the ith summand in the partition. Since the boxes are indistinguishable, we may arrange them such that the number of balls they contain is nonincreasing: box 1 gets at least as many balls as box 2, and so forth.

Cell 10 is similar, but now we don't have to use all the boxes, so we might use fewer than k summands. Thus the answer is $\sum_{i=0}^{k} p_i(n)$. Once again, starting from p_0 is only necessary to make the formula hold when $n = 0$. (Think this through if you aren't used to this sort of thing!)

21.2   Problems

(1) In combinatorics, a *word* is a sequence of letters from some alphabet, regardless of whether or not it forms a word in some human language. Determine the number of n-letter words from the usual a–z alphabet, such that the letters are nondecreasing lexicographically (so for instance an "a" can come before a "b," but not after). The word "bccghhi" is one such example of a permissible 7-letter word.

(2) Find the number of ordered triples (A, B, C), where $A, B, C \subseteq [n]$, $A \cup B \cup C = [n]$, and $A \cap B \cap C = \varnothing$.

(3) Recall that an *equivalence relation* on a set X is a binary relation $\sim$ such that

- $x \sim x$ for all $x \in X$,
- if $x \sim y$, then $y \sim x$,
- if $x \sim y$ and $y \sim z$, then $x \sim z$.

An *equivalence class* is a nonempty set $S \subseteq X$ such that for all $s, t \in S$, we have $s \sim t$.

(a) Determine the number of equivalence relations on $[n]$ consisting of exactly k equivalence classes.

(b) Now, declare two equivalence relations $\sim_1$ and $\sim_2$ on $[n]$ to be equivalent if there is a permutation σ of $[n]$ such that $x \sim_1 y$ if and only if $\sigma(x) \sim_2 \sigma(y)$. Show that this is an equivalence relation on equivalence relations. How many equivalence classes of equivalence relations with exactly k equivalence classes are there? (Make sure you understand exactly what this means. Try some examples until you are confident that you get the question.)

Chapter 22

Counting labeled trees

22.1   Labeled trees

The last thing we will do in this part of the book is to count trees. There are many proofs of the associated theorem, but we will only look at one of them. But consider looking into some of the others, as it is instructive to see several notably different proofs of the same theorem. Several especially beautiful proofs can be found in [AZ18, Chapter 33].

Definition 22.1. Let V be a set of size n. A *tree* on V is a connected graph with vertex set V having $n - 1$ edges.

There are many other equivalent definitions of a tree, such as a connected graph without cycles, or a connected graph such that removing any edge disconnects the graph.

Let us suppose that our vertices come with labels $1, 2, \ldots, n$. We would like to know how many trees there are with this vertex set. See Figure 22.1 for $n = 1, 2, 3, 4$.

It is not clear what the pattern is supposed to be, starting with 1, 1, 3, 16, so let us compute more examples. Instead of drawing all the pictures (there are a lot!), let us try to work more systematically. To do that, we first forget about the labels and look at the *shape* (or isomorphism type) of the graph, and then count how many distinct labelings there are with that shape.

To see what we mean, let us consider first the case of $n = 4$. There are two basic shapes of the graph: either there can be one vertex that connects to all the rest of them, as in the first graph for $n = 4$, or it can be a path of length 4, as in the fifth graph for $n = 4$. For the first type, there are 4 ways of labeling the central vertex, so there are 4 graphs of this type. For

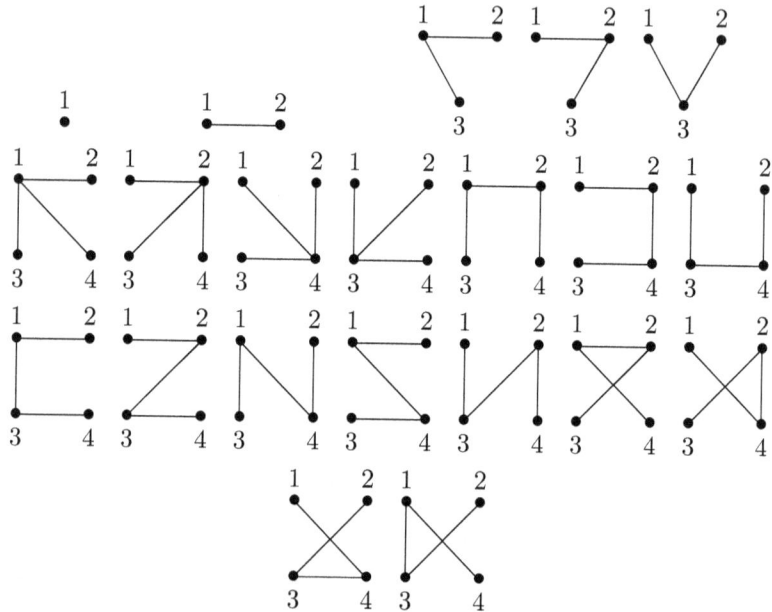

Figure 22.1. 1 tree on 1 vertex, 1 tree on 2 vertices, 3 trees on 3 vertices, 16 on 4 vertices.

the second type, we have $\binom{4}{2} = 6$ ways of labeling the two ends, and then we have two vertices that can go next to the "starting" vertex, for a total of 12. Thus we have $4 + 12 = 16$ labeled trees on four vertices.

What are the possible shapes for $n = 5$? Since 5 is a small number, it is not so difficult to enumerate all of them: there are only 3, as shown in Figure 22.2.

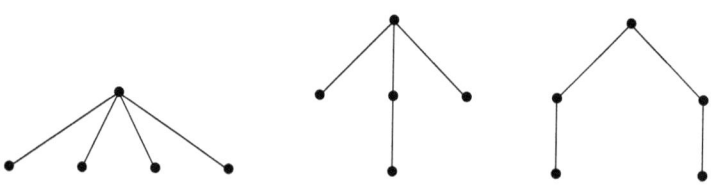

Figure 22.2. Three isomorphism types of trees on five vertices.

Now let's count them. For the first type, the central vertex can take on 5 different labels, and then we're done, so there are 5 labeled trees of this shape. For the second type, there are again 5 choices for the central vertex, then 4 for the vertex below it, then 3 for the vertex below that, and then we're done, so there are 60 labelings of this tree. For the last type, there are 5 choices for the first vertex, then 4 for the second, then 3 for the third, then 2 for the fourth, then 1 for the last, but the graph has a symmetry: if we label them 1,2,3,4,5, in that order, then we get the same tree as if we had labeled them 5,4,3,2,1. So we have to divide by 2, to get 60 again. In total, there are $5 + 60 + 60 = 125$ labeled trees on 5 vertices.

It might still not be clear what the pattern is from the sequence 1, 1, 3, 16, 125, so it's good to do one more example, with $n = 6$. There are six isomorphism types of trees, as shown in Figure 22.3. There are a total of 1296 labeled trees on 6 vertices.

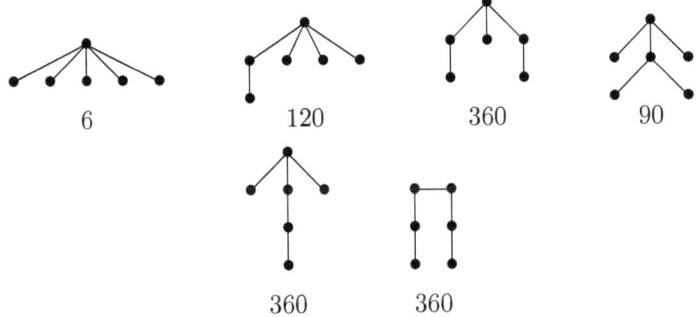

Figure 22.3. There are six isomorphism types of trees on six vertices, and $6 + 120 + 360 + 90 + 360 + 360 = 1296$ labeled trees.

It is now reasonable to note that the sequence 1, 1, 3, 16, 125, 1296 appears to be n^{n-2}. And indeed, it is.

Theorem 22.2 (Cayley). *There are n^{n-2} labeled trees on n vertices.*

In order to give a bijective proof of Cayley's Theorem, we need to find a set that "obviously" has n^{n-2} elements. The most obvious such set is the set of sequences $(a_1, \ldots, a_{n-2})$, where each $a_i \in [n]$. So, we want to construct such a sequence out of a labeled tree, and conversely, we want to construct a labeled tree out of each such sequence. The sequence associated to the tree is known as the *Prüfer sequence*.

Definition 22.3. Let T be a tree. A vertex $v \in T$ is said to be a *leaf* if it has exactly one neighbor.

Note that if T is a tree with at least two vertices, then T has some leaves. (At least two, in fact.)

Proof of Theorem 22.2. Starting with a labeled tree T on n vertices, we produce its Prüfer sequence of $n - 2$ numbers from 1 to n—possibly with repetition—as follows. For $1 \leq k \leq n - 2$, we find the leaf v remaining on the tree with the smallest label. Since v is a leaf, it has a unique neighbor w. We let the kth element of the Prüfer sequence be the label of w. Then we delete v from the tree and move on to the next step. See Figure 22.4 for the steps in the construction.

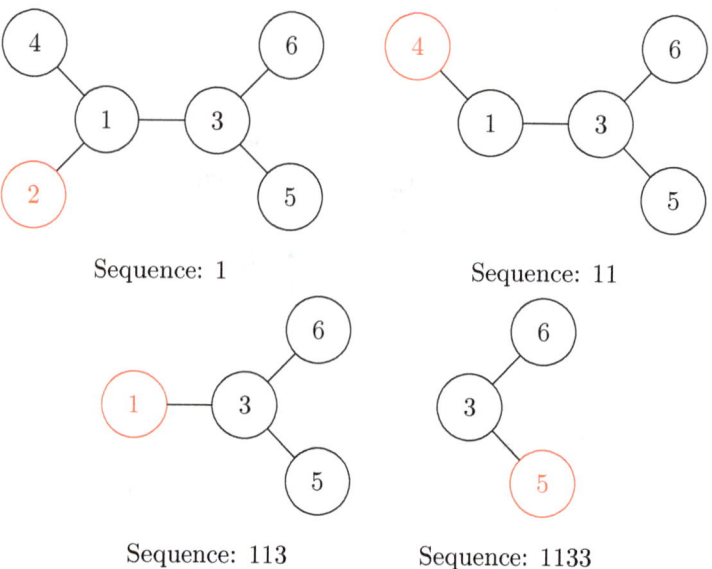

Figure 22.4. Constructing the Prüfer sequence for a tree on six vertices.

Really, we shouldn't stop the construction after the $(n - 2)^{\text{nd}}$ step, but we should continue on with the $(n - 1)^{\text{st}}$ step. However, the label at the $(n - 1)^{\text{st}}$ step is always n. The reason is that, since there are always at least two leaves in a tree with $k \geq 2$ vertices, we never remove the vertex labeled n. At the $(n - 1)^{\text{st}}$ step, there are exactly two vertices remaining, one of which has label n. Thus we must remove the other one, and the next term in the Prüfer sequence is n. So, since the last term is always the same, we may freely omit it without any loss of information.

So, now we know how to go from a labeled tree to a Prüfer sequence. Next, we must explain how to go in the other direction: we need to start with a Prüfer sequence and use it to generate a labeled tree. Let us think about how to reverse the process. Given a sequence, what is the label of the first vertex removed? Since it is removed, it had better not be part of the sequence. Furthermore, since we remove leaves starting with the one with the smallest label, the first vertex removed *must* be the smallest number *not* in the Prüfer sequence. Furthermore, we know what it is connected to, namely the first term of the Prüfer sequence. Now delete the first term of the Prüfer sequence and remove the vertex just added from the list of vertices under consideration, and continue the process, as shown in Figure 22.5.

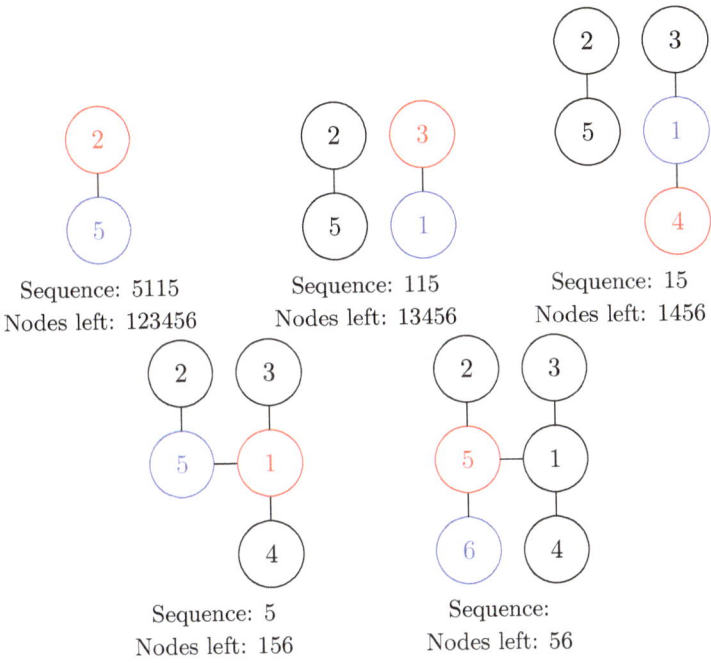

Figure 22.5. Reconstructing the labeled tree from the Prüfer sequence 5115. At each stage, the red node is the leaf that would be deleted in the construction of the Prüfer sequence, and the blue node is its neighbor.

When doing the last step, remember that the sequence is supposed to end with an n, but we omit it since that is automatic. However, we need to remember it when we reconstruct the tree from the sequence, in order to draw the final edge of the tree.

Now that we have both directions of our construction, it remains to check that these operations are inverses: if we start with a tree, construct its Prüfer sequence, then build a tree from the sequence, we end up with the original tree. (It might not look the same in terms of how we draw it, but all the vertices have the same neighbors.) Conversely, if we start with a sequence, build a tree, then reconstruct the Prüfer sequence, we end up with the original sequence. We leave it as an exercise to check this! ∎

22.2   Further reading

There are several proofs of Cayley's Theorem, some generalizing in interesting ways or connecting to seemingly different parts of combinatorics. Several of them can be found in [AZ18].

One combinatorial object that appears to have nothing to do with labeled trees, but secretly does, is the class of *parking functions*. Suppose there are n cars on a one-way street, and there are n parking spaces on the street. The driver of car i prefers spot $f(i)$, so $1 \leq f(i) \leq n$ for all i with $1 \leq i \leq n$. The driver of car i will park in spot $f(i)$ if it is available, and will otherwise park in the first spot available after $f(i)$, so driver i parks in spot j where j is the smallest available parking spot $\geq f(i)$. If there are no more available spots after spot $f(i)$, then the driver gives up and goes home. We call the function f a *parking function* if all the drivers are able to park.

Theorem 22.4. *The number of parking functions of length n is $(n+1)^{n-1}$.*

Proof. Let us modify the road so that it is now a circular road with $n + 1$ parking spots, labeled 1 through $n+1$, but still only n cars. We also permit the driver to prefer spot $n + 1$. Now, because the road is circular and there are at least as many parking spaces as drivers, everyone will always be able to park regardless of preferences, so there are $(n + 1)^n$ functions that allow everyone to park.

After everyone has parked, there will always be exactly one empty spot. A circular parking function is an ordinary parking function if and only if, after everyone has parked, the empty space is the one labeled $n + 1$. But in the case of circular parking functions, all the spots are symmetric, so the proportion of functions leaving the $n + 1$ spot empty is $\frac{1}{n+1}$. Thus the number of ordinary parking functions is $\frac{1}{n+1}(n + 1)^n = (n + 1)^{n-1}$. ■

Note that the number of parking functions of length n is equal to the number of labeled trees on $n + 1$ vertices. Can you find a bijection between these two sets? For more on parking functions, see [Sta20].

22.3 Problems

(1) How many labeled trees are there on n vertices with no edge between vertices 1 and 2?

(2) Prüfer sequences are allowed to have repeated elements. Describe the trees for which their Prüfer sequences have no repeated elements. (For example, (1,3,5,2,7), but not (1,3,1,5,4).)

(3) Characterize the trees whose Prüfer sequences are constant (i.e. $(a_1, a_2, \ldots, a_{n-2})$ with $a_1 = a_2 = \cdots = a_{n-2}$).

(4) A *forest* is a collection of disjoint trees. A *rooted forest* is a forest in which one vertex in each tree is chosen to be a distinguished vertex known as the root.

   (a) Prove that the number of rooted trees on n labeled vertices is n^{n-1}.

   (b) Prove that the number of rooted forests on n labeled vertices is $(n + 1)^{n-1}$. Can you prove this using a variant on Prüfer sequences? What about without them, directly from Cayley's formula for trees?

PART 3
Analysis

Chapter 23

Countable and uncountable sets

23.1 Sizes of sets

Given a set S with finitely many elements, we can assign it a *cardinality*, which is just a fancy term for the number of elements it contains. For example, the set

$$S = \{a, 50, 8\pi - 9, \clubsuit, \triangle\}$$

has cardinality 5. Two finite sets with the same number of elements have the same cardinality. However, it's not immediately obvious how to tell whether two infinite sets have the same number of elements (or cardinality), or even what that means. So the first order of business is to give a precise definition for what it means for two sets, which might be infinite, to have the same cardinality.

Definition 23.1. Two sets S and T are said to have the same *cardinality* if there is a bijective function $f : S \to T$.

Recall the definition of a bijective function:

Definition 23.2. Let S and T be two sets, and let $f : S \to T$ be a function.

- We say that f is *surjective* or *onto* if, for every $t \in T$, there is some $s \in S$ such that $f(s) = t$.
- We say that f is *injective* or *one-to-one* if, whenever $s \neq s'$, we have $f(s) \neq f(s')$.
- We say that f is *bijective* if it is both injective and surjective.

See Figure 23.1 for pictures of surjective and injective functions.

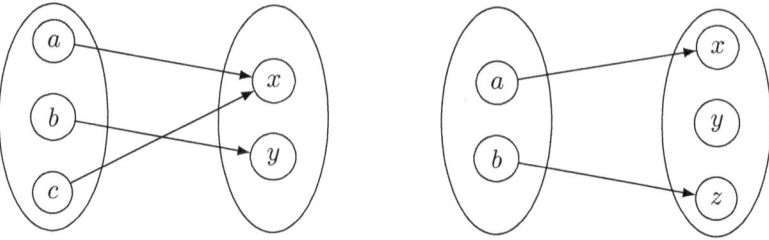

Figure 23.1. Left: a surjective function. Right: an injective function.

The key point here is that *if there is a bijection between two sets S and T, then they have the same number of elements.* Recall also that the existence of an inverse function is equivalent to being a bijection:

Theorem 23.3. *Let S and T be sets, and let $f : S \to T$ be a function. Then f is a bijection if and only if there is a function $g : T \to S$ such that $g \circ f(s) = s$ for all $s \in S$, and $f \circ g(t) = t$ for all $t \in T$.*

All this is pretty logical: two sets S and T have the same size (or cardinality) if we can take the elements of S and match them up with the elements of T, so that nothing gets skipped or repeated.

Here is another collection of basic results about bijections (and injections and surjections) that will be helpful later on.

Proposition 23.4. *Let R, S, T be three sets, and suppose that $f : R \to S$ and $g : S \to T$ are functions.*

(1) If f and g are both surjective, then so is $g \circ f : R \to T$.
(2) If f and g are both injective, then so is $g \circ f$.
(3) If f and g are both bijective, then so is $g \circ f$.

Proof.

(1) Let $t \in T$ be arbitrary. We must show that there is some $r \in R$ such that $g \circ f(r) = t$. Since g is surjective, there is some $s \in S$ such that $g(s) = t$. Since f is surjective, there is some $r \in R$ such that $f(r) = s$. For this r, we have

$$(g \circ f)(r) = g(f(r)) = g(s) = t.$$

Since t was arbitrary, this shows that $g \circ f$ is surjective.

(2) Suppose that $g \circ f(r) = g \circ f(r')$. Since g is injective, we have $f(r) = f(r')$. Since f is injective, we have $r = r'$. Thus $g \circ f$ is injective.

(3) Since $g \circ f$ is both surjective and injective, it is bijective. ∎

Of course, by induction, these results hold for chains of more than two (sur/in/bi)jective functions as well.

23.2　Countable sets

A popular infinite set to compare with other sets is the set $\mathbb{N}$ of positive integers, i.e. $\mathbb{N} = \{1, 2, 3, \ldots\}$. (Some people use $\mathbb{N}$ to mean the *nonnegative* integers. We shall soon see that, for the purposes of counting, it doesn't matter.)

Definition 23.5. A set S is said to be *countable* if it is in bijection with $\mathbb{N}$, i.e. if there is a bijective function $f : \mathbb{N} \to S$ (or equivalently $g : S \to \mathbb{N}$).[1] An infinite set S is said to be *uncountable* if it is not in bijection with $\mathbb{N}$.

Let's see some examples of countable sets.

Example. The set $\mathbb{Z}_{\geq 0}$ of *nonnegative* integers is countable. To see this, let's define a function $f : \mathbb{N} \to \mathbb{Z}_{\geq 0}$ and then prove it's a bijection. We'll take $f(n) = n - 1$, so $f(1) = 0$, $f(2) = 1$, and so forth. We claim that this is a bijection. We can check this either by showing that it's both injective and surjective, or by finding an inverse. Both methods are pretty easy in this case; we'll do it by writing down an inverse $g : \mathbb{Z}_{\geq 0} \to \mathbb{N}$. Our inverse is $g(n) = n + 1$. For any $n \in \mathbb{N}$, we have $g(f(n)) = g(n-1) = (n-1) + 1 = n$, and similarly for any $n \in \mathbb{Z}_{\geq 0}$, we have $f(g(n)) = f(n+1) = (n+1) - 1 = n$. Both compositions leave us back where we started, so f and g are inverses and thus f is a bijection. It follows that $\mathbb{Z}_{\geq 0}$ is countable.

This is already a little bit surprising, because $\mathbb{Z}_{\geq 0}$ contains $\mathbb{N}$ as a proper subset. Thus we learn that adding an element to an infinite set doesn't necessarily make it bigger. In fact, one occasionally used definition of an infinite set is a set S that has a bijection with one of its proper subsets.

Example. The set $\mathbb{Z}$ of all integers is countable. We have to define a function $f : \mathbb{N} \to \mathbb{Z}$ and then show that it is bijective. Again, we'll prove that it's bijective by constructing an inverse. This is easier to do by drawing a picture than by writing down an actual formula (although the latter can be

---

[1]Some people use the term "countable" to include finite sets. We will not do that, but be aware that that convention also exists and is quite common.

done by anyone sufficiently committed to the cause). Here's our bijection. The picture makes it clear that it's actually a bijection, since the inverse just goes the other way.

$$
\begin{array}{ccccccccccc}
\mathbb{N}: & 1 & 2 & 3 & 4 & 5 & 6 & 7 & 8 & 9 & \cdots \\
& \updownarrow & \updownarrow & \updownarrow & \updownarrow & \updownarrow & \updownarrow & \updownarrow & \updownarrow & \updownarrow & \\
\mathbb{Z}: & 0 & 1 & -1 & 2 & -2 & 3 & -3 & 4 & -4 & \cdots
\end{array}
$$

It's clear that this is a bijection, because all the integers appear in the bottom row: after 0, we alternate writing down the next positive number and the next negative number, and we never run out of numbers on either row. Furthermore, every positive integer in the top row is paired with some integer in the bottom row, and vice versa. Thus this map is a bijection, so $\mathbb{Z}$ is countable.

Theorem 23.6. *The set $\mathbb{Q}$ of all rational numbers is countable.*

Proof. We have to produce a bijection from $\mathbb{N}$ to $\mathbb{Q}$. There are ways of writing down precise formulae,[2] but an easy way to see this is with a picture. Instead of starting with *all* the rationals, let's start with just the positive ones, and then we'll see how to modify the argument to handle all rationals. To do this, we note that the rationals are just numbers $\frac{a}{b}$ where a and b are relatively prime, so we can identify them with some subset of the ordered pairs (a, b) of integers. We'll put the rational number $\frac{a}{b}$ at the point (a, b) in the plane. We need to skip the pairs like $(3, 6)$ that represent rational numbers we'll already have counted. A picture of the beginning of our bijection is shown in Figure 23.2. The bijection is given by

$$
f(1) = 1, \quad f(2) = 2, \quad f(3) = \frac{1}{2}, \quad f(4) = \frac{1}{3}, \quad f(5) = 3,
$$

$$
f(6) = 4, \quad f(7) = \frac{3}{2}, \quad f(8) = \frac{2}{3}, \quad f(9) = \frac{1}{4},
$$

and so on. In general, $f(n)$ is what we get when we start from 1 and follow along in the diagram for $n - 1$ arrows. Note that we simply skip over any fractions that aren't in reduced form.

To see that this map is a bijection from $\mathbb{N}$ to the positive rationals, we'll check that it's injective and surjective. First, it's injective because $f(m) \neq f(n)$ whenever $m \neq n$: the result of tracing $m - 1$ arrows is not

---

[2] I encourage you to look up the Stern–Brocot tree (see for instance [GKP94]) or the Calkin–Wilf tree (see [CW00]) for ways of doing this. They are quite elementary, but it would take us a bit off-topic to give an adequate treatment of either one.

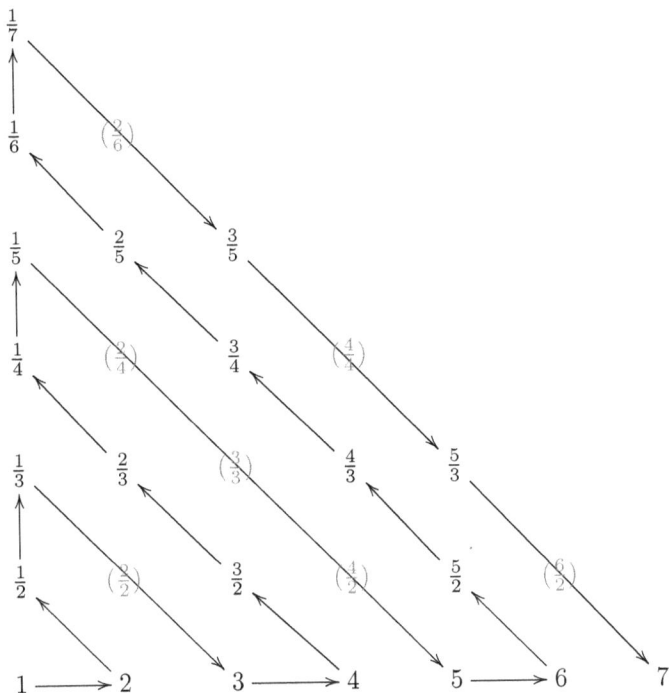

Figure 23.2. A bijection between $\mathbb{N}$ and $\mathbb{Q}_{>0}$.

the same as the result of tracing $n-1$ arrows. Furthermore, it's surjective because every positive rational number $\frac{m}{n}$ is at the point (m,n) in the plane, so we eventually get to it on our diagonal sweep through the line $x+y = m+n$ that it lies on.

Now, we really want to show that $\mathbb{Q}$ is countable—not just that the *positive* rationals are countable. One slick way of doing this is to make use of the fact that $\mathbb{Z}$ is countable, so it's just as good to produce a bijection between $\mathbb{Z}$ and $\mathbb{Q}$. To do this, consider our bijection f above from $\mathbb{N}$ to the positive rationals. Extend it to $f : \mathbb{Z} \to \mathbb{Q}$ by setting $f(0) = 0$ and $f(-n) = -f(n)$. Now f is a bijection from $\mathbb{Z}$ to $\mathbb{Q}$, by the same reasoning as before. We have a chain of bijections

$$\mathbb{N} \to \mathbb{Z} \to \mathbb{Q},$$

and the composition of bijections is a bijection by Proposition 23.4. Thus we have proven that $\mathbb{Q}$ is countable. ∎

23.3   Uncountable sets

It might feel surprising that $\mathbb{Q}$ is countable, since it seems so much bigger than $\mathbb{N}$. At this point, one might wonder if maybe all infinite sets have the same cardinality: if they are all in bijection with $\mathbb{N}$. However, this is not the case, as Georg Cantor, the first mathematician to study set theory seriously, proved in one of the classic proofs in all of mathematics.

Theorem 23.7. *The set $\mathbb{R}$ of real numbers is uncountable.*

Remark 23.8. At this point, we do not yet have a formal definition of the real numbers. In future chapters of this book, we will see two formal definitions of the real numbers, constructing them out of just the integers or the rationals. For now, we'll think of the real numbers as decimal expansions, noting that $0.9999\cdots = 1$ and other similar things. Decimal expansions are unique except for the caveat that a number ending with all 9's can be turned into a terminating decimal. That will be good enough for our purposes at the moment.

Proof. Let $(0,1) = \{x \in \mathbb{R} : 0 < x < 1\}$. We will show that if $f : \mathbb{N} \to (0,1)$ is any function, then f fails to be surjective, hence is not a bijection. (See problem 5 for why it is sufficient to work with $(0,1)$ rather than all of $\mathbb{R}$.) We begin by making a table of values of f, say

$$f(1) = .d_{11}d_{12}d_{13}d_{14}d_{15}d_{16}d_{17}d_{18}\ldots$$
$$f(2) = .d_{21}d_{22}d_{23}d_{24}d_{25}d_{26}d_{27}d_{28}\ldots$$
$$f(3) = .d_{31}d_{32}d_{33}d_{34}d_{35}d_{36}d_{37}d_{38}\ldots$$
$$f(4) = .d_{41}d_{42}d_{43}d_{44}d_{45}d_{46}d_{47}d_{48}\ldots$$
$$f(5) = .d_{51}d_{52}d_{53}d_{54}d_{55}d_{56}d_{57}d_{58}\ldots$$
$$f(6) = .d_{61}d_{62}d_{63}d_{64}d_{65}d_{66}d_{67}d_{68}\ldots$$
$$f(7) = .d_{71}d_{72}d_{73}d_{74}d_{75}d_{76}d_{77}d_{78}\ldots$$
$$f(8) = .d_{81}d_{82}d_{83}d_{84}d_{85}d_{86}d_{87}d_{88}\ldots$$
$$\vdots$$

Here, each d_{ij} is a digit from 0 to 9. Note that i and j range over the positive integers, so we eventually encounter digits with labels like $d_{12,44}$, where at this point we must use a comma to separate the indices for the sake of clarity. We will now produce a real number $x \in (0,1)$ that is definitely *not* $f(n)$ for any n, from which we can conclude that f is not surjective, hence not bijective. To do this, consider the number y, which consists of the first digit (after the decimal point) of the first number, followed by the second digit of the second number, then the third digit of the third number, and so forth; in other words, the digits marked in red in the table. That is,

$$y = .d_{11}d_{22}d_{33}d_{44} \cdots$$

Now, in order to obtain x from y, we change each digit of y to some digit from 1–8 (so as to avoid problems with terminating decimals or numbers that end with infinitely many 9's) that is different from the original digit in that position in y. The resulting number is x. Note that for every positive integer k, the kth digits of x and y are different.

Is the number x of the form $f(n)$ for some n? We know that $x \neq f(1)$, because their first digits are different. Similarly, $x \neq f(2)$, because their second digits are different. Likewise, $x \neq f(3)$, because their third digits are different. In general, $x \neq f(n)$, because their n^{th} digits are different. Thus f is not surjective, hence not a bijection. ∎

So, the cardinality of $\mathbb{R}$ is not the same as the cardinality of $\mathbb{N}$. It is now time to give names to the cardinalities of the positive integers and the real numbers:

Definition 23.9. The cardinality of $\mathbb{N}$ is called $\aleph_0$ (pronounced "aleph-zero" or "aleph-null"). The cardinality of $\mathbb{R}$ is called $\mathfrak{c}$, or "the continuum."

Another name for $\mathfrak{c}$ is $2^{\aleph_0}$, since it is also the cardinality of binary strings of length $\aleph_0$. (You will prove this in problem 8.) Note that there are 2^n binary strings of length n, so this notation makes sense.

23.4 Problems

(1) (a) Prove that the union of two countable sets is countable. More generally, prove that any finite union of countable sets is countable. (The sets might not be disjoint, so figure out how to deal with that potential issue.)

(b) Is a countable union of countable sets necessarily countable? Prove it or give a counterexample.

(2) Prove the following two infinitary analogues of the pigeonhole principle:

(a) If n is a positive integer, $A_1, A_2, \ldots, A_n$ are sets, and $\bigcup_{i=1}^{n} A_i$ is an infinite set, then there is some i with $1 \leq i \leq n$ such that A_i is an infinite set.

(b) If n is a positive integer, $A_1, A_2, \ldots, A_n$ are sets, and $\bigcup_{i=1}^{n} A_i$ is an uncountable set, then there is some i with $1 \le i \le n$ such that A_i is an uncountable set.

(3) Let S and T be nonempty sets, and let $f : S \to T$ be a function.

    (a) Prove that f is injective if and only if there is a function $g : T \to S$ such that $g \circ f(s) = s$ for all $s \in S$.

    (b) Prove that f is surjective if and only if there is a function $g : T \to S$ such that $f \circ g(t) = t$ for all $t \in T$.

(4) What is the cardinality of $\mathbb{Q} \times \mathbb{Q}$, the set of ordered pairs (a, b) where both a and b are rational? Prove that your answer is correct.

(5) Prove that $\mathbb{R}$ has the same cardinality as the open interval $(0, 1)$.

(6) Suppose we try to use the diagonalization proof that $\mathbb{R}$ is uncountable to prove that $\mathbb{Q}$ is also uncountable. This can't work, because $\mathbb{Q}$ is countable. But where does the proof break down?

(7) Let A be an uncountable set and $B \subseteq A$ a countable subset. Show that A and $A \setminus B$ have the same cardinality. (The set $A \setminus B$ is the set of elements of A that are not in B.)

(8) Prove that the set of binary strings of length $\aleph_0$ (or functions from $\mathbb{N}$ to $\{0, 1\}$) has cardinality $\mathfrak{c}$.

(9) Find an explicit bijection between binary strings of length $\aleph_0$ and ternary (i.e. the digits are from $\{0, 1, 2\}$) strings of length $\aleph_0$.

Chapter 24

Set-theoretic odds and ends

24.1 The Cantor–Schröder–Bernstein Theorem

Sometimes, we would like to be able to say something about the relative sizes of various cardinals. For example, if we have two sets S and T such that there is a bijection between S and T, then S and T have the same cardinality. But what happens if there is only an injection, or only a surjection? It seems that if there is an injective function from S to T, then we should be able to conclude that the cardinality of S is less than or equal to the cardinality of T. Similarly, if there is a surjection from S to T, then we should be able to conclude that the cardinality of S is greater than or equal to the cardinality of T. A good sanity check, to make sure that all of this has some sort of internal consistency, is the theorem of Cantor, Schröder, and Bernstein (sometimes only a proper subset of those names are used).

Theorem 24.1 (Cantor–Schröder–Bernstein). *Let S and T be two sets. Suppose that there is an injective function $f : S \to T$ and another injective function $g : T \to S$. Then there is a bijection $h : S \to T$.*

Proof. Assume that S and T are disjoint; if not, relabel the elements of one of them (say T) to make them disjoint. For any $s \in S$ (and similarly for $t \in T$), we can form a chain of elements in $S \cup T$ containing s, as follows:

$$\cdots \to f^{-1}(g^{-1}(s)) \to g^{-1}(s) \to s \to f(s)$$
$$\to g(f(s)) \to f(g(f(s))) \to g(f(g(f(s)))) \to \cdots .$$

By $g^{-1}(s)$, for instance, we mean the unique element $t \in T$ such that $g(t) = s$. In the backward direction, the chain may terminate at some point. For instance, we will not be able to continue backward from s if

263

there is no $t \in T$ such that $g(t) = s$. Note that every element in $S \cup T$ appears in exactly one chain of this type.

In terms of these chains, we now define $h(s)$ for each $s \in S$. We have three possibilities:

- The chain stops at some point to the left, and the first element is in S. In this case, define $h(s) = f(s)$.

- The chain stops at some point to the left, and the first element is in T. In this case, define $h(s) = g^{-1}(s)$. (Note that this makes sense, because the first element in the chain is in T, so we can always go backward from any element of S, including s itself.)

- The chain never stops to the left. This means that either it hits new elements each time without ever getting stuck, or it goes in a loop. Either way, we set $h(s) = f(s)$.

We claim that this is a bijection. To see this, we'll construct its inverse function $k : T \to S$, by setting $k(t) = f^{-1}(t)$, $g(t)$, and $f^{-1}(t)$, respectively, in the three cases. This inverse function is well-defined in the cases where $k(t) = f^{-1}(t)$, because the chain either stops at an element of S or never stops. Either way, for every such t, $f^{-1}(t)$ is defined. ∎

Since this proof is a bit complicated, let's take a look at a simple example.

Example. Consider the functions $f : \mathbb{N} \to \mathbb{N}$ and $g : \mathbb{N} \to \mathbb{N}$, given by $f(n) = n + 1$ and $g(n) = n + 2$ for all n. These are both injections, so we can use the proof of the Cantor–Schröder–Bernstein Theorem to produce a bijection. Of course, it is very easy to find a bijection from $\mathbb{N}$ to $\mathbb{N}$ (like $h(n) = n$), but let's find the one that the proof produces out of f and g.

Let $h : \mathbb{N} \to \mathbb{N}$ be the bijection produced from f and g. Let's compute $h(12)$. To do this, we have to look at the chain containing 12, $f(12)$, $g(f(12))$, and so forth, as well as the preimages $g^{-1}(12)$, $f^{-1}(g^{-1}(12))$, and so forth. The construction of h involves the preimages, so let's compute those. We'll use red for the left $\mathbb{N}$ and blue for the right $\mathbb{N}$, so that we don't get confused since both sets (S and T as in the notation of the theorem) are the same. In the notation of the proof, the red numbers are in S and the blue numbers are in T; see Figure 24.1. We have

$$g^{-1}(12) = 10,$$
$$f^{-1}(g^{-1}(12)) = 9,$$
$$g^{-1}(f^{-1}(g^{-1}(12))) = 7,$$
$$f^{-1}(g^{-1}(f^{-1}(g^{-1}(12)))) = 6,$$
$$g^{-1}(f^{-1}(g^{-1}(f^{-1}(g^{-1}(12))))) = 4,$$
$$f^{-1}(g^{-1}(f^{-1}(g^{-1}(f^{-1}(g^{-1}(12)))))) = 3,$$
$$g^{-1}(f^{-1}(g^{-1}(f^{-1}(g^{-1}(f^{-1}(g^{-1}(12))))))) = 1,$$

and at that point we have to stop, because $f^{-1}(1)$ does not exist. Since the chain stops with a blue element, i.e. something in T, this means that $h(12) = g^{-1}(12)$, i.e. $h(12) = 10$. We can use the same logic, but starting with an arbitrary number rather than 12, to show that

$$h(n) = \begin{cases} n - 2 & \text{if } n \equiv 0 \pmod{3}, \\ n + 1 & \text{if } n \equiv 1, 2 \pmod{3}. \end{cases}$$

We've just constructed the Cantor–Schröder–Bernstein bijection in this case! It's a little weirder than the identity bijection, but it's a perfectly good one nonetheless.

We can use the Cantor–Schröder–Bernstein Theorem to make the proof that $\mathbb{Q}$ is countable a little bit easier. It was slightly complicated by the fact that we had to skip over the rational numbers that weren't in lowest terms. Let's use the Cantor–Schröder–Bernstein Theorem to explain how to avoid doing that. As before, we'll only prove that the positive rationals are countable, since we can use the same trick as before to extend to the case of all rationals.

Theorem 24.2. $\mathbb{Q}$ *is countable.*

Proof 2 of Theorem 24.2. We will construct injections $f : \mathbb{N} \to \mathbb{Q}_{>0}$ and $g : \mathbb{Q}_{>0} \to \mathbb{N}$. (As we discussed in Chapter 23, in order to show that $\mathbb{Q}$ is countable, it suffices to show that $\mathbb{Q}_{>0}$ is countable.) First, let $f(n) = n$; this is obviously injective. There are many ways of constructing a suitable g, but here's one that illustrates a useful point. Instead of writing rational numbers $\frac{a}{b}$ where a and b are in base 10, write them in base 9, so that a and b contain only digits from 0 to 8. Now let $g(\frac{a}{b}) = a9b$, i.e. replace the / with a 9 and then read off in base 10. (For instance, $g(\frac{1}{3}) = 193$.) This is an injective function from $\mathbb{Q}_{>0}$ to $\mathbb{N}$. Now that we have injective functions

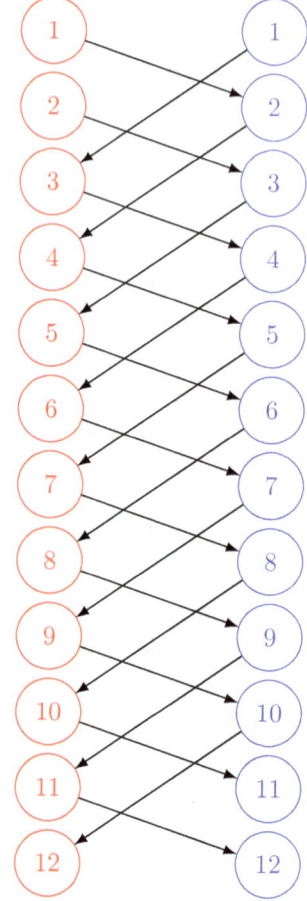

Figure 24.1. A Cantor–Schröder–Bernstein bijection in action: $h(n)$ depends on where the chain of backward arrows staring from n ends.

both ways between $\mathbb{N}$ and $\mathbb{Q}_{>0}$, the Cantor–Schröder–Bernstein Theorem guarantees that there must be a bijection! ∎

The cool thing about this approach is that we didn't have to say exactly what the bijection is. In theory, we could unpack the proof of Cantor–Schröder–Bernstein to figure it out, but why bother when we know that there must be a bijection? Also worth noting is that neither of the two injections is anywhere close to being a bijection. This is a frequent feature of bijection proofs using Cantor–Schröder–Bernstein: there is no need to do the theorem's work for it by creating a bijection or near-bijection yourself. Instead, just find a simple injection each way and allow the theorem to do its magic.

The Cantor–Schröder–Bernstein Theorem also allows us to define the notion of inequality among cardinalities.

Definition 24.3. Let S and T be two sets, with cardinalities $|S|$ and $|T|$, respectively. We say that $|S| \leq |T|$ if there is an injective function $f : S \to T$.

A restatement of the Cantor–Schröder–Bernstein Theorem, then, is that if $|S| \leq |T|$ and $|T| \leq |S|$, then $|S| = |T|$. The notation makes that seem obvious, but remember that the definitions don't instantly imply it because the notation $\leq$ means something new. It's only after we have proven the Cantor–Schröder–Bernstein Theorem that we can guarantee it.

24.2   Power sets

So far, we have only seen two cardinalities of infinite sets: $\aleph_0$ and $\mathfrak{c}$. It is natural to wonder whether there are others. In fact, there are many, many, many others—a (very) infinite number of other cardinalities. One way of proving this is by using power sets.

Definition 24.4. Let S be a set. The *power set* of S, denoted $\mathscr{P}(S)$, is the set of all subsets of S.

Example. Suppose that $S = \{1, 2, 3\}$. Then

$$\mathscr{P}(S) = \{\varnothing, \{1\}, \{2\}, \{3\}, \{1, 2\}, \{1, 3\}, \{2, 3\}, \{1, 2, 3\}\}.$$

Note that it contains 8 elements. More generally, if S is a finite set of cardinality n, then $\mathscr{P}(S)$ is also a finite set, with cardinality 2^n.

In Chapter 23, we used Cantor's diagonalization argument to prove that the real numbers are uncountable—and, in fact, that the cardinality of the real numbers is greater than that of the natural numbers. A similar argument can be used to show that, if S is any set at all, then the cardinality of $\mathscr{P}(S)$ is greater than that of S. It's clear that the cardinality of $\mathscr{P}(S)$ must be at least as large as that of S, because for every $s \in S$, $\mathscr{P}(S)$ contains a set $\{s\}$. Note that when $S = \varnothing$, then $|S| = 0$ and $|\mathscr{P}(S)| = 1$, so it's even true of the empty set.

Theorem 24.5. *For any set S, we have $|\mathscr{P}(S)| > |S|$.*

Proof. We must show that there is no bijection $f : S \to \mathscr{P}(S)$. Suppose that $f : S \to \mathscr{P}(S)$ is any function. We will show that it fails to be surjective. To do this, we create a subset $T \subseteq S$ that cannot be in the image of f, and we do this by building T one element at a time: for every $s \in S$, we will decide whether or not $s \in T$. For each $s \in S$, we have two cases: either $s \in f(s)$ or $s \notin f(s)$. If $s \in f(s)$, then $s \notin T$. On the other hand, if $s \notin f(s)$, then $s \in T$. In terms of a concise expression, we have

$$T = \{s \in S : s \notin f(s)\}.$$

We claim that T is not $f(s)$ for any $s \in S$. For any $s \in S$, $f(s) \neq T$, because $s \in T$ if and only if $s \notin f(s)$, so $f(s)$ and T differ on whether or not they contain s. Thus T is not $f(s)$ for any $s \in S$, so f is not surjective. ∎

One interesting consequence is that there is no largest cardinality. Given any set S, we can always find some other set T such that $|T| > |S|$, and indeed $T = \mathscr{P}(S)$ is one possibility. There are lots of other choices as well, such as $\mathscr{P}(\mathscr{P}(S))$.

24.3   The Cantor set

Now is as good a time as any to introduce a remarkable subset of the interval $[0, 1]$, known as the *Cantor set*. Here is how we construct it: Let C_0 be the interval $[0, 1]$. For each n, C_n is a union of closed intervals, and we form C_{n+1} from C_n by removing the open middle third of each interval in C_n. So, to obtain C_1 from C_0, we start with C_0, which is $[0, 1]$ and hence just a single interval, and remove the middle third of it, which is $(\frac{1}{3}, \frac{2}{3})$. Thus we are left with $C_1 = [0, \frac{1}{3}] \cup [\frac{2}{3}, 1]$. To form C_2 from C_1, we remove the middle third of each of the two intervals in C_1. The middle third of $[0, \frac{1}{3}]$ is $(\frac{1}{9}, \frac{2}{9})$, and the middle third of $[\frac{2}{3}, 1]$ is $(\frac{7}{9}, \frac{8}{9})$. Thus we are left with $C_2 = [0, \frac{1}{9}] \cup [\frac{2}{9}, \frac{1}{3}] \cup [\frac{2}{3}, \frac{7}{9}] \cup [\frac{8}{9}, 1]$.

We can draw a picture of what each iteration C_n looks like, as shown in Figure 24.2.

The Cantor set is the intersection of all the C_n's, i.e. $C = \bigcap_{n=0}^{\infty} C_n$. It is natural to wonder whether there is anything left: every iteration, we remove a bunch of points, so that we end up with shorter and shorter intervals, so it might be the case that C is just the empty set. However, it is not. For example, 0 and 1 are in the Cantor set, because they are clearly in each C_n. Furthermore, all the endpoints of the intervals created are in C, such as $\frac{1}{3}$ and $\frac{2}{9}$.

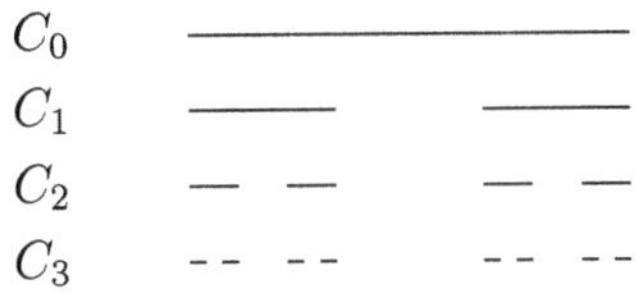

$$
\begin{array}{ll}
C_0 & \text{\rule{4cm}{0.4pt}} \\
C_1 & \text{\rule{1.5cm}{0.4pt}} \qquad \text{\rule{1.5cm}{0.4pt}} \\
C_2 & - \ - \qquad - \ - \\
C_3 & -- \ \ -- \qquad -- \ \ -- \\
\end{array}
$$

Figure 24.2. Iterations of the Cantor set construction.

But that's not all! In fact, we can describe the elements of the Cantor set very concretely. To do this, we have to use base-3 (also known as *ternary*) notation instead of base-10 notation for the real numbers. For example, $\frac{2}{9}$ in base-3 notation is 0.02.

When we remove the middle third of the full interval $[0, 1]$ to obtain C_1, we remove all the numbers whose first base-3 digit is a 1. (Well, except for $\frac{1}{3} = .1$, but we get away with this because there is another base-3 representation for $\frac{1}{3}$, namely $0.0222222\ldots$ This is the same phenomenon as $0.999999\ldots = 1$ in usual base-10 notation.) So, we can say that C_1 contains all points in $[0, 1]$ that have a base-3 representation without a 1 in the first digit. Similarly, when passing from C_1 to C_2, we remove all numbers whose base-3 expansions have a 1 in the second digit. Similarly, when passing from C_{n-1} to C_n, we remove all numbers whose base-3 expansions have a 1 in the nth digit. Thus C_n consists of those numbers whose base-3 expansions do not have any 1's in the first n digits. It follows that C consists of those numbers whose base-3 expansions do not have any 1's at all. For example, $0.0202020202\ldots \in C$. One can check that this number is $\frac{1}{4}$. There are also irrational numbers in C, which are sequences of 0's and 2's that do not eventually repeat.

Theorem 24.6. *The Cantor set has cardinality* $\mathfrak{c}$.

Proof. We'll use the Cantor–Schröder–Bernstein Theorem, knowing that $[0, 1]$ has cardinality $\mathfrak{c}$. First, we have an injective function $f : C \to [0, 1]$ because every point in C is already in $[0, 1]$; that is $f(x) = x$. Now, we must construct an injective function $g : [0, 1] \to C$. For this, we shall fix a binary representation of each $y \in [0, 1]$. (Some numbers have more than one; choose either one, say the one with finitely many nonzero digits.) If $y = \sum_{n=1}^{\infty} \frac{b_n}{2^n}$ where $b_n \in \{0, 1\}$, then we define

$$
g(y) = g\left(\sum_{n=1}^{\infty} \frac{b_n}{2^n}\right) = \sum_{n=1}^{\infty} \frac{2b_n}{3^n}.
$$

In other words, replace each 1 with a 2 and then read the number off in base 3. This is an injective (but not bijective!) function $g : [0,1] \to C$. Since we have injective functions both ways, C and $[0,1]$ have the same cardinality by the Cantor–Schröder–Bernstein Theorem.   ∎

Example. If $y = 0.10010101$ in binary, then $g(y) = 0.20020202$, considered as a number in base 3.

24.4   Problems

(1) (a) Prove that if S and T are two nonempty sets, then there is an injective function $f : S \to T$ if and only if there is a surjective function $g : T \to S$.

   (b) Prove the *dual* of the Cantor–Schröder–Bernstein Theorem: if S and T are two sets such that there are surjections $f : S \to T$ and $g : T \to S$, then there is a bijection $h : S \to T$.

(2) A complex number α is said to be *algebraic* if it is a root of a nonzero polynomial with integer coefficients. For example, $\sqrt[4]{3}$ is algebraic, because it is a root of the polynomial $x^4 - 3$. What is the cardinality of the set of algebraic numbers? Prove that your answer is correct.

(3) Use the Cantor–Schröder–Bernstein Theorem to prove that the sets $[0,1]$ and $(0,1)$, the closed and open intervals, have cardinality $\mathfrak{c}$.

(4) Describe an explicit bijection between $[0,1]$ and $(0,1)$. (You can unpack the proof of Cantor–Schröder–Bernstein if you'd like, after giving injections both ways. But there are easier ways to do this.)

(5) Suppose that $|S| = |T|$. Prove that $|\mathscr{P}(S)| = |\mathscr{P}(T)|$. More generally, prove that if $|S| \le |T|$, then $|\mathscr{P}(S)| \le |\mathscr{P}(T)|$.

(6) Prove that $|\mathscr{P}(\mathbb{N})| = \mathfrak{c}$.

(7) Prove that the closed unit square $[0,1] \times [0,1]$ has cardinality $\mathfrak{c}$.

(8) What is the cardinality of the set of functions from the integers to the integers?

(9) Prove that the Cantor set C does not contain any open intervals (a,b) where $a < b$.

(10) (a) A *figure-8* consists of a pair of circles in the plane which are tangent, and neither one is inside the other. The circles may be of any size. The circles consist of just the boundary curve, not the interior. Is it possible to place uncountably many figure-8's in the plane such that no two of them touch? You can place them inside each other.

(b) It is clearly possible to place uncountably many circles in the plane such that no two touch: just make them all concentric. Is it still possible to place uncountably many circles in the plane such that no two touch or are concentric?

Chapter 25

Equivalence classes of pairs

25.1 From $\mathbb{N}$ to $\mathbb{Z}$

The first numbers we encounter as kids are the positive integers $\mathbb{N}$. We consider those to be fundamental objects: we are given them for free.[1] All other numbers have to be defined in terms of them. Our goal, over the course of this chapter and the next chapter, is to give a construction of the real numbers starting only from the positive integers.[2]

The first step along the way is to construct all the integers from just the positive integers. How do we get from the positive integers to all the integers? By means of subtraction: every integer—whether positive, negative, or zero—can be written as a difference of two positive integers. So, that will be our *definition* of an integer: an integer is the difference between two positive integers.

However, there is a slight problem with this definition: we can write an integer as a difference of two positive integers in many ways. We want $3 - 5$ and $7 - 9$ to be the same integer, even though it is written in different ways. To this end, we recall from Chapter 5 the notion of an equivalence relation.

Definition 25.1. Let X be a set, and let $\sim$ be a relation on X, i.e. a subset of the set $X \times X = \{(x, y) : x, y \in X\}$. We say that $\sim$ is an *equivalence relation* if the following properties hold:

Reflexivity: $x \sim x$ for all $x \in X$.

---

[1]There are constructions of the positive integers too though, most notably using the Peano axioms; see for instance [Fef05, Chapter 3].

[2]We will eventually discuss two constructions, with the second one coming in Chapter 28.

Symmetry: If $x \sim y$, then $y \sim x$.
Transitivity: If $x \sim y$ and $y \sim z$, then $x \sim z$.

When we have an equivalence relation $\sim$ on a set X, the set $[x] = \{y \in X : y \sim x\}$ is called an *equivalence class*.

In the terminology of equivalence relations and equivalence classes, what we really want to say about integers is that they are equivalence classes of pairs of positive integers, under a certain equivalence relation. When talking about only positive integers, subtraction isn't always defined, so let's instead phrase everything in terms of addition alone.

Consider the set $\mathbb{N} \times \mathbb{N}$ of pairs of positive integers, where a typical element has the form (m, n). Put a relation $\sim$ on $\mathbb{N} \times \mathbb{N}$ by saying that $(a, b) \sim (c, d)$ if $a + d = b + c$.

Proposition 25.2. $\sim$ *is an equivalence relation.*

Proof. We must check the three properties that define an equivalence relation.

Reflexivity: We must show that $(a, b) \sim (a, b)$. We have $a + b = b + a$, so this is true.
Symmetry: We must show that if $(a, b) \sim (c, d)$, then $(c, d) \sim (a, b)$. Note that $(a, b) \sim (c, d)$ means that $a + d = b + c$, whereas $(c, d) \sim (a, b)$ means that $c + b = d + a$. But these are the same.
Transitivity: Suppose that $(a, b) \sim (c, d)$ and $(c, d) \sim (e, f)$. This means that $a + d = b + c$ and $c + f = d + e$. Adding these two equations together, we get $a + d + c + f = b + c + d + e$; and, after canceling the $c + d$ on both sides, we get $a + f = b + e$, which means that $(a, b) \sim (e, f)$. This completes the proof of transitivity and thus that $\sim$ is an equivalence relation. ∎

Let us now define the integers.

Definition 25.3. An *integer* is an equivalence class of $\mathbb{N} \times \mathbb{N}$ under the equivalence relation $\sim$, as described above. We write $\mathbb{Z}$ for the set of all the integers.

For instance, the integer we usually call -2 is the equivalence class of $(3, 5)$, which we'll write as $[3, 5]$ when we need to refer to it in terms of its equivalence class. In general, we'll denote the equivalence class of (m, n) by $[m, n]$.[3]

So far, we have only defined the integers as a *set*. But usually we want more out of the integers than just their structure as a set. For instance, we'd like to add, subtract, and multiply them. We have to explain how to do those things in terms of our new formalism. In case you're wondering why we have to do all this when we already know how to do arithmetic on integers, keep in mind that this is practice for when we define objects that we have a less clear understanding of. Some examples of this appear in the problem set.

Let's define addition of integers.

Definition 25.4. Let $[a, b]$ and $[c, d]$ be two integers. Their sum is defined to be

$$[a, b] + [c, d] = [a + c, b + d].$$

This agrees with our usual notion, since $[a, b]$ is supposed to mean $a - b$, and $[c, d]$ is supposed to mean $c - d$. So $(a - b) + (c - d) = (a + c) - (b + d)$, which is just what the definition says.

Actually, there is a problem with Definition 25.4, even though it looks okay at a glance. The problem is that it is not immediately clear that it is *well-defined*, i.e. that we always get the same answer when adding the same two integers. How can this be? Well, suppose that $[a_1, b_1] = [a_2, b_2]$ and $[c_1, d_1] = [c_2, d_2]$. Is it necessarily the case that $[a_1, b_1] + [c_1, d_1] = [a_2, b_2] + [c_2, d_2]$? It had better be, since we're adding the same two integers, just writing them a bit differently. However, this isn't guaranteed anywhere in our definition. So that's something that we have to check by hand. Let's do it!

Proposition 25.5. *Suppose that* $[a_1, b_1] = [a_2, b_2]$ *and* $[c_1, d_1] = [c_2, d_2]$. *Then* $[a_1 + c_1, b_1 + d_1] = [a_2 + c_2, b_2 + d_2]$.

Proof. We have to check that

$$(a_1 + c_1) + (b_2 + d_2) = (b_1 + d_1) + (a_2 + c_2),$$

---

[3]Note that, if we stick to our usual notation of $[x]$ for the equivalence class of x, then we would have to write $[(m, n)]$ instead. However, this is needlessly cumbersome notation, so we'll stick to the more nimble $[m, n]$.

given that $a_1 + b_2 = b_1 + a_2$ and $c_1 + d_2 = d_1 + c_2$. We have

$$
\begin{aligned}
(a_1 + c_1) + (b_2 + d_2) &= (a_1 + b_2) + (c_1 + d_2) \\
&= (b_1 + a_2) + (d_1 + c_2) \\
&= (b_1 + d_1) + (a_2 + c_2),
\end{aligned}
$$

as desired.   ■

Similarly, let us define multiplication.

Definition 25.6. Let $[a, b]$ and $[c, d]$ be two integers. Their product is defined to be

$$
[a, b][c, d] = [ac + bd, ad + bc].
$$

Why is this the definition? Because

$$
(a - b)(c - d) = ac - bc - ad + bd = (ac + bd) - (ad + bc).
$$

Once again, we have the same issue, which is that we need to check that it is well-defined, i.e. that if $[a_1, b_1] = [a_2, b_2]$ and $[c_1, d_1] = [c_2, d_2]$, then $[a_1, b_1][c_1, d_1] = [a_2, b_2][c_2, d_2]$. The argument here is similar to that in the proof of Proposition 25.5 but involves a lengthier calculation.

You might think we also need to define subtraction. However, subtraction can already be defined in terms of addition and multiplication. For integers written the usual way, we have $x - y = x + (-1)y$. In terms of equivalence classes of pairs, we have $[a, b] - [c, d] = [a, b] + [1, 2][c, d]$, since $[1, 2]$ is the equivalence class version of -1.

If one were being even more thorough at this point, one would check that addition and multiplication of pairs satisfy the sort of properties that we expect. For example, $[1, 1] + [a, b] = [a, b]$, $[1, 1][a, b] = [1, 1]$, $[a, b] + [c, d] = [c, d] + [a, b]$, $[a, b]([c, d] + [e, f]) = [a, b][c, d] + [a, b][e, f]$. In problem 1, you will decode these properties and see which standard properties they are referring to!

From now on, we'll drop this annoying notation and write integers the way we usually do, except when we need to refer directly to the elements of the pair. For example, we'll write 0 instead of $[1, 1]$ and -1 instead of $[1, 2]$.

25.2   From $\mathbb{Z}$ to $\mathbb{Q}$

At this point, we have the integers defined rigorously. So it's time to take the next step up the ladder and define the rational numbers. The rational numbers are just the fractions: $\frac{a}{b}$, where a and b are integers, and $b \neq 0$. But, just like in the case of the integers, these things aren't uniquely defined, and so we need to use equivalence relations. The equivalence relation for rationals is similar to, but not quite the same as, the one for integers.

Consider the set $\mathbb{Z} \times (\mathbb{Z} \setminus \{0\})$ of pairs of integers, where the second one is nonzero. Put a relation $\equiv$ on it by saying that $(a, b) \equiv (c, d)$ if $ad = bc$. (Recall that the equivalence relation used to construct $\mathbb{Z}$ from $\mathbb{N}$ was $a + d = b + c$, so this is just a multiplicative version of what we did to construct the integers.) We think of (a, b) as being the fraction $\frac{a}{b}$.

Proposition 25.7. $\equiv$ *is an equivalence relation.*

You will prove this in problem 2.

Definition 25.8. A *rational number* is an equivalence class of $\mathbb{Z} \times (\mathbb{Z} \setminus \{0\})$ under the equivalence relation $\equiv$. We write $\mathbb{Q}$ for the set of all the rational numbers.

In order to distinguish these equivalence classes from the other ones, we'll write $\langle a, b \rangle$ for the equivalence class of (a, b). Remember that $\langle a, b \rangle$ is supposed to be $\frac{a}{b}$.

Just like with the integers, we want to do more with the rationals than just consider them as a set: we want to do arithmetic. That is, we want to be able to add, subtract, multiply, and divide them. You can probably already predict what is coming, but here are the definitions of these operations:

Definition 25.9. Let $\langle a, b \rangle$ and $\langle c, d \rangle$ be two rational numbers.

- Their *sum* is

$$\langle a, b \rangle + \langle c, d \rangle = \langle ad + bc, bd \rangle.$$

- Their *difference* is

$$\langle a, b \rangle - \langle c, d \rangle = \langle ad - bc, bd \rangle.$$

- Their *product* is

$$\langle a, b \rangle \langle c, d \rangle = \langle ac, bd \rangle.$$

- Assuming that $c \neq 0$, their *quotient* is

$$\langle a, b \rangle \div \langle c, d \rangle = \langle ad, bc \rangle.$$

Just like in the case of integers, we shouldn't be satisfied with these definitions until we have checked that they are well-defined, i.e. they do not depend on the choice of representatives of their respective equivalence classes. The verification of this is similar to the case of the integers, but we'll just move on rather than spending more time on this particular tedious issue.

25.3   The ordering on $\mathbb{Z}$ and $\mathbb{Q}$

So far, we have discussed the arithmetic of $\mathbb{Z}$ and $\mathbb{Q}$: how to add (and do other operations on) integers and rational numbers, directly in terms of their ordered pair representations. But there is another crucial aspect of these number systems: the ordering. That is, given two integers or rational numbers, there is a way of telling when one of them is greater than or less than the other.

Before we explain how to do that directly in terms of the pairs, let's make sure we understand precisely what we expect out of an ordering relation. Just as you would expect, we use $<$ and $\leq$ as our ordering relations. They are very similar to each other: $<$ means $\leq$ but not $=$, i.e. $a < b$ is equivalent to $a \leq b$ and $a \neq b$.

Definition 25.10. A *totally ordered set* is a set S together with a relation $\leq$ (written $x \leq y$) satisfying the following properties:

Reflexivity: $x \leq x$ for all $x \in S$.
Antisymmetry: if $x \leq y$ and $y \leq x$, then $x = y$.
Transitivity: if $x \leq y$ and $y \leq z$, then $x \leq z$.
Comparability: if $x, y \in S$, then either $x \leq y$ or $y \leq x$.[4]

We call $\leq$ an *ordering* or a *total ordering*.

To define the order on $\mathbb{Z}$ and $\mathbb{Q}$, we must build up from the order on $\mathbb{N}$, so let's start by defining that.

Definition 25.11. Let $x, y \in \mathbb{N}$. We say that $x < y$ if there exists some $z \in \mathbb{N}$ such that $x + z = y$. We say that $x \leq y$ if either $x < y$ or $x = y$.

It's obvious[5] that this is an ordering on $\mathbb{N}$. We'll now use it to build up an ordering on $\mathbb{Z}$.

---

[4]Recall the convention on "or" in mathematics: we allow the possibility that both $x \leq y$ and $y \leq x$, in which case we have $x = y$ by antisymmetry.
[5]Well, maybe: If we were going to do this properly, we would start with an actual definition of $\mathbb{N}$, so that we could check these properties on that construction. But since we skipped that part, we'll have to skip checking that this is an ordering as well.

The ordering on $\mathbb{Z}$ is similar to the ordering on $\mathbb{N}$. For integers x, y, we say that $x \leq y$ if there is some $z \in \mathbb{N} \cup \{0\}$ such that $x + z = y$.

Proposition 25.12. *The relation $\leq$ on $\mathbb{Z}$ is an ordering.*

Proof. Let's check that it satisfies the four properties of an ordering.

Reflexivity: $x + 0 = x$, so $x \leq x$.
Antisymmetry: Suppose that $x + a = y$ and $y + b = x$, where $a, b \in \mathbb{N} \cup \{0\}$.
 Then we have $x + a + b = x$, so $a + b = 0$. Thus $x = x$.
Transitivity: Suppose $x + a = y$ and $y + b = z$. Then $x + a + b = z$, so $x \leq z$.
Comparability: Let $x, y \in \mathbb{Z}$. Then $x - y \in \mathbb{Z}$, so we can write $x - y = [a, b]$
 for some $a, b \in \mathbb{N}$. Thus $x - y = a - b$. If $a \geq b$, then $a - b \in \mathbb{N} \cup \{0\}$,
 and $y + (a - b) = x$, so $y \leq x$. If $b \geq a$, then $b - a \in \mathbb{N} \cup \{0\}$,
 and $x + (b - a) = y$, so $x \leq y$. Either way, we have either $x \leq y$ or
 $y \leq x$. ∎

We won't prove it right now (instead, we'll save it for problem 3), but the ordering is compatible with arithmetic operations. For example, if $a \leq b$ and $c \leq d$, then $a + c \leq b + d$. Similarly, the ordering acts as you expect with respect to the other arithmetical operations. For instance, if $0 \leq a \leq b$ and $0 \leq c \leq d$, then $0 \leq ac \leq bd$. There is a lot to check, of course, but it is all straightforward, following along the lines of what we have already done.

Now we need to move on to define the ordering on $\mathbb{Q}$. We have to do that in terms of the ordering on $\mathbb{Z}$ only. Here's one way of defining it:

Definition 25.13. Let $x, y \in \mathbb{Q}$. We say that $x \leq y$ if there are integers $a \in \mathbb{N}$ and $b \in \mathbb{N} \cup \{0\}$ such that $ax + b = ay$.

You will easily recognize that, in normal parlance, this is the same as saying that $y - x = \frac{b}{a} \geq 0$, so this is the same as the ordering relation you are used to. But let's check that it actually is an ordering.

Proposition 25.14. *The relation $\leq$ is an ordering on $\mathbb{Q}$.*

Proof. Again, we just have to check the four properties.

Reflexivity: If $y = x$, then we may take $a = 1$ and $b = 0$.
Antisymmetry: Suppose $ax + b = ay$ and $cy + d = cx$. Multiplying the
 first equation by c and the second equation by a, we have $acx + bc =$
 acy and $acy + ad = acx$. Substituting $acx + bc$ for acy in the second

equation, we get $acx + bc + ad = acx$. Thus $bc + ad = 0$. Since $a, c > 0$ and $b, d \geq 0$, it follows that $b, d = 0$, so $ax = ay$ and $cy = cx$. Thus $x = y$.

Transitivity: Suppose $x \leq y$ and $y \leq z$. Then we have $a, c \in \mathbb{N}$ and $b, d \in \mathbb{N} \cup \{0\}$ such that $ax + b = ay$ and $cy + d = cz$. Multiplying the first equation by c and the second by a, we get $acx + bc = acy$ and $acy + ad = acz$. Thus we have $acx + bc + ad = acz$. Letting $a' = ac$ and $b' = bc + ad$, we have $a'x + b' = a'z$, where $a' \in \mathbb{N}$ and $b' \in \mathbb{N} \cup \{0\}$. Thus $x \leq z$ by definition.

Comparability: Suppose $x = \langle p, q \rangle$ and $y = \langle r, s \rangle$. We can rewrite them as $x = \langle ps, qs \rangle$ and $y = \langle qr, qs \rangle$, where $qs \in \mathbb{N}$. If $ps \leq qr$, then we can take $a = qs$ and $b = qr - ps$, so that $ax + b = ay$. If $qr \leq ps$, then we can take $a = qs$ and $b = ps - qr$, so that $ay + b = ax$. Thus either $x \leq y$ or $y \leq x$. ∎

25.4 Problems

(1) Explain what the following statements about integers mean, in common parlance:

    (a) $[1, 1] + [a, b] = [a, b]$.
    (b) $[1, 1][a, b] = [1, 1]$.
    (c) $[a, b] + [c, d] = [c, d] + [a, b]$.
    (d) $[a, b]([c, d] + [e, f]) = [a, b][c, d] + [a, b][e, f]$.

(2) Prove Proposition 25.7.

(3) Prove, directly from the definitions, that if $a, b, c, d \in \mathbb{Z}$, and $a \leq b$ and $c \leq d$, then $a + c \leq b + d$.

(4) Prove, directly from the definitions, that if a and b are integers with $a, b < 0$, then $ab > 0$.

(5) Prove that for every $x \in \mathbb{Q}$, there is exactly one $y \in \mathbb{Q}$ such that $x + y = 0$. (Here, 0 means $\langle 0, 1 \rangle$.)

(6) Prove that for every $x \in \mathbb{Q}$ other than $\langle 0, 1 \rangle$, there is exactly one $y \in \mathbb{Q}$ such that $xy = 1$, where $1 = \langle 1, 1 \rangle$.

(7) The set $\mathbb{Z}[\frac{1}{2}]$, known as the *dyadic rationals*, consists of all rational numbers whose denominator is a power of 2. So, for example, $-\frac{7}{4} \in \mathbb{Z}[\frac{1}{2}]$, but $\frac{1}{3} \notin \mathbb{Z}[\frac{1}{2}]$.

    (a) Explain how to construct $\mathbb{Z}[\frac{1}{2}]$ from $\mathbb{Z}$ using a similar construction to the one used to construct $\mathbb{Q}$ from $\mathbb{Z}$. Prove that anything you claim to be an equivalence relation actually is one.

(b) Can we add, subtract, multiply, and divide elements of $\mathbb{Z}[\frac{1}{2}]$? (Don't try to divide by 0.) Which ones work, and which ones don't? Give definitions for these operations, and check whether or not they work.

(8) Suppose we mimic the construction used to construct $\mathbb{Q}$ from $\mathbb{Z}$, but we apply it to $\mathbb{Q}$ instead of to $\mathbb{Z}$. That is, we consider equivalence classes of pairs $(a, b) \in \mathbb{Q} \times (\mathbb{Q} \setminus \{0\})$, under the equivalence relation $(a, b) \sim (c, d)$ if $ad = bc$. Which familiar structure do we end up with in this way?

(9) Let $i = \sqrt{-1}$, and define $\mathbb{Z}[i] = \{a + bi : a, b \in \mathbb{Z}\}$. (The set $\mathbb{Z}[i]$ is called the set of *Gaussian integers*.) Describe, as best you can, the set of equivalence classes of $\mathbb{Z}[i] \times (\mathbb{Z}[i] \setminus \{0\})$ under the equivalence relation $(a, b) \sim (c, d)$ if $ad = bc$.

Chapter 26

Dedekind cuts

26.1 Dedekind cuts

In Chapter 25, we saw a technique for constructing $\mathbb{Z}$ from $\mathbb{N}$, and then $\mathbb{Q}$ from $\mathbb{Z}$. Now we want to take the next step and construct $\mathbb{R}$ from $\mathbb{Q}$. There are two standard ways of doing this, and we'll discuss both of them in this book. The way we'll do it in this chapter uses Dedekind cuts of rationals.

The idea of Dedekind cuts is as follows. Take the real number line, and pick some real number α on it. Now take a pair of scissors and cut the number line at α. We end up with two rays, one consisting of all the numbers less than α, and the other consisting of all the numbers greater than α. (What happens to α itself? We can make a convention for which set it ends up in, or we can suppose that it disappears after the cut is made. The actual convention chosen doesn't matter, as long as we're consistent about it. We'll choose to put it in the upper set.) The two sets of numbers we have left uniquely determine α: what we're saying is that the number α itself carries exactly the same information as knowing which numbers are less than α, and which ones are greater than α.

However, we want to build the real numbers out of only the rationals, rather than using all the reals. That's okay: we don't need information about *all* the numbers less than α (or greater than α); it's enough to know which *rational* numbers are less than α and which ones are greater than α. That's the basic idea of Dedekind cuts.

Now let's do it formally.

Definition 26.1. A *Dedekind cut* of the rational numbers is an ordered pair (A, B), where A and B are subsets of $\mathbb{Q}$, with the following properties:

- Both A and B are nonempty.
- $A \cup B = \mathbb{Q}$.
- If $a \in A$ and $b \in B$, then $a < b$.
- A contains no greatest element: if $a \in A$, then there is some $c \in A$ with $a < c$.

What all this is saying is that both A and B are rays in $\mathbb{Q}$, with A pointing to the left and B pointing to the right. While A is not allowed to have a maximal element, B is allowed to have a minimal element. Sometimes it does, and sometimes it doesn't.

The idea is that given a real number α, we can construct a Dedekind cut (A_α, B_α) out of it, by letting $A_\alpha = \{r \in \mathbb{Q} : r < \alpha\}$ and $B_\alpha = \{r \in \mathbb{Q} : r \geq \alpha\}$. For example, from $\sqrt{2}$, we construct its Dedekind cut to be $(A_{\sqrt{2}}, B_{\sqrt{2}})$ where $A_{\sqrt{2}} = \mathbb{Q}_{\leq 0} \cup \{r \in \mathbb{Q} : r^2 < 2\}$ and $B_{\sqrt{2}} = \{r \in \mathbb{Q}_{>0} : r^2 > 2\}$. (See Figure 26.1.) However, this sort of thing doesn't make sense in generality yet, because we don't rigorously know what the real numbers are yet. So instead, we'll go backward: we'll *define* the real numbers to be the Dedekind cuts.

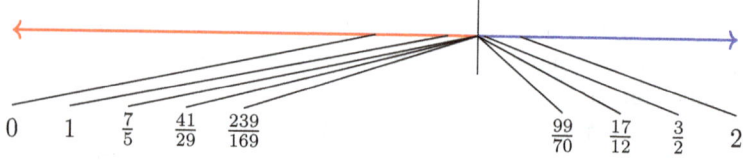

$$0 \quad 1 \quad \tfrac{7}{5} \quad \tfrac{41}{29} \quad \tfrac{239}{169} \qquad\qquad \tfrac{99}{70} \quad \tfrac{17}{12} \quad \tfrac{3}{2} \quad 2$$

Figure 26.1. The Dedekind cut corresponding to $\sqrt{2}$. The set $A_{\sqrt{2}}$ is drawn in red, and the set $B_{\sqrt{2}}$ is drawn in blue.

Definition 26.2. A *real number* is a Dedekind cut of the rationals. We write $\mathbb{R}$ for the set of all real numbers.

This means that each real number is an ordered pair, where each element in the pair is a subset of the rational numbers. This is a bit clumsy, and eventually we'll drop all of that, but it's all we have to work with at the moment, so we'll use it to define all the structure on $\mathbb{R}$.

26.2   The structure of $\mathbb{R}$

The first bit of structure we'll want to define is that of inequalities. Given two real numbers α and β, what does it mean to say that $\alpha < \beta$?

Remember that we have to do this in terms of Dedekind cuts, since that's all we have available to us. We'll actually define $\leq$ instead of $<$, since it's slightly cleaner, and then we'll say that $\alpha < \beta$ if $\alpha \leq \beta$ and $\beta \not\leq \alpha$.

Definition 26.3. Let $\alpha = (A_\alpha, B_\alpha)$ and $\beta = (A_\beta, B_\beta)$ be two real numbers. We say that $\alpha \leq \beta$ if $A_\alpha \subseteq A_\beta$. We say that $\alpha < \beta$ if $\alpha \leq \beta$ and $\beta \not\leq \alpha$.

This is logical: $\alpha \leq \beta$ if every rational number $x < \alpha$ is also $< \beta$. But so far it's just a definition; we don't know that this definition has any of the usual good properties of inequalities, i.e. that $\leq$ makes $\mathbb{R}$ into a totally ordered set. So we have to prove them. Let's just do two of the properties required of a totally ordered set: transitivity and comparability.

Proposition 26.4.

(1) If $\alpha \leq \beta$ and $\beta \leq \gamma$, then $\alpha \leq \gamma$.
(2) If $\alpha, \beta \in \mathbb{R}$, then either $\alpha \leq \beta$ or $\beta \leq \alpha$.

Proof.

(1) Suppose $\alpha \leq \beta$ and $\beta \leq \gamma$. Then $A_\alpha \subseteq A_\beta$ and $A_\beta \subseteq A_\gamma$. Thus $A_\alpha \subseteq A_\gamma$, so $\alpha \leq \gamma$.
(2) If not, then $A_\alpha \not\subseteq A_\beta$, and $A_\beta \not\subseteq A_\alpha$. This means that there is some $x \in A_\alpha \setminus A_\beta$, and some $y \in A_\beta \setminus A_\alpha$. Thus $x \neq y$ and, by the total ordering of $\mathbb{Q}$, we either have $x \leq y$ or $y \leq x$. Let us suppose that $x \leq y$. Since $x \notin A_\beta$ and $A_\beta \cup B_\beta = \mathbb{Q}$, we must have $x \in B_\beta$. Since $y \in A_\beta$, we have $y < x$ by Definition 26.3, contradicting our assumption that $x \leq y$. We get a similar contradiction if we instead assume that $y \leq x$. Thus it must be the case that either $A_\alpha \subseteq A_\beta$, i.e. $\alpha \leq \beta$, or $A_\beta \subseteq A_\alpha$, i.e. $\beta \leq \alpha$. ∎

Next, we want to define arithmetic on $\mathbb{R}$: How do we add, subtract, multiply, and divide real numbers? Addition is easy, so we'll start with that.

Definition 26.5. Let $\alpha = (A_\alpha, B_\alpha)$ and $\beta = (A_\beta, B_\beta)$ be two real numbers. We define $\alpha + \beta$ to be the ordered pair $(A_{\alpha+\beta}, \mathbb{Q} \setminus A_{\alpha+\beta})$, where $A_{\alpha+\beta} = \{r + s : r \in A_\alpha, s \in A_\beta\}$.

Note the careful wording in this definition: $\alpha + \beta$ is defined to be an ordered pair, not a real number (or a Dedekind cut). And that is because we have not yet proven that it is always a Dedekind cut. So that's something we have to prove.

286 Transition to Proofs

Proposition 26.6. *If $\alpha, \beta \in \mathbb{R}$, then $\alpha + \beta \in \mathbb{R}$.*

Proof. We must check all the properties in the definition.

- The first property to check is that $A_{\alpha+\beta}$ and $\mathbb{Q} \setminus A_{\alpha+\beta}$ are both nonempty. Since A_α and A_β are both nonempty, we can find some $r \in A_\alpha$ and $s \in A_\beta$. Note that $r, s \in \mathbb{Q}$, so we already know how to evaluate $r + s$. By definition, $r + s \in A_{\alpha+\beta}$, so $A_{\alpha+\beta}$ is nonempty. To show that $\mathbb{Q} \setminus A_{\alpha+\beta}$ is nonempty, let $b_\alpha \in B_\alpha$ and $b_\beta \in B_\beta$ be arbitrary elements. For any $a_\alpha \in A_\alpha$ and $a_\beta \in A_\beta$, we have $a_\alpha < b_\alpha$ and $a_\beta < b_\beta$, so $a_\alpha + a_\beta < b_\alpha + b_\beta$. Thus $b_\alpha + b_\beta \notin A_{\alpha+\beta}$, so $\mathbb{Q} \setminus A_{\alpha+\beta}$ is nonempty.
- Next, we need to check that $A_{\alpha+\beta} \cup (\mathbb{Q} \setminus A_{\alpha+\beta}) = \mathbb{Q}$. This is clear, because $\mathbb{Q} \setminus A_{\alpha+\beta}$ consists exactly of those elements of $\mathbb{Q}$ not in $A_{\alpha+\beta}$.
- The next thing to check is that if $a \in A_{\alpha+\beta}$ and $b \in \mathbb{Q} \setminus A_{\alpha+\beta}$, then $a < b$. Suppose $a \in A_{\alpha+\beta}$ and $b \in \mathbb{Q} \setminus A_{\alpha+\beta}$. Let us suppose that $a = a_\alpha + a_\beta$, where $a_\alpha \in A_\alpha$ and $a_\beta \in A_\beta$. Since $b \notin A_{\alpha+\beta}$, we have $b = a_\alpha + b_\beta$ for some $b_\beta \in B_\beta$. (In this case, $b_\beta = b - a_\alpha$, which must be in B_β, for otherwise b would be in $A_{\alpha+\beta}$.) Since (A_β, B_β) is a Dedekind cut, $a_\beta < b_\beta$, so $a = a_\alpha + a_\beta < a_\alpha + b_\beta = b$, as desired.
- Finally, we must prove that $A_{\alpha+\beta}$ has no greatest element. Let $a = a_\alpha + a_\beta \in A_{\alpha+\beta}$, where $a_\alpha \in A_\alpha$ and $a_\beta \in A_\beta$. Since α and β are real numbers, A_α and A_β have no greatest elements, so there exist $c_\alpha \in A_\alpha$ and $c_\beta \in A_\beta$ with $a_\alpha < c_\alpha$ and $a_\beta < c_\beta$. Thus we have

$$a = a_\alpha + a_\beta < c_\alpha + c_\beta \in A_{\alpha+\beta},$$

so a is not the greatest element in $A_{\alpha+\beta}$. Since a was arbitrary, we conclude that $A_{\alpha+\beta}$ has no greatest element. ∎

We might like to define multiplication similarly: $\alpha\beta = (A_{\alpha\beta}, \mathbb{Q} \setminus A_{\alpha\beta})$ where $A_{\alpha\beta} = \{rs : r \in A_\alpha, s \in A_\beta\}$. However, this is not a Dedekind cut. So, we need to be more careful. The problem is that multiplication and inequalities don't always play nicely together: if $a < a'$ and $b < b'$, we cannot conclude that $ab < a'b'$. However, if we also know that a and b are positive, then we can conclude that $ab < a'b'$. Thus we start by dealing with multiplication only of positive real numbers.

Definition 26.7. A real number α is said to be *positive* (resp. *nonnegative*) if $\alpha > 0$ (resp. $\alpha \geq 0$). A real number β is said to be *negative* (resp. *nonpositive*) if $\beta < 0$ (resp. $\beta \leq 0$).

Definition 26.8. Let $\alpha = (A_\alpha, B_\alpha)$ and $\beta = (A_\beta, B_\beta)$ be two positive real numbers. Then we define their product $\alpha\beta$ to be the pair $(A_{\alpha\beta}, \mathbb{Q} \setminus A_{\alpha\beta})$, where

$$A_{\alpha\beta} = \{r \in \mathbb{Q} : r \leq 0\} \cup \{a_\alpha a_\beta : a_\alpha \in A_\alpha, a_\beta \in A_\beta, a_\alpha > 0, a_\beta > 0\}.$$

As in the case of addition, we ought to check that $(A_{\alpha\beta}, \mathbb{Q} \setminus A_{\alpha\beta})$ is actually a Dedekind cut. This is similar to the case of addition, so we'll omit the somewhat tedious details.

The full definition of multiplication is more complicated, because we have several cases where each of α and β can be positive, negative, or zero. One way of dealing with this to start by defining multiplication by -1. This, at least, is easy: if $\alpha = (A, B)$, then $-\alpha = (C, D)$, where C consists of the negatives of all elements of B, except for the smallest element of B if there is one, and $D = \mathbb{Q} \setminus C$. Note that if $\alpha > 0$, then $-\alpha < 0$. (Of course, strictly speaking, this also needs to be checked, but we'll omit this verification.) With this in place, we can now define multiplication in all cases, other than the one we've already done:

Definition 26.9.

- If $\alpha = 0$, then $\alpha\beta = 0$.
- If $\beta = 0$, then $\alpha\beta = 0$.
- If $\alpha < 0$ and $\beta > 0$, then $\alpha\beta = -((-\alpha)\beta)$.
- If $\alpha > 0$ and $\beta < 0$, then $\alpha\beta = -(\alpha(-\beta))$.
- If $\alpha < 0$ and $\beta < 0$, then $\alpha\beta = (-\alpha)(-\beta)$.

Subtraction is now easy, since $\alpha - \beta = \alpha + (-\beta)$. It will still take some work to define division using Dedekind cuts, but we will skip the definition of division here.

26.3 The least upper bound property

Dedekind cuts are built around ordering of rational (and hence real) numbers. Thus they tend to be helpful for proving properties of real numbers that have to do with ordering. One of the most important properties of real numbers is the least upper bound property.

Definition 26.10. A subset $S \subseteq \mathbb{R}$ is said to be *bounded above* if there is some real number N such that for all $\alpha \in S$, $\alpha \leq N$.

In terms of the number line, that means that S doesn't contain numbers arbitrarily far to the right.

Definition 26.11. Given a subset $S \subseteq \mathbb{R}$, an *upper bound* is some real number N such that $\alpha \leq N$ for all $\alpha \in S$.

Clearly, being bounded above is equivalent to the existence of some upper bound. But might there be a *best* possible upper bound? That is, if N is an upper bound for S and $N' > N$, then N' is also an upper bound, but it's worse. The smaller the upper bound, the better it is. Does a subset $S \subseteq \mathbb{R}$ which is bounded above have a lowest upper bound?

Theorem 26.12 (Least Upper Bound Property). *Let $S \subseteq \mathbb{R}$ be bounded above and nonempty. Then S has a least upper bound.*

Definition 26.13. If $S \subseteq \mathbb{R}$ is bounded above and nonempty, then its least upper bound is called the *supremum* of S and is denoted $\sup(S)$. If S is empty, then its supremum is defined to be $-\infty$, whereas if S is not bounded above, then its supremum is defined to be ∞.

The least upper bound property is something special about $\mathbb{R}$. There is no corresponding property in $\mathbb{Q}$, for instance: the set $S = \{s \in \mathbb{Q} : s^2 < 2\}$ has no least upper bound in $\mathbb{Q}$. That is because $\mathbb{Q}$ has a "gap" at $\sqrt{2}$. The least upper bound property is essentially saying that $\mathbb{R}$ has no gaps.

Proof of Theorem 26.12. Let $S \subseteq \mathbb{R}$ be bounded above and nonempty. For each $s \in S$, let us write (A_s, B_s) for the Dedekind cut corresponding to s. Let us define a pair (C, D), where

$$C = \bigcup_{s \in S} A_s,$$

and $D = \mathbb{Q} \setminus C$. Note that C consists of all rational numbers less than some element of S, so D consists of all rational numbers that are not less than some element of S, i.e. those rational numbers that are greater than or equal to all elements of S. We must show the following things:

- $\alpha = (C, D)$ is a Dedekind cut.
- α is an upper bound for S.
- α is the *least* upper bound for S.

We start by showing that α is a Dedekind cut. The first thing to check is that C is nonempty. Since S is nonempty and A_s is nonempty for each $s \in S$, C is also nonempty. Next, we must check that D is nonempty. Since we are assuming that S is bounded above, any $r \in \mathbb{Q}$ greater than every $s \in S$ is in D. Thus D is also nonempty. Next, we must show that if $c \in C$ and $d \in D$, then $c < d$. To do this, we note that D consists of all rational numbers that are greater than or equal to every element of S, whereas C consists of all rational numbers that are less than some element of S. Thus if $c \in C$, there is some $s \in S$ such that $c < s$, while if $d \in D$, then $d \geq s$. Thus $c < d$. Finally, we must show that C has no greatest element. If it did have a greatest element, say c, then c would lie in some A_s, which would mean that A_s has a greatest element. But this would contradict (A_s, B_s) being a Dedekind cut. Thus C has no greatest element. We have now checked all the properties needed to guarantee that $\alpha = (C, D)$ is a Dedekind cut and hence a real number.

Next, we must show that α is an upper bound for S. If $d \in D$ and $s \in S$, then $s \leq d$. Thus every element of D is greater than or equal to every element of S, so $\alpha \geq s$ for every $s \in S$ as well. Thus α is an upper bound for S.

Finally, we must show that α is the *least* upper bound for S. Suppose that $\beta \in \mathbb{R}$ is an upper bound for S, say with $\beta = (E, F)$ as a Dedekind cut. Since we are assuming β to be an upper bound for S, this means that for every $s \in S$, we have $(A_s, B_s) \leq (E, F)$. Thus $A_s \subseteq E$ for all $s \in S$, or $\bigcup_{s \in S} A_s \subseteq E$. But $C = \bigcup_{s \in S} A_s$. Thus $C \subseteq E$, which means that $\alpha \leq \beta$. Thus α is less than or equal to any upper bound for S, and is therefore the least upper bound. ∎

26.4 Problems

(1) Suppose $\alpha = (A, B)$ is given as a Dedekind cut. In terms of A and B, how can you tell if α is rational or irrational?

(2) Prove that there is no smallest positive real number. (Note that you must do this directly in terms of Dedekind cuts.)

(3) Prove that if α and β are any two real numbers with $\alpha < \beta$, then there is a rational number r with $\alpha < r < \beta$.

(4) Prove that if α and β are any two real numbers with $\alpha < \beta$, then there is an irrational number s with $\alpha < s < \beta$.

(5) Prove that there exists a countable set that contains an uncountable collection of nested subsets. (A collection of subsets is said to be *nested* if for any two subsets A and B, either $A \subseteq B$ or $B \subseteq A$.)

(6) When we defined addition of real numbers in terms of Dedekind cuts, we defined $\alpha + \beta$ to be $(A_{\alpha+\beta}, \mathbb{Q} \setminus A_{\alpha+\beta})$ rather than $(A_{\alpha+\beta}, B_{\alpha+\beta})$, where $B_{\alpha+\beta} = \{b_\alpha + b_\beta : b_\alpha \in B_\alpha, b_\beta \in B_\beta\}$. Why does the latter definition not work?

(7) Prove that there is a $\sqrt{2} \in \mathbb{R}$, by showing directly from the definition of multiplication that there is some cut $x = (A, B) \in \mathbb{R}$ such that $x \cdot x = 2$. (You will need to use the least upper bound property.)

(8) Prove that if $x = (A, B)$ is a Dedekind cut, then $x = \sup(A)$.

(9) Prove the *greatest lower bound property*: if $S \subseteq \mathbb{R}$ is bounded below and nonempty, then S has a greatest lower bound. We call this greatest lower bound the *infimum* of S, and we denote it $\inf(S)$. (Note: you can mimic the proof of the least upper bound property, making the necessary changes, which might require a small amount of care due to the asymmetry of A and B in the definition of a Dedekind cut. Or you can prove this directly from the least upper bound property.)

Chapter 27

Sequences in $\mathbb{Q}$

27.1   The idea of completions

We now know how to construct $\mathbb{R}$ in terms of Dedekind cuts: this is one of two standard constructions of $\mathbb{R}$ out of $\mathbb{Q}$. We'll now move toward the other main construction, in terms of equivalence classes of Cauchy sequences. Before we define Cauchy sequences formally, let's see an example of what the idea is.

Think of a real number, say π. How would you describe it to someone? You can just say it's π, but if ey is not familiar with that particular number, then that description won't be of much use to em. You can also explain it in terms of its geometric properties, saying something about the circumference and diameter of a circle. But that's very specific to π and doesn't work for other numbers.

Instead, you can describe π to someone, at least approximately, by telling em a bunch of its digits. You might tell em that $\pi = 3.14159265358979\ldots$ What this means is that we have a sequence of numbers that gets closer and closer to π, so that by going far enough out in the sequence, we get as close as desired to π. The sequence here consists of the following sequence of truncations of the decimal expansion of π, namely

$$3, 3.1, 3.14, 3.141, 3.1415, 3.14159, \ldots$$

If someone wants to know what π is to within any error, say 10^{-100}, then ey can figure that out by going far enough in the sequence.

The idea, then, is that we can describe *any* real number by giving an infinite sequence of rational numbers that approximate it better and better. However, not all sequences of rational numbers describe real numbers. For example, the sequence $1, 2, 3, 4, 5, \ldots$ does not describe any real number, because the terms stay far apart from each other. So the real numbers can only be described by sequences of rational numbers that get close together as we go far out in the sequence.

27.2   Cauchy sequences

Definition 27.1. A *sequence* of rational numbers is a function $f : \mathbb{N} \to \mathbb{Q}$. Typically if f is a sequence, we write $a_1, a_2, a_3, \ldots$, where $a_n = f(n)$.

Example. Suppose $f(n) = \frac{1}{n}$. Then the sequence is $1, \frac{1}{2}, \frac{1}{3}, \frac{1}{4}, \ldots$

Remark 27.2. Sometimes it's convenient to start the sequence with a_0 instead of a_1, or perhaps somewhere else. Nothing much changes when we do that, so we'll start with a_0 or a_1, or anything else, as it is convenient at the moment.

Remark 27.3. A sequence doesn't have to be describable by a nice formula. For instance, there is a sequence $1, \frac{5}{3}, -\frac{81}{19}, \frac{991}{4096}, \ldots$, and in fact infinitely many of them depending on what goes into the ellipsis. We'd have trouble writing down a concise formula for a_n, but the definition of a sequence doesn't require that we be able to do so.

As mentioned above, not all sequences get close together. There are two possible intuitive notions of what it means for terms of a sequence to get close together, but only one of them behaves well. The first possibility is that we could say that the sequence $a_1, a_2, a_3, \ldots$ gets close together if, for large enough n, $|a_n - a_{n+1}|$ can be made as small as possible. However, this is not a very good idea, as we shall see when we discuss infinite series in Chapter 32.

A much better idea is to say that the terms of a sequence $a_1, a_2, a_3, \ldots$ get close together if, once we go out far enough in the sequence, *all* the future terms are close to each other. That means that not only do two consecutive terms need to be close, but *any* two terms far enough out need to be close to each other.

Somewhat more precisely, this means that if we choose *any* positive number ε, then once we go sufficiently far in the sequence, any two terms are within ε of each other. This is the notion of a *Cauchy sequence*. Here is the precise definition:

Definition 27.4. Let $a_1, a_2, a_3, \ldots$ be a sequence in $\mathbb{Q}$. We say that this sequence is a *Cauchy sequence* if, for every $\varepsilon > 0$, there is some $N > 0$ (depending on ε) such that if $m, n > N$, then $|a_m - a_n| < \varepsilon$.

Example. Consider the sequence $a_n = \frac{1}{n}$. Let us check that this sequence is Cauchy. Choose some $\varepsilon > 0$. We must find some N such that if $m, n > N$, then $|a_m - a_n| < \varepsilon$. This means that we need a way of selecting an N in terms of ε. In this case, we can write down a fairly simple formula, but this isn't always possible. Here, we note that

$$|a_m - a_n| = \left| \frac{1}{m} - \frac{1}{n} \right| < \max\left(\frac{1}{m}, \frac{1}{n} \right) < \frac{1}{N},$$

whenever $m, n > N$. We need to choose N such that $|a_m - a_n| < \varepsilon$, so it suffices to choose an N such that $\frac{1}{N} < \varepsilon$, or $N > \frac{1}{\varepsilon}$. When that happens, we have

$$|a_m - a_n| = \left| \frac{1}{m} - \frac{1}{n} \right| < \max\left(\frac{1}{m}, \frac{1}{n} \right) < \frac{1}{N} < \varepsilon,$$

which is just what we want.

Example. Consider the sequence $a_n = (-1)^n$, so that the sequence goes $-1, 1, -1, 1, -1, 1, \ldots$ This sequence is not Cauchy. For instance, take $\varepsilon = 1$. (In fact, even $\varepsilon = 2$ presents the same problem. However, there are no bonus points for finding the largest possible ε that presents a problem.) Then there is no N such that whenever $m, n > N$, we have $|a_m - a_n| < \varepsilon$. The reason is that, regardless of how large N is, we can find $m, n > N$ such that $a_m = 1$ and $a_n = -1$, so that $|a_m - a_n| = 2$. Thus we can *never* go far enough in the sequence so that all the terms a_m, a_n with $m, n > N$ are of distance less than 1 from each other.

See Figure 27.1 for a picture of a Cauchy sequence.

Figure 27.1. A Cauchy sequence. Darker points represent a_n for larger values of n.

Example. Consider our sequence $3, 3.1, 3.14, 3.141, \ldots$ approximating π, so that $a_1 = 3$, $a_2 = 3.1$, $a_3 = 3.14$, and so forth. This sequence is Cauchy. To see this, let's suppose that $m > n$. Then

$$|a_m - a_n| < 10^{1-n}.$$

Thus if we pick any $\varepsilon > 0$ and then choose an N such that $10^{1-N} < \varepsilon$, then whenever $m, n > N$, we have $|a_m - a_n| < \varepsilon$.

27.3   Convergent sequences

Given a sequence of rational numbers $a_1, a_2, a_3, \ldots$, we would like to know whether it is getting closer and closer to some number. In the case of the sequence $3, 3.1, 3.14, 3.141, \ldots$, this sequence seems to be getting close to π. But there's a problem here: we're trying to use this sequence to *create* π, since at this point we haven't yet defined real numbers other than rationals. (Well, we did using Dedekind cuts, but we want to start from scratch and not use that, since this is a second, completely independent, construction of $\mathbb{R}$.) So it does not make sense at this point to say that that sequence converges to π. Furthermore, since π is irrational,[1] it doesn't make much sense to say that the sequence converges to anything at all. Clearly, we need a precise definition here to cut through some of the complications.

Definition 27.5. Let $a_1, a_2, a_3, \ldots$ be a sequence of rational numbers, and let s be a rational number. We say that $a_1, a_2, a_3, \ldots$ *converges* to s and write

$$\lim_{n \to \infty} a_n = s$$

if, for any $\varepsilon > 0$, there is an $N > 0$ such that if $n > N$, we have $|a_n - s| < \varepsilon$. If there is some $s \in \mathbb{Q}$ such that $\lim_{n \to \infty} a_n = s$, then we say that the sequence *converges*. We call this value of s the *limit* of the sequence.

Note that we require that $s \in \mathbb{Q}$, because that's all we know at the moment. Later on, we will redo much of this to allow sequences of arbitrary real numbers converging to a real limit, but that only makes sense once we have defined the real numbers. So, for now, we're stuck with sequences of rational numbers and rational limits.

Let's see some examples of convergent sequences.

Example. Let $a_n = \frac{1}{n}$. We claim that this is a convergent sequence, and that its limit is 0. To see this, we have to show that for every $\varepsilon > 0$, there is some $N > 0$ such that whenever $n > N$, we have $|a_n - 0| = |\frac{1}{n}| < \varepsilon$. Note that $|\frac{1}{n}| = \frac{1}{n} < \frac{1}{N}$ if $n > N$, so it suffices to choose an $N \geq \frac{1}{\varepsilon}$, or $\frac{1}{N} \leq \varepsilon$. Indeed, when $N \geq \frac{1}{\varepsilon}$ and $n > N$, we have

$$|a_n - 0| = \left| \frac{1}{n} \right| = \frac{1}{n} < \frac{1}{N} \leq \varepsilon,$$

which is just what we needed. Thus we have shown that

$$\lim_{n \to \infty} \frac{1}{n} = 0.$$

---

[1] This is not an easy theorem at all, but one we will take for granted for the sake of this discussion; see [Niv47] for a proof.

Intuitively, it seems that Cauchy sequences and convergent sequences are trying to capture a similar idea, so there ought to be some sort of connection between them. And indeed there is!

Proposition 27.6. *Convergent sequences are Cauchy.*

The point is that if x and y are two numbers that are both close to the same number s, then x and y are also close to each other. In order to prove Proposition 27.6 by making this idea rigorous, we need the triangle inequality.

Theorem 27.7 (Triangle Inequality). *If $x, y \in \mathbb{R}$ (or perhaps just $\mathbb{Q}$ at the moment), then*

$$|x - y| \le |x| + |y|.$$

Or, replacing x by $x - z$ and y by $y - z$,

$$|x - y| = |x - z - y + z| \le |x - z| + |y - z|.$$

The triangle inequality is obvious with a moment's thought, so we won't bother to prove it.

Proof of Proposition 27.6. Let $a_1, a_2, a_3, \ldots$ be a convergent sequence, with $\lim_{n \to \infty} a_n = s$. We must show that this sequence is Cauchy. Pick an $\varepsilon > 0$. Since $\lim_{n \to \infty} a_n = s$, there is some N such that whenever $n > N$, we have $|a_n - s| < \frac{\varepsilon}{2}$.[2] We will show that, for the same N, if $m, n > N$, then $|a_m - a_n| < \varepsilon$. To see this, we use the triangle inequality: if $m, n > N$, then $|a_m - s| < \frac{\varepsilon}{2}$ and $|a_n - s| < \frac{\varepsilon}{2}$, so

$$|a_m - a_n| \le |a_m - s| + |a_n - s| < \frac{\varepsilon}{2} + \frac{\varepsilon}{2} = \varepsilon,$$

as desired.   ∎

On the other hand, some Cauchy sequences are not convergent, because while the terms of the sequence get closer and closer to each other, they don't get arbitrarily close to some fixed rational number. Observe:

---

[2]Why did we pick $\frac{\varepsilon}{2}$ rather than ε? This is a standard trick that we'll see over and over again in analysis. You'll see shortly why that's just what we need to make the proof work out. Sometimes we'll need $\frac{\varepsilon}{3}$ or $\frac{\varepsilon}{4}$ instead of $\frac{\varepsilon}{2}$. Often, when you start working on a proof, you won't know what you need to divide ε by, so you can just keep everything as ε at first, and then when you have a bound of $k\varepsilon$ at the end, just divide all the ε's by k.

Example. Consider our sequence $3, 3.1, 3.14, 3.141, \ldots$ for π. As discussed above, this sequence is Cauchy. But it isn't convergent in $\mathbb{Q}$. If it were, it would have to converge to some *rational* number s. Let's take one and see why it doesn't converge to it; for illustrative purposes, let's try $s = 3$. Why doesn't the sequence converge to 3? We have to pick a sufficiently small ε, say $\varepsilon = .1$. Is it true that there is some N such that whenever $n > N$, we have $|a_n - 3| < .1$? No, because there are lots of terms of the sequence that are ≥ 3.1. (In fact, all but the first.) If we had instead tried $s = 3.14$, then we could take $\varepsilon = .001$, and it isn't true that all sufficiently far out terms in the sequence satisfy $|a_n - 3.14| < .001$, because lots of the terms are ≥ 3.141.

Definition 27.8. A set is said to be *complete* if every Cauchy sequence converges.

As the example shows, $\mathbb{Q}$ is not complete, because we have examples of Cauchy sequences in $\mathbb{Q}$ that do not converge in $\mathbb{Q}$. Our goal in the next chapter will be to create a *completion* of $\mathbb{Q}$: adding in numbers to make all Cauchy sequences converge. The resulting set is $\mathbb{R}$; this will be our second construction of $\mathbb{R}$ from $\mathbb{Q}$.

One thing that we swept under the rug—but that you might believe especially in light of the example—is that limits, if they exist, are unique. That is, if $a_1, a_2, a_3, \ldots$ is a sequence and $\lim_{n \to \infty} a_n = s$ and $\lim_{n \to \infty} a_n = t$, then $s = t$. This isn't built into the definition, because the definition of convergence involved first picking the limiting value s. But the terminology and notation strongly suggest that limits are unique. And this isn't hard to prove. Once again, the triangle inequality is the star of the show.

Proposition 27.9. *A sequence $a_1, a_2, a_3, \ldots$ of rational numbers has at most one limit.*

Proof. Suppose there were two limits, say s and t, such that $\lim_{n \to \infty} a_n = s$ and $\lim_{n \to \infty} a_n = t$. Suppose that $s \neq t$. Then let $\varepsilon = \frac{|s-t|}{2}$, which is positive because $s \neq t$. Since $\lim_{n \to \infty} a_n = s$, we can find some N_1 such that $|a_n - s| < \varepsilon$ whenever $n > N_1$. Similarly, we can find some N_2 such that $|a_n - t| < \varepsilon$ whenever $n > N_2$. Let $N = \max(N_1, N_2)$. This means that whenever $n > N$, we have both $|a_n - s| < \varepsilon$ and $|a_n - t| < \varepsilon$. Thus for all $n > N$, we have

$$2\varepsilon = |s - t| \leq |s - a_n| + |t - a_n| < \varepsilon + \varepsilon = 2\varepsilon.$$

But that means we have $2\varepsilon < 2\varepsilon$, which is false. Thus we have a contradiction to our assumption that $s \neq t$. This shows that if the sequence has two limits, then those limits are equal. In other words, there's only at most one limit. ∎

27.4 Problems

(1) Which of the following sequences in $\mathbb{Q}$ are Cauchy sequences? Which ones converge in $\mathbb{Q}$? For those that converge, determine their limits. In all cases, prove that your answers are correct.

(a) $a_n = 1 - \frac{1}{n^3}$ (b) $a_n = \frac{2n^2}{n^2+1}$ (c) $a_n = \frac{n^2}{n+6}$

(2) Let $a \in \mathbb{Q}$. Prove that the sequence $a_n = a$ for all n is a Cauchy sequence. Does it converge to some rational number? If so, which one?

(3) Find an example of a sequence a_n in $\mathbb{Q}$ such that infinitely many of the a_n's are 0, but the sequence does not converge to 0.

(4) If a sequence a_n of rational numbers is Cauchy, and $a_n = 0$ for infinitely many values of n, does a_n necessarily converge to 0? Prove or find a counterexample.

(5) If a_n is a convergent sequence in $\mathbb{Q}$, is $b_n = \lfloor a_n \rfloor$ necessarily convergent? Prove or find a counterexample.

(6) A *subsequence* of a sequence $a_1, a_2, a_3, \ldots$ in $\mathbb{Q}$ is another sequence $a_{n_1}, a_{n_2}, a_{n_3}, \ldots$, where $1 \leq n_1 < n_2 < n_3 < \cdots$. If a_n is convergent, is any subsequence also necessarily convergent? Prove that your answer is correct.

(7) Find a sequence a_n that is not convergent, but that has a convergent subsequence. Prove that your answer is correct.

(8) Let a_n be a sequence. Suppose that, for every k, there is a subsequence of a_n containing a_k that is convergent. Is a_n necessarily convergent? Prove or find a counterexample.

(9) Let a_n be a sequence. Suppose that there is some rational number r such that for every k, there is a subsequence of a_n containing a_k that converges to r. Is a_n necessarily convergent? Prove or find a counterexample.

Interlude: Proofs using ε

In this interlude, we'll go through a few more examples of proofs using ε and δ, and how to write them well. Use these proofs as models for how to write your own.

Proposition 27.10. *Let $a_n = 2 + \frac{n^2+5}{n^3-4}$. Then $\lim_{n\to\infty} a_n = 2$.*

Proof. Let $\varepsilon > 0$ be arbitrary. We have
$$|2 - a_n| = \left|\frac{n^2 + 5}{n^3 - 4}\right|.$$
For $n \geq 2$, we have $n^3 - 4 \geq \frac{n^3}{2}$, and for $n \geq 3$, we have $n^2 + 5 < 2n^2$. Thus for $n \geq 3$, we have
$$|2 - a_n| = \left|\frac{n^2 + 5}{n^3 - 4}\right| \leq \frac{2n^2}{n^3/2} = \frac{4}{n}.$$
Let $N = \max(\frac{4}{\varepsilon}, 3)$. Then for all $n > N$, we have
$$|2 - a_n| \leq \frac{4}{n} < \frac{4}{N} < \varepsilon.$$
Thus $\lim_{n\to\infty} a_n = 2$, as claimed. ∎

Proposition 27.11. *Let $a_n = 2 + \frac{n^2+5}{n^3-4}$. Then the sequence a_n is Cauchy.*

This follows from Proposition 27.10, because all convergent sequences are Cauchy, but let's prove that the sequence is Cauchy directly, just to get another example of how to write that sort of proof.

Proof. Let $\varepsilon > 0$ be arbitrary. Let $N = \max(\frac{8}{\varepsilon}, 3)$. As in the proof of Proposition 27.10, for $n \geq 3$, we have $|2 - a_n| \leq \frac{4}{n}$, so if $m, n \geq N$, we have
$$|a_m - a_n| = |(a_m - 2) - (a_n - 2)| \leq |a_m - 2| + |a_n - 2| \leq \frac{4}{m} + \frac{4}{n} < \frac{\varepsilon}{2} + \frac{\varepsilon}{2} = \varepsilon,$$
as desired. ∎

Chapter 28

A second construction of $\mathbb{R}$

28.1 Equivalence classes of Cauchy sequences

In Chapter 27, we introduced Cauchy sequences in $\mathbb{Q}$ and noted that not all Cauchy sequences in $\mathbb{Q}$ converge. Our goal is to fill in the gaps in $\mathbb{Q}$ so as to make all Cauchy sequences converge. We will get $\mathbb{R}$ by filling in all these gaps.

Consider, as usual, our Cauchy sequence for π: 3, 3.1, 3.14, 3.141, ... This Cauchy sequence doesn't converge in $\mathbb{Q}$, so if we're going to construct some set where all Cauchy sequences converge, we need to add in a number for this one. The problem is that there are other Cauchy sequences that we also want to converge to π; for instance 3, $\frac{22}{7}$, $\frac{333}{106}$, $\frac{355}{113}$, $\frac{103993}{33102}$, ...[1] So we don't just want to add one number for each nonconvergent Cauchy sequence in $\mathbb{Q}$, or else we'll have many different π's, one for each Cauchy sequence that is supposed to converge to π. Instead, we want to make sure that they all converge to the *same* π.

In order to do that, we need to put an equivalence relation on the set of all Cauchy sequences in $\mathbb{Q}$. The equivalence relation is as follows:

Definition 28.1. Let $a = (a_1, a_2, a_3, \ldots)$ and $b = (b_1, b_2, b_3, \ldots)$ be two Cauchy sequences in $\mathbb{Q}$. We say that a and b are *equivalent* (and write $a \sim b$, at least for the moment) if, for every $\varepsilon > 0$, there is some N such that whenever $n > N$, we have $|a_n - b_n| < \varepsilon$.

Example. Let $a = (a_1, a_2, a_3, \ldots)$ be the usual Cauchy sequence for π,

---

[1] Where did these come from? They are the continued fraction convergents for π. Each of them is the best rational approximation for π with that denominator or smaller. See for instance [NZM91, Chapter 7] for a detailed look at continued fractions and their importance in approximations.

namely $a_1 = 3$, $a_2 = 3.1$, $a_3 = 3.14$, and so on. Let $b = (b_1, b_2, b_3, \ldots)$ be the Cauchy sequence that rounds up rather than down: $b_1 = 4$, $b_2 = 3.2$, $b_3 = 3.15$, $b_4 = 3.142$, $b_5 = 3.1416$, and so forth. Let's check that $a \sim b$. Note that $|a_1 - b_1| = 1$, $|a_2 - b_2| = .1$, $|a_3 - b_3| = .01$, and in general, $|a_n - b_n| = 10^{1-n}$. For any $\varepsilon > 0$, we can find some N such that $10^{1-n} < \varepsilon$ whenever $n > N$ (and in particular, when $n = N + 1$). Thus $a \sim b$.

There are two especially important types of equivalent Cauchy sequences.

Example. Suppose that a and b are eventually equal, i.e. there exists some N such that $a_n = b_n$ whenever $n > N$. Then $a \sim b$. The reason is simple: whenever $n > N$, we have $|a_n - b_n| = 0$, which is less than ε for any $\varepsilon > 0$. Note that in this case, we can always choose the same N, regardless of the value of ε.

Example. Suppose that a and b are shifts of each other. This means that there is some integer k (which we'll assume to be positive, or else we can just switch the roles of a and b) such that $b_n = a_{n+k}$ for all n (or even all sufficiently large n). Then $a \sim b$. Let's check. Pick some $\varepsilon > 0$. We must find an N such that whenever $n > N$, we have $|a_{n+k} - a_n| < \varepsilon$. Now, since a is a Cauchy sequence, there is some N such that whenever $m, n > N$, we have $|a_m - a_n| < \varepsilon$. Now, if $n > N$, then both n and $n + k$ are greater than N, so in particular we have $|a_n - a_{n+k}| < \varepsilon$. But that's just what we need to conclude that $a \sim b$.

There is another characterization of equivalence of Cauchy sequences that is sometimes useful. This could serve as the definition just as easily as the one in Definition 28.1.

Proposition 28.2. *Two Cauchy sequences $a = (a_1, a_2, a_3, \ldots)$ and $b = (b_1, b_2, b_3, \ldots)$ are equivalent if and only if for every $\varepsilon > 0$, there exists an N such that whenever $m, n > N$, we have $|a_m - b_n| < \varepsilon$.*

This is a good opportunity to get more familiar with $\frac{\varepsilon}{2}$ tricks.

Proof. Suppose that for all $\varepsilon > 0$, there exists an N such that whenever $m, n > N$, we have $|a_m - b_n| < \varepsilon$. Then, letting $m = n$, we have that whenever $n > N$, $|a_n - b_n| < \varepsilon$, which is the definition of equivalence of a and b.

Now let's do the other direction. Suppose that $a \sim b$. Pick an $\varepsilon > 0$. We must show that there exists an N such that whenever $m, n > N$, we have $|a_m - b_n| < \varepsilon$. Since $a \sim b$, there is some N_1 such that whenever $n > N_1$, we have $|a_n - b_n| < \frac{\varepsilon}{2}$. And since a is Cauchy, there is some N_2 such that whenever $m, n > N_2$, we have $|a_m - a_n| < \frac{\varepsilon}{2}$. Now, let $N = \max(N_1, N_2)$. Thus whenever $m, n > N$, we have both $|a_n - b_n| < \frac{\varepsilon}{2}$ and $|a_m - a_n| < \frac{\varepsilon}{2}$. Thus whenever $m, n > N$, we have

$$|a_m - b_n| \le |a_m - a_n| + |a_n - b_n| < \frac{\varepsilon}{2} + \frac{\varepsilon}{2} = \varepsilon,$$

as desired. $\blacksquare$

We haven't checked that $\sim$ is an equivalence relation yet, so it's time to do that.

Proposition 28.3. $\sim$ *is an equivalence relation.*

Proof. **Reflexivity:** We must show that $a \sim a$. This means we must check that for every $\varepsilon > 0$, there is some N such that whenever $n > N$, we have $|a_n - a_n| < \varepsilon$. Any N will do, since $|a_n - a_n| = 0 < \varepsilon$.

Symmetry: We must check that if $a \sim b$, then $b \sim a$. Suppose that $a \sim b$, and let $\varepsilon > 0$. There is some N such that whenever $n > N$, we have $|a_n - b_n| < \varepsilon$, and we must show that there is some N such that whenever $n > N$, we have $|b_n - a_n| < \varepsilon$. Since $|a_n - b_n| = |b_n - a_n|$, the same N will do.

Transitivity: We must check that if $a \sim b$ and $b \sim c$, then $a \sim c$. Pick some $\varepsilon > 0$. There is some N_1 such that whenever $n > N_1$, we have $|a_n - b_n| < \frac{\varepsilon}{2}$. Also, there is some N_2 such that whenever $n > N_2$, we have $|b_n - c_n| < \frac{\varepsilon}{2}$. Let $N = \max(N_1, N_2)$. Thus whenever $n > N$, we have

$$|a_n - c_n| \le |a_n - b_n| + |b_n - c_n| < \frac{\varepsilon}{2} + \frac{\varepsilon}{2} = \varepsilon,$$

as desired. $\blacksquare$

Now that we know that $\sim$ is an equivalence relation, we can talk about the equivalence classes of Cauchy sequences.

Definition 28.4. A *real number* is an equivalence class of Cauchy sequences of rational numbers with respect to the equivalence relation $\sim$. We write $\mathbb{R}$ for the set of all real numbers. Given a Cauchy sequence a in $\mathbb{Q}$, we write $[a]$ for its equivalence class, which is a real number.

That is our second construction of $\mathbb{R}$ from $\mathbb{Q}$. Observe the differences between our two constructions, the one coming from Dedekind cuts and the one coming from Cauchy sequences. The Dedekind cut construction is based on the fact that $\mathbb{Q}$ (and $\mathbb{R}$) is an *ordered* set: we fill in the gaps in the ordering. The Cauchy sequence construction is based on the notion of distance.

While these two constructions may seem similar in the case of $\mathbb{R}$, you should keep in mind that there are many other structures that have a notion of ordering but not distance, or a notion of distance but not ordering. So it's good to be familiar with both constructions, since they generalize in very different ways.[2]

Remark 28.5. At some point, one ought to check that the two definitions of $\mathbb{R}$ actually agree, in some sense. We explain how this is done in the Interlude following this chapter. We will use properties of $\mathbb{R}$ coming both from Dedekind cuts (such as the least upper bound property) as well as properties of $\mathbb{R}$ coming from Cauchy sequences (especially completeness) in the future, wherever convenient. We won't mention whether we're thinking of the Dedekind cut model of $\mathbb{R}$ or the Cauchy sequence model of $\mathbb{R}$ because they define the same object.

Note that $\mathbb{R}$ contains $\mathbb{Q}$. Since $\mathbb{R}$ is made out of $\mathbb{Q}$, it seems plausible that the $\mathbb{Q}$ sitting inside of $\mathbb{R}$ is somehow noticeable. And this is true. Given a rational number q, there is an obvious Cauchy sequence representing q, namely the sequence $(q, q, q, q, \ldots)$: the constant sequence.

28.2   Arithmetic in $\mathbb{R}$

We have now defined $\mathbb{R}$ as a set, but we want to do arithmetic in $\mathbb{R}$. How do we add, subtract, multiply, and divide real numbers? Given the construction, we had better say how to do this directly in terms of Cauchy sequences.

Definition 28.6. Let α and β be real numbers. Suppose that $\alpha = [a]$ and $\beta = [b]$, where $a = (a_1, a_2, a_3, \ldots)$ and $b = (b_1, b_2, b_3, \ldots)$.

- We define $\alpha + \beta$ to be $[a+b]$, where $a+b = (a_1+b_1, a_2+b_2, a_3+b_3, \ldots)$.

---

[2]I won't dwell on this much, but let me just say that there is a type of structure called a *partially ordered set* (or *poset* for short), and you can perform a Dedekind cut sort of completion, known as the *Dedekind–MacNeille completion*, on an arbitrary poset; see [Mac37]. Similarly, there is a type of structure called a *metric space* that admits a sort of distance, and you can perform a completion (in this case just called the completion) of an arbitrary metric space, using equivalence classes of Cauchy sequences; see [Pug15, §2.10]. I encourage you to look into these topics if this sort of thing excites you.

- We define $\alpha - \beta$ to be $[a - b]$, where $a - b = (a_1 - b_1, a_2 - b_2, a_3 - b_3, \ldots)$.
- We define $\alpha\beta$ to be $[ab]$, where $ab = (a_1 b_1, a_2 b_2, a_3 b_3, \ldots)$.
- Suppose that $b_n \neq 0$ for all n, and that $\beta \neq 0$. Then we define $\frac{\alpha}{\beta}$ to be $[\frac{a}{b}]$, where $\frac{a}{b} = (\frac{a_1}{b_1}, \frac{a_2}{b_2}, \frac{a_3}{b_3}, \ldots)$.

Recall that when we define $\mathbb{Z}$ from $\mathbb{N}$, and $\mathbb{Q}$ from $\mathbb{Z}$, we did something very similar. Also recall that there was a bit of trouble with the analogous definition, because there were a bunch of things we needed to check. The same is true here. For instance, we need to check that $a + b$ is actually a Cauchy sequence (and similarly for the others). Then we need to check that it is well-defined. That amounts to checking that if $a_1 \sim a_2$ and $b_1 \sim b_2$, then $a_1 + b_1 \sim a_2 + b_2$. Then we have to do the same thing for the others. We'll do both of these just for addition, leaving it as an exercise for you to do the rest on your own. (See problem 3 for multiplication.)

Proposition 28.7. *If $a = (a_1, a_2, a_3, \ldots)$ and $b = (b_1, b_2, b_3, \ldots)$ are Cauchy sequences, then $a + b$ is also a Cauchy sequence.*

Proof. Let $\varepsilon > 0$. Since a is Cauchy, there is some N_1 such that whenever $m, n > N_1$, we have $|a_m - a_n| < \frac{\varepsilon}{2}$. Similarly, since b is Cauchy, there is some N_2 such that whenever $m, n > N_2$, we have $|b_m - b_n| < \frac{\varepsilon}{2}$. Let $N = \max(N_1, N_2)$. Then whenever $m, n > N$, we have

$$
\begin{aligned}
|(a_m + b_m) - (a_n + b_n)| &= |(a_m - a_n) + (b_m - b_n)| \\
&\leq |a_m - a_n| + |b_m - b_n| \\
&< \frac{\varepsilon}{2} + \frac{\varepsilon}{2} \\
&= \varepsilon,
\end{aligned}
$$

as desired. ∎

Proposition 28.8. *If $a \sim c$ and $b \sim d$, then $a + b \sim c + d$.*

Proof. Let us write $a = (a_1, a_2, a_3, \ldots)$ and similarly for the others. Let $\varepsilon > 0$. We must show that there is some N such that if $n > N$, then $|(a_n + b_n) - (c_n + d_n)| < \varepsilon$. Since $a \sim c$, there is some N_1 such that if $n > N_1$, then $|a_n - c_n| < \frac{\varepsilon}{2}$. Similarly, since $b \sim d$, there is some N_2 such that if $n > N_2$, then $|b_n - d_n| < \frac{\varepsilon}{2}$. Let $N = \max(N_1, N_2)$. Then whenever

$n > N$, we have

$$|(a_n + b_n) - (c_n + d_n)| = |(a_n - c_n) + (b_n - d_n)|$$
$$\leq |a_n - c_n| + |b_n - d_n|$$
$$< \frac{\varepsilon}{2} + \frac{\varepsilon}{2}$$
$$= \varepsilon,$$

as desired.   ∎

The last two propositions combine to show that addition of real numbers is well-defined. Similar arguments show that subtraction, multiplication, and division (but not by zero) are well-defined as well.

So far, our absolute value function $|\cdot|$ only takes rational arguments, because we've only just constructed $\mathbb{R}$ and haven't ported everything over yet. So we need to define the absolute value function on all of $\mathbb{R}$ to make future steps.

Definition 28.9. Let $\alpha \in \mathbb{R}$ be represented by a Cauchy sequence $a = (a_1, a_2, a_3, \ldots)$. Then $|\alpha|$ is the real number represented by the Cauchy sequence $|a| = (|a_1|, |a_2|, |a_3|, \ldots)$.

Of course, just like everything else, we need to check that this definition makes sense, i.e. that $|a|$ is indeed a Cauchy sequence, and that $|\alpha|$ is independent of the choice of Cauchy sequence that represents α. These are similar to previous proofs, so we leave one of these checks as an exercise. (See problem 2.)

28.3   Cauchy sequences in $\mathbb{R}$

So far, we have only discussed sequences and Cauchy sequences in $\mathbb{Q}$. Now that we have defined $\mathbb{R}$, we can go back and redo everything we have done for sequences in $\mathbb{R}$. Since we can add and subtract real numbers, as well as take their absolute values, we can define Cauchy sequences in $\mathbb{R}$.

Definition 28.10. Let $a_1, a_2, a_3, \ldots$ be a sequence in $\mathbb{R}$. We say that this sequence is a *Cauchy sequence* if, for every $\varepsilon > 0$, there is some $N > 0$ (depending on ε) such that if $m, n > N$, then $|a_m - a_n| < \varepsilon$.

If this definition looks familiar, it should: it's exactly the same definition we saw in Chapter 27 for Cauchy sequences in $\mathbb{Q}$, except with $\mathbb{Q}$ replaced with $\mathbb{R}$. We can also talk about convergence in $\mathbb{R}$:

Definition 28.11. Let $a_1, a_2, a_3, \ldots$ be a sequence of real numbers, and let s be a real number. We say that $a_1, a_2, a_3, \ldots$ *converges* to s and write

$$\lim_{n \to \infty} a_n = s$$

if, for every $\varepsilon > 0$, there is an $N > 0$ such that if $n > N$, we have $|a_n - s| < \varepsilon$. If there is some $s \in \mathbb{R}$ such that $\lim_{n \to \infty} a_n = s$, then we say that the sequence *converges*. We call this value of s the *limit* of the sequence.

Again, this is the same definition as before, but with $\mathbb{Q}$ (or rational) replaced with $\mathbb{R}$ (or real).

Now, recall that we constructed $\mathbb{R}$ by filling in the gaps formed by sequences in $\mathbb{Q}$. We might be concerned that we have to do that process again with $\mathbb{R}$. Just because all Cauchy sequences in $\mathbb{Q}$ converge to real numbers, that doesn't necessarily mean that all Cauchy sequence in $\mathbb{R}$ also converge to real numbers, because suddenly we have a lot more Cauchy sequences lying around. However, we're in luck: all Cauchy sequences in $\mathbb{R}$ also converge in $\mathbb{R}$. Recall that we say that a set S is *complete* if all Cauchy sequences in S converge to something in S. In other words, we are claiming the following:

Theorem 28.12. $\mathbb{R}$ *is complete.*

Proof. We must show that every Cauchy sequence in $\mathbb{R}$ converges in $\mathbb{R}$. Consider a Cauchy sequence $\alpha = (\alpha_1, \alpha_2, \alpha_3, \ldots)$ in $\mathbb{R}$. We will show that α is equivalent to a Cauchy sequence in $\mathbb{Q}$. This suffices to prove that $\mathbb{R}$ is complete, because we already know that every Cauchy sequence in $\mathbb{Q}$ converges to a real number. Thus we need to find a sequence of rational numbers close to the α_i's.

Each α_i is an element of $\mathbb{R}$, i.e. an equivalence class of Cauchy sequences in $\mathbb{Q}$. For each α_i, choose a Cauchy sequence $a_i = (a_{i1}, a_{i2}, a_{i3}, \ldots)$ in the equivalence class α_i. For each i, the Cauchy sequence a_i converges to α_i in $\mathbb{R}$, so there is some $N(i) > 0$, depending on i, such that $|a_{in} - \alpha_i| < \frac{1}{i}$ whenever $n \geq N(i)$.

Now, consider the sequence $b = (b_1, b_2, b_3, \ldots)$ in $\mathbb{Q}$ (and hence $\mathbb{R}$), where $b_i = a_{i,N(i)}$. We claim that b is equivalent to α. By definition, this means that for any $\varepsilon > 0$, there is some $N > 0$ such that whenever $n > N$, then $|\alpha_n - b_n| < \varepsilon$. By the construction of b_n, we have

$$|\alpha_i - b_i| = |\alpha_i - a_{i,N(i)}| < \frac{1}{i}.$$

Choose $N > \frac{1}{\varepsilon}$, so that $\frac{1}{N} < \varepsilon$. Then whenever $n > N$, we have

$$|a_n - b_n| < \frac{1}{n} < \varepsilon.$$

This shows that b is equivalent to α. Since b is a Cauchy sequence in $\mathbb{Q}$, it converges to a number in $\mathbb{R}$. Thus α also converges to a number in $\mathbb{R}$. Since α is an arbitrary Cauchy sequence in $\mathbb{R}$, this shows that all Cauchy sequences in $\mathbb{R}$ converge, so $\mathbb{R}$ is complete. ∎

28.4   Further reading

We have seen two constructions of the real numbers in this book: the one using Dedekind cuts and the one using equivalence classes of Cauchy sequences. While these are the most popular constructions, they aren't the only ones. Let's take a brief look at another construction, called the *Eudoxus reals*.

Definition 28.13. A function $f : \mathbb{Z} \to \mathbb{Z}$ is called an *almost homomorphism* if there is some integer C such that for all $m, n \in \mathbb{Z}$, $|f(m+n) - f(m) - f(n)| \leq C$.

Example. For any real number α, the function $f_\alpha : \mathbb{Z} \to \mathbb{Z}$ defined by $f_\alpha(n) = \lfloor \alpha n \rfloor$ is an almost homomorphism.

Definition 28.14. We say that two almost homomorphisms f and g are *almost equal* if there is some integer B such that $|f(n) - g(n)| \leq B$ for all integers n.

Let $\mathcal{F}$ be the set of almost homomorphisms. The relation $\sim$ on $\mathcal{F}$ defined by $f \sim g$ if f and g are almost equal is an equivalence relation. Let $\mathcal{R} = \mathcal{F}/\sim$ be the set of equivalence classes under this equivalence relation.

Definition 28.15. The set $\mathcal{R}$ of equivalence classes is called the set of *Eudoxus reals*.

The Eudoxus reals is another construction of the real numbers, since it turns out that for any almost homomorphism f, there is a unique real number α such that $f \sim f_\alpha$, so we identify the $\sim$-equivalence class of f with the real number α.

We can define the arithmetic operations on $\mathcal{R}$ directly in terms of almost homomorphisms. Given two almost homomorphisms f and g, we define $[f] + [g]$ to be $[f + g]$, where $f + g$ is the function $(f + g)(n) = f(n) + g(n)$, which can be checked to be another almost homomorphism. Similarly, we define $[f] \cdot [g]$ to be $[fg]$, where fg is the function $(fg)(n) = f(n)g(n)$, which is again an almost homomorphism. We can also define the ordering: $[f] < [g]$ means that there exists an integer N such that $f(n) < g(n)$ for all $n > N$.

See [Art04] for a survey of the Eudoxus reals, as well as [Wei15] for a survey of various constructions of the real numbers.

28.5   Problems

(1) Below are several Cauchy sequences in $\mathbb{Q}$. Which ones are equivalent to which other ones? Prove that your answer is correct.

   (a) $a_n = 0$.
   (b) $b_n = 1$.
   (c) $c_n = \frac{1}{n}$.
   (d) $d_n = \frac{1}{n^2}$.
   (e) $e_1 = 2$, $e_{n+1} = \frac{1}{2}(e_n + \frac{1}{e_n})$ for $n \geq 1$.
   (f) $f_n = \frac{n^3}{n^3 - 5n + 9}$.

(2) Check that the definition of $|\alpha|$ given in Definition 28.9 is well-defined, i.e. independent of the choice of Cauchy sequence representing α.

(3) Prove that the product of two Cauchy sequences, as defined in Definition 28.6, is Cauchy.

(4) A subset $S \subseteq \mathbb{R}$ is said to be *dense* if, for every $\alpha \in \mathbb{R}$ and every $\varepsilon > 0$, there is some $s \in S$ such that $|\alpha - s| < \varepsilon$. Which of the following subsets of $\mathbb{R}$ are dense? In each case, prove that your answer is correct.

   (a) $\mathbb{Q}$.
   (b) The set $\mathbb{Z}[\frac{1}{2}]$ of dyadic rationals, i.e. rational numbers $\frac{a}{b}$ where b is a power of 2.
   (c) $\mathbb{Z}$.
   (d) The positive real numbers.
   (e) $\{x^2 : x \in \mathbb{Q}\} \cup \{-x^2 : x \in \mathbb{Q}\}$.
   (f) Numbers of the form $q\pi$, where $q \in \mathbb{Q}$.

(5) Does there exist a finite subset of $\mathbb{R}$ which is dense? Find such a set or prove that none can exist.

(6) Prove that a subset $S \subseteq \mathbb{Q}$ is dense if and only if, for every $\alpha \in \mathbb{R}$, there is a Cauchy sequence with terms in S representing α (i.e. every Cauchy sequence in $\mathbb{Q}$ is equivalent to a Cauchy sequence in S).

(7) Prove that if $\alpha \geq 0$ is a real number, then there is a real number β such that $\beta^2 = \alpha$, using Cauchy sequences.

(8) Let F_n denote the nth Fibonacci number, and let $a_n = \frac{F_{n+1}}{F_n}$.

(a) Prove that a_n is a Cauchy sequence in $\mathbb{Q}$.

(b) Which real number does it correspond to, in common parlance? Prove that your answer is correct.

Interlude: The real numbers as the unique complete ordered field

The goal of this interlude is to explain why the various construction of the real numbers, such as the one based on Dedekind cuts on $\mathbb{Q}$ and the one based on equivalence classes of Cauchy sequences, as well as more exotic constructions like the Eudoxus reals, all end up constructing the same object, and even what that means.

First, let's discuss what it means for two constructions of $\mathbb{R}$ to be the same. Any construction of the real numbers involves a set of numbers, together with operations of addition and multiplication, as well as the relation of $<$. Let's say we have two such sets, namely $(\mathcal{R}, +, \cdot, <)$ and $(\mathscr{R}, \oplus, \odot, \prec)$. We say that they are *isomorphic* if there exists a bijection $f : \mathcal{R} \to \mathscr{R}$ such that for all $a, b \in \mathcal{R}$, $f(a+b) = f(a) \oplus f(b)$, $f(ab) = f(a) \odot f(b)$, and $a < b$ if and only if $f(a) \prec f(b)$. In other words, f is a bijection that maps all the relevant structure of $\mathcal{R}$ onto analogous structure for $\mathscr{R}$.

It is possible, but somewhat awkward, to describe a bijection directly from the Dedekind cut reals to the Cauchy sequence reals. But a better way to do this is to define certain axioms that mimic the properties of the real numbers, and then show that, up to isomorphism, there is only one structure satisfying all those axioms. Since the Dedekind cut reals and the Cauchy sequence reals both satisfy all the axioms, they must then be isomorphic.

Definition 28.16. A *field* is a set F together with two binary operations, $+$ and $\cdot$, satisfying the following properties:

Associativity of addition: For all $a, b, c \in F$, $a + (b + c) = (a + b) + c$.
Commutativity of addition: For all $a, b \in F$, $a + b = b + a$.
Additive identity: There is an element $0 \in F$ such that $a + 0 = a$ for all $a \in F$.

Additive inverses: For all $a \in F$, there exists an element $-a \in F$ such that $a + (-a) = 0$.

Associativity of multiplication: For all $a, b, c \in F$, $a(bc) = (ab)c$.

Commutativity of multiplication: For all $a, b \in F$, $ab = ba$.

Multiplicative identity: There exists an element $1 \in F$, with $1 \neq 0$, such that for all $a \in F$, $a \cdot 1 = a$.

Multiplicative inverses: For all $a \in F \setminus \{0\}$, there exists an $a^{-1} \in F \setminus \{0\}$ such that $a \cdot (a^{-1}) = 1$.

Distributive law: For all $a, b, c \in F$, $a(b + c) = ab + ac$.

Example. The rational numbers $\mathbb{Q}$ form a field under the usual operations of addition and multiplication. The integers $\mathbb{Z}$ do not, because some nonzero integers, such as 2, do not have multiplicative inverses in the integers: there is no integer b such that $2b = 1$.

Definition 28.17. An *ordered field* is a field $(F, +, \cdot)$ together with a relation $<$ on F satisfying the following order axioms:

- $<$ is trichotomous: For all $a, b \in F$, exactly one of $a < b$, $a = b$, and $b < a$ holds.
- $<$ is transitive: If $a < b$ and $b < c$, then $a < c$.
- If $a < b$, then for all $c \in F$, we have $a + c < b + c$.
- For all $a, b, c \in F$, if $a < b$ and $0 < c$, then $ac < bc$.

Definition 28.18. An ordered field $(F, +, \cdot, <)$ is said to be *complete* if it satisfies the least upper bound property: if a nonempty subset $S \subseteq F$ is bounded above, i.e. if there exists a $B \in F$ such that $a \leq B$ for all $a \in S$, then S has a least upper bound in F, i.e. there is a smallest $x \in F$ such that $a \leq x$ for all $a \in S$.

Theorem 28.19. *Up to isomorphism, there is a unique complete ordered field.*

We'll omit a few details in the proof below in the interest of keeping this section reasonably short.

Proof. The existence follows from our constructions since (for instance) the Dedekind cut reals form a complete ordered field. So, it remains to show uniqueness.

Suppose $(F, +, \cdot, <)$ and $(K, \oplus, \odot, \prec)$ are two complete ordered fields. Because F is a field, it contains an element 1. By adding 1 to itself repeatedly, we can construct any positive integer, so F contains a copy of the positive integers. Together with their negatives and zero, they form a copy of all the integers inside of F. By taking quotients of two of them, we can in fact find a copy of $\mathbb{Q}$ inside of F. Let us call it $\mathcal{Q}$. Similarly, we have a copy of $\mathbb{Q}$ inside of K, which we'll call $\mathscr{Q}$.

We now begin to define $f : F \to K$, starting by defining it on $\mathcal{Q}$. Each element of $\mathcal{Q}$ corresponds to a rational number, and f sends the copy of a rational number q in $\mathcal{Q}$ to the copy of the same rational number in $\mathscr{Q}$. This gives us a bijection from $\mathcal{Q}$ to $\mathscr{Q}$.

We must now extend the bijection from all of F to all of K. Let $\alpha \in F$ be arbitrary, and let $A_\alpha = \{q \in \mathcal{Q} : q < \alpha\}$. Let $B_\alpha = \{f(q) : q \in A_\alpha\}$. Then B_α is a nonempty subset of $\mathscr{Q}$ (hence of K). Since K is complete, B_α has a least upper bound $\sup B_\alpha$, so we define $f(\alpha)$ to be $\sup B_\alpha$. Note that f is well-defined because if $\alpha, \beta \in F$ with $\alpha < \beta$, then there is a rational number between α and β, so A_α and A_β are not the same set.

We must now show that f is an isomorphism. To begin, we must prove that it is a bijection. To begin, we show that if $\alpha < \beta$, then $f(\alpha) \prec f(\beta)$. Find two rational numbers $q, r \in \mathcal{Q}$ with $\alpha < q < r < \beta$. Then because $\alpha < q$, we have $B_\alpha \subseteq B_q$, so $f(\alpha) \preceq f(q)$, and similarly $f(r) \preceq f(\beta)$. Since f is an order-preserving bijection on the rationals, we have $f(q) \prec f(r)$. Thus $f(\alpha) \prec f(\beta)$. Since F is ordered, this implies that f is injective.

We must now show that f is surjective. Let $\beta \in K$. We must find an $\alpha \in F$ such that $f(\alpha) = \beta$. Let $D_\beta = \{q \in \mathscr{Q} : q \prec \beta\}$, and let $C_\beta = \{f^{-1}(q) : q \in D_\beta\} \subseteq \mathcal{Q}$. Then C_β is nonempty and bounded above, so it has a least upper bound $\sup C_\beta$. One can check that $f(\sup C_\beta) = \beta$. Thus f is an order-preserving bijection.

It remains to check that f preserves the arithmetical operations, i.e. that $f(\alpha + \beta) = f(\alpha) \oplus f(\beta)$ and $f(\alpha\beta) = f(\alpha) \odot f(\beta)$ for all $\alpha, \beta \in F$. We leave this as an exercise. ∎

Chapter 29

Limits of sequences

29.1   Limits of sequences

Now that we have constructed $\mathbb{R}$ (a second time), it's time to go back and redo what we did with sequences in $\mathbb{Q}$, only this time for sequences in $\mathbb{R}$. We start by recalling the definition of a limit of a sequence in $\mathbb{Q}$.

Definition 29.1. Let $a_1, a_2, a_3, \ldots$ be a sequence of rational numbers, and let s be a rational number. We say that $a_1, a_2, a_3, \ldots$ *converges* to s and write

$$\lim_{n \to \infty} a_n = s$$

if, for any $\varepsilon > 0$, there is an $N > 0$ such that if $n > N$, we have $|a_n - s| < \varepsilon$. If there is some $s \in \mathbb{Q}$ such that $\lim_{n \to \infty} a_n = s$, then we say that the sequence *converges*. We call this value of s the *limit* of the sequence.

The real-number version is almost identical; we only have to replace "rational" with "real" (and $\mathbb{Q}$ with $\mathbb{R}$) everywhere, which gives the following definition.

Definition 29.2. Let $a_1, a_2, a_3, \ldots$ be a sequence of real numbers, and let s be a real number. We say that $a_1, a_2, a_3, \ldots$ *converges* to s and write

$$\lim_{n \to \infty} a_n = s$$

if, for any $\varepsilon > 0$, there is an $N > 0$ such that if $n > N$, we have $|a_n - s| < \varepsilon$. If there is some $s \in \mathbb{R}$ such that $\lim_{n \to \infty} a_n = s$, then we say that the sequence *converges*. We call this value of s the *limit* of the sequence.

While the definitions look essentially identical, there is a major difference in how we should think of these two things. In the first case, it is fairly tricky for sequences in $\mathbb{Q}$ to converge in $\mathbb{Q}$. Not only must the terms get close to each other, but they also have to get close to some fixed rational number. On the other hand, in the case of $\mathbb{R}$, if we have a sequence whose terms get close to each other, they *automatically* get close to some fixed real number. This is because $\mathbb{R}$ is complete, as we showed in Chapter 28. We summarize this as follows:

Proposition 29.3. *A sequence in $\mathbb{R}$ is convergent if and only if it is a Cauchy sequence.*

Example. Consider the sequence $a_n = \frac{n-1}{n}$, so that the sequence begins $0, \frac{1}{2}, \frac{2}{3}, \frac{3}{4}, \frac{4}{5}, \ldots$ We claim that this sequence converges in $\mathbb{R}$. To check just this, we don't need to find the limit; all we need to do is to show that it is Cauchy. So let's do that. Let us look at $|a_m - a_n|$. We have

$$
\begin{aligned}
|a_m - a_n| &= \left| \frac{m-1}{m} - \frac{n-1}{n} \right| \\
&= \left| \left(1 - \frac{1}{m} \right) - \left(1 - \frac{1}{n} \right) \right| \\
&= \left| \frac{1}{n} - \frac{1}{m} \right| \\
&\leq \frac{1}{n} + \frac{1}{m}
\end{aligned}
$$

by the triangle inequality. (We can get better bounds, but this one is good enough for our purposes.) Now we must show that for any $\varepsilon > 0$, there is some N such that whenever $m, n > N$, then $|a_m - a_n| < \varepsilon$. If $m, n > N$, then we have

$$
|a_m - a_n| \leq \frac{1}{n} + \frac{1}{m} < \frac{2}{N},
$$

so if we select N such that $\frac{2}{N} \leq \varepsilon$, i.e. $N \geq \frac{2}{\varepsilon}$, then we're in good shape: we have

$$
|a_m - a_n| < \frac{2}{N} \leq \varepsilon,
$$

which is just what we need to conclude that the sequence is Cauchy.

Now, we can already conclude that the sequence converges to some real number. But we might want to determine exactly what the limit is: we'd like to compute $\lim_{n \to \infty} \frac{n-1}{n}$. From looking at the first few terms, we can guess that the limit is 1, so let's verify that. Pick an arbitrary $\varepsilon > 0$. We need to explain why there is some N such that if $n > N$, then $\left| \frac{n-1}{n} - 1 \right| < \varepsilon$. In order to do this, it's helpful to express $\left| \frac{n-1}{n} - 1 \right|$ better: this is just $\frac{1}{n}$. So in other words, we need to explain why there is some N such that whenever

$n > N$, we have $\frac{1}{n} < \varepsilon$. Pick N such that $N \geq \frac{1}{\varepsilon}$, or equivalently $\varepsilon \geq \frac{1}{N}$. Then whenever $n > N$, we have

$$\frac{1}{n} < \frac{1}{N} \leq \varepsilon,$$

as desired.

In this case, the sequence is actually a sequence of rational numbers that converges to a rational number. But with the Cauchy sequence argument, we weren't able to guarantee that it would converge to a rational number, only to a real number. In practice, most of the examples we'll look at actually converge to rational numbers rather than irrational numbers, because describing irrational numbers and sequences that converge to them is generally more complicated. There are, of course, plenty of sequences of reals or rationals converging to irrational numbers, and some of them can even be expressed in terms of relatively straightforward formulae. For instance, consider the sequence

$$a_n = \sum_{k=1}^{n} \frac{(-1)^{k-1}}{2k-1},$$

so that $a_1 = 1$, $a_2 = 1 - \frac{1}{3}$, $a_3 = 1 - \frac{1}{3} + \frac{1}{5}$, $a_4 = 1 - \frac{1}{3} + \frac{1}{5} - \frac{1}{7}$, and so forth. This sequence is Cauchy (can you see how to prove this?), and it converges to $\frac{\pi}{4}$. This is the celebrated *Gregory–Leibniz series*; see for instance [Roy90]. But proving that directly from the definition seems like it would be quite complicated.[1] So instead we use other methods to prove such things, rather than doing them directly using the definition.

Sometimes, we will investigate the convergence of sequences in subsets of $\mathbb{R}$ rather than $\mathbb{R}$ itself. Let $S \subseteq \mathbb{R}$, and let $a_1, a_2, a_3, \ldots$ be a sequence of numbers in S. It might seem that nothing changes—after all, a sequence in S is a sequence in $\mathbb{R}$, so all our rules still apply. However, we're likely to ask slightly different questions about sequences in S and their convergence. In particular, if $a_1, a_2, a_3, \ldots$ is a Cauchy sequence of numbers in S, then we know it converges to a number in $\mathbb{R}$, but does it converge to a number in S? We have already seen this issue come up when $S = \mathbb{Q}$, which is why we had to construct $\mathbb{R}$ in the first place: Cauchy sequences in $\mathbb{Q}$ do not necessarily converge to numbers in $\mathbb{Q}$, only to numbers in $\mathbb{R}$.

Similarly, if S is an open interval like $(0, 1)$, meaning $(0, 1) = \{x \in \mathbb{R} : 0 < x < 1\}$, then Cauchy sequences in S do not necessarily converge in S.

---

[1] If you want to try this, you first have to decide what the precise definition of π is, something that is already a nontrivial task.

For example, consider the sequence where $a_n = \frac{1}{n+1}$, so the sequence starts $\frac{1}{2}, \frac{1}{3}, \frac{1}{4}, \frac{1}{5}, \ldots$ This sequence converges to 0 in $\mathbb{R}$, but of course $0 \notin (0,1)$.

On the other hand, it might seem that if S is a *closed* interval $[a, b] = \{x \in \mathbb{R} : a \leq x \leq b\}$, we can't pull off that sort of trick, and indeed that's right.

Theorem 29.4. *Let S be a closed interval $[c, d]$ in $\mathbb{R}$. Then any Cauchy sequence in S converges to a number in S.*

Proof. Let $a_1, a_2, a_3, \ldots$ be a Cauchy sequence in S. Thus it converges to some number $s \in \mathbb{R}$; we need to show that $s \in S$. Since $a_i \geq c$ for all i, we have $\lim_{n \to \infty} a_n \geq c$ as well, so $s \geq c$. (We haven't proven this yet, but you will do so in problem 4.) Similarly, since $a_i \leq d$ for all i, we have $\lim_{n \to \infty} a_n \leq d$, so $s \leq d$. Thus $s \in [c, d]$, as desired. ∎

29.2   Subsequences

A *subsequence* of a sequence is just a sequence formed out of some of the terms of a sequence. Typically they come in the same order as they do in the original sequence, so we'll usually assume that without mentioning it. For example, suppose a_n is a sequence. Then $b_n = a_{n^2}$ is a subsequence of a_n. The subscripts of the terms chosen don't necessarily have a convenient formula like n^2; sometimes they might be specified by properties of the a_n's rather than directly by a formula. For example, maybe b_n is some a_k such that $a_k < \frac{1}{k}$. If we don't know a precise formula for the a_k's (perhaps because we're trying to prove a general theorem), then we just have to describe a subsequence in some less direct manner.

Subsequences can be useful for checking that sequences do not converge. The reason is as follows:

Proposition 29.5. *Let $a = (a_1, a_2, a_3, \ldots)$ be a sequence in $\mathbb{R}$, and let $b = (b_1, b_2, b_3, \ldots)$ and $c = (c_1, c_2, c_3, \ldots)$ be two subsequences. If b and c both converge, but*

$$\lim_{n \to \infty} b_n \neq \lim_{n \to \infty} c_n,$$

then a does not converge.

Proof. Suppose that $\lim_{n \to \infty} b_n = s$ and $\lim_{n \to \infty} c_n = t$, where $s \neq t$. We will show that a_n is not a Cauchy sequence. Let $\varepsilon = \frac{|s-t|}{3}$. Since b_n converges to s, there is some N_1 such that whenever $n > N_1$, we have $|b_n - s| < \varepsilon$. Similarly, there is some N_2 such that whenever $n > N_2$, we have $|c_n - t| < \varepsilon$. Let $N = \max(N_1, N_2)$. Now, for any $n > N$, we have $|b_n - s| < \varepsilon$ and $|c_n - t| < \varepsilon$, so by the triangle inequality, we have

$$|s - t| \le |s - b_n| + |b_n - c_n| + |c_n - t|,$$

or

$$|b_n - c_n| \ge |s - t| - |s - b_n| - |c_n - t|.$$

Since $|s - t| = 3\varepsilon$ and $|s - b_n|$ and $|c_n - t|$ are both less than ε, we have

$$|b_n - c_n| > 3\varepsilon - \varepsilon - \varepsilon = \varepsilon.$$

Since b_n and c_n can be made to be arbitrarily far out terms of the sequence a_n, we have shown that there is no N such that if $m, n > N$, then $|a_m - a_n| < \varepsilon$. Thus a_n is not Cauchy and thus does not converge. ∎

Note that in this proof, we didn't have to conjure up some potential number that a_n could converge to, since we sidestepped all that by showing that the sequence isn't Cauchy. This is often a convenient thing to do.

While sequences might or might not converge, we can often say something about the existence of subsequences that converge. This is the content of the Bolzano–Weierstraß Theorem, one of the cornerstone results of basic real analysis.

Definition 29.6. A subset $S \subseteq \mathbb{R}$ is said to be *bounded* if there is some positive real number B such that $|x| \le B$ for all $x \in S$.

Theorem 29.7 (Bolzano–Weierstraß). *Let $a = (a_1, a_2, a_3, \dots)$ be a bounded sequence in $\mathbb{R}$. Then a contains a convergent subsequence.*

Proof. Let us suppose that a is bounded by B, so that $|a_n| \le B$ for all n. We will construct a convergent subsequence. First, consider the subintervals $[-B, 0]$ and $[0, B]$. Since there are infinitely many a_n's, at least one of these two subintervals must contain infinitely many a_n's; if both of them do, choose one of them. Call this subinterval I_1. From this subinterval I_1 with infinitely many a_n's, choose one of them to be b_1, say a_{n_1}. Now, split I_1 into two closed subintervals of equal length. Again, one of them has infinitely many terms of the sequence; call this subinterval I_2. Pick some $n_2 > n_1$ such that $a_{n_2} \in I_2$, and let $b_2 = a_{n_2}$. Continue on this way, bisecting I_{n-1} and choosing one of the halves with infinitely many terms of the sequence to be I_n, and then construct $b_n \in I_n$.

We claim that $b = (b_1, b_2, b_3, \dots)$ is a convergent subsequence. It suffices to prove that b is a Cauchy sequence. Note that the length of I_n is $\frac{B}{2^{n-1}}$.

Now let $\varepsilon > 0$ be arbitrary. We must find some N such that whenever $m, n > N$, we have $|b_m - b_n| < \varepsilon$. Choose N such that $\frac{B}{2^{N-1}} < \varepsilon$. Then if $m, n > N$, b_m and b_n are both in I_N, so $|b_m - b_n| \le \frac{B}{2^{N-1}} < \varepsilon$. Thus b is Cauchy and hence converges in $\mathbb{R}$. ∎

29.3   Limit laws

One way to make it easier to compute limits is to show that they behave well with respect to basic arithmetic operations. For example, if we have two sequences $a = (a_1, a_2, a_3, \ldots)$ and $b = (b_1, b_2, b_3, \ldots)$, then we can form a new sequence $a + b = (a_1 + b_1, a_2 + b_2, a_3 + b_3, \ldots)$. We should hope that $\lim_{n\to\infty}(a_n + b_n) = \lim_{n\to\infty} a_n + \lim_{n\to\infty} b_n$, at least if both limits on the right exist. This is indeed true, as are other related limit properties.

Proposition 29.8. *Let $a = (a_1, a_2, a_3, \ldots)$ and $b = (b_1, b_2, b_3, \ldots)$ be sequences, and let $c \in \mathbb{R}$. Suppose that $\lim_{n\to\infty} a_n$ and $\lim_{n\to\infty} b_n$ both exist. Then*

- $\lim_{n\to\infty}(a_n + b_n) = \lim_{n\to\infty} a_n + \lim_{n\to\infty} b_n$.
- $\lim_{n\to\infty}(c a_n) = c \cdot \lim_{n\to\infty} a_n$.
- $\lim_{n\to\infty}(a_n b_n) = (\lim_{n\to\infty} a_n)(\lim_{n\to\infty} b_n)$.
- $\lim_{n\to\infty} \frac{a_n}{b_n} = (\lim_{n\to\infty} a_n)/(\lim_{n\to\infty} b_n)$, *as long as $\lim_{n\to\infty} b_n \neq 0$. (We may wish to exclude the case in which any $b_n = 0$ due to an issue with the left side, although it's also possible to define what we mean by a limit when a finite number of the terms involve division by zero.)*

The point here is that often, we only need to know limits of certain simple sequences (which we must compute with bare hands), and then we can combine them in order to compute a limit that would be challenging to compute directly from the definition. For simplicity, we'll prove only the first of the limit laws. The proof follows the pattern of several others we have already seen.

Proof. Let $s = \lim_{n\to\infty} a_n$ and $t = \lim_{n\to\infty} b_n$. We will prove that $\lim_{n\to\infty}(a_n + b_n) = s + t$. Let $\varepsilon > 0$. We must find an N such that whenever $n > N$, we have $|a_n + b_n - s - t| < \varepsilon$. Since $\lim_{n\to\infty} a_n = s$, we can find an N_1 such that whenever $n > N_1$, we have $|a_n - s| < \frac{\varepsilon}{2}$. Similarly, we can find an N_2 such that whenever $n > N_2$, we have $|b_n - t| < \frac{\varepsilon}{2}$. Let $N = \max(N_1, N_2)$. Then whenever $n > N$, we have

$$|a_n + b_n - s - t| \le |a_n - s| + |b_n - t| < \frac{\varepsilon}{2} + \frac{\varepsilon}{2} = \varepsilon,$$

as desired.   ∎

Example. We have already shown that $\lim_{n\to\infty} \frac{n-1}{n} = 1$. By shifting by one, it follows that $\lim_{n\to\infty} \frac{n}{n+1} = 1$ as well. Thus, by our limit laws, we can compute $\lim_{n\to\infty} \frac{n^2-1}{n^2}$ without doing any more work with ε's:

$$\lim_{n\to\infty} \frac{n^2-1}{n^2} = \lim_{n\to\infty} \frac{(n-1)/n}{n/(n+1)} = \frac{\lim_{n\to\infty} \frac{n-1}{n}}{\lim_{n\to\infty} \frac{n}{n+1}} = \frac{1}{1} = 1.$$

While this may not seem very exciting at first, remember that it saved us from having to do yet another tedious calculation straight from the definition. So, once we have built up a database of known limits, it becomes much easier to work out future ones.

29.4 Problems

(1) Do the following sequences converge or diverge? Give a proof. If they converge, find their limits and prove your answer is correct using the definition of a limit.

(a) $a_n = \frac{3n+4}{5n+7}$.

(b) $a_n = \frac{n}{n^2+3}$.

(c) $a_n = \frac{n^2-2}{n+1}$.

(d) $a_n = \frac{(-1)^n}{n}$.

(e) $a_n = \frac{\sin(n)}{n}$.

(f) $a_n = \frac{\cos(n)}{n}$.

(g) $a_n = n - \sqrt{n(n-1)}$.

(2) Use limit laws to help you evaluate the limits of the following sequences. You may still need to use the definition of limits to work out some parts.

(a) $a_n = (1 + \frac{n^2}{n^2+3})(3 + \frac{n}{n^2+6})$. (b) $a_n = \frac{\sin^3(n)+\cos(n)}{n^3+3n^2+1}$.

(3) Let $a = (a_1, a_2, a_3, \ldots)$ be a sequence, and let $b = (b_1, b_2, b_3, \ldots)$ be a subsequence of a. Prove that if $\lim_{n\to\infty} a_n = s$, then $\lim_{n\to\infty} b_n = s$ as well. (Part of what you need to show is that b converges. Then you need to show that it has the same limit as a does.)

(4) Suppose that $a_1, a_2, a_3, \ldots$ is a sequence such that $\lim_{n\to\infty} a_n = s$. Suppose furthermore that, for some real number C, we have $a_n < C$ for all n. Prove that $s \leq C$. Why can't we conclude that $s < C$? Prove the same result if we only assume that $a_n \leq C$ for all n.

(5) Prove that every convergent sequence is bounded. Find an example to show that a bounded sequence need not be convergent.

(6) A sequence $a = (a_1, a_2, a_3, \ldots)$ is said to be *monotonically increasing* if $a_1 \leq a_2 \leq a_3 \leq \cdots$ and *monotonically decreasing* if $a_1 \geq a_2 \geq a_3 \geq \cdots$. We say that a is *monotone* if it is either monotonically increasing or monotonically decreasing. Prove that every sequence in $\mathbb{R}$ contains a monotone subsequence.

(7) Prove that a monotone sequence is convergent if and only if it is bounded.

(8) A sequence $a_1, a_2, a_3, \ldots$ is said to be *contractive* if there exists some r with $0 < r < 1$ such that $|a_{n+2} - a_{n+1}| \leq r|a_{n+1} - a_n|$ for all $n \geq 1$.

(a) Prove that any contractive sequence is Cauchy.

(b) Let $a_1 = 1$, and define $a_{n+1} = \frac{1}{3+a_n}$ for all $n \geq 1$. Prove that the sequence $a_1, a_2, a_3, \ldots$ is contractive (and hence convergent). What is its limit? Prove that your answer is correct.

Chapter 30

Functions and continuity

30.1 Limits of functions

So far, we have investigated limits of sequences. There is a parallel notion of limits for functions. While we can consider functions more generally, we will restrict our attention to functions defined on an open interval $(a, b) \subseteq \mathbb{R}$. We will allow a to be $-\infty$ and b to be $+\infty$, so as to allow all of $\mathbb{R}$ (or a ray in $\mathbb{R}$) as a possible domain of functions. All our functions will take values in $\mathbb{R}$, i.e. $f(x) \in \mathbb{R}$ for all $x \in (a, b)$.

When dealing with sequences, we said that a sequence $a_1, a_2, a_3, \ldots$ converges to some $s \in \mathbb{R}$ if, for large enough values of n, a_n is close to s. When working with functions, we are interested in the behavior of $f(x)$ as x gets close to some number y, while not actually being equal to y. That is, changing the value of $f(y)$, but not of any other number, does not change the limit: the limit is only concerned with nearby numbers. The statement

$$\lim_{x \to y} f(x) = s$$

means that $f(x)$ is always as close as desired to s, provided that x is sufficiently close to y but not equal to it.

Here is the formal definition of a limit of functions:

Definition 30.1. Let $f : (a, b) \to \mathbb{R}$ be a function, and let $y \in (a, b)$. We say that the *limit* of $f(x)$ as x approaches y is s, and write

$$\lim_{x \to y} f(x) = s,$$

if, for all $\varepsilon > 0$, there is a $\delta > 0$ such that whenever $0 < |x - y| < \delta$, we have $|f(x) - s| < \varepsilon$.

That's a bit of a mouthful, so let's unpack it a bit. The statement $|f(x) - s| < \varepsilon$ means that $f(x)$ has to be close to s: the number ε measures

323

how close we're required to get. If $\lim_{x \to y} f(x) = s$, this means that, regardless of how close we insist that $f(x)$ is to s, we can make this happen: just choose a small enough (positive) value of δ, and for then all values of x within δ of y other than y itself, $f(x)$ is within ε of s.

The definition of limits of functions is completely analogous to the case of limits of sequences. But now we have to consider *all* numbers close to y, rather than simply the terms in some sequence. Recall that the definition of a limit of a sequence doesn't have a δ in it, but it has an N instead. Think of δ as being $\frac{1}{N}$.

Let's compute some examples directly from the definition.

Example. Consider the function $f : \mathbb{R} \to \mathbb{R}$ defined by

$$f(x) = \begin{cases} x + 2 & x \neq 2, \\ 7 & x = 2. \end{cases}$$

What is $\lim_{x \to 2} f(x)$? You might think it's 7, because $f(2) = 7$. However, this is wrong! Instead, $\lim_{x \to 2} f(x) = 4$, because the definition of a limit does not see what happens *at* $x = 2$, only what happens *near* $x = 2$. So, let's check straight from the definition. In order to show that $\lim_{x \to 2} f(x) = 4$, we must show that, for every $\varepsilon > 0$, there is a $\delta > 0$ such that whenever $0 < |x - 2| < \delta$, we have $|f(x) - 4| < \varepsilon$. For any $\delta > 0$, whenever $0 < |x - 2| < \delta$, we have $f(x) = x + 2$, so $\lim_{x \to 2} f(x)$ must be the same as $\lim_{x \to 2}(x + 2)$. In order to show that the latter is 4, we must show that for any $\varepsilon > 0$, there is a $\delta > 0$ such that whenever $0 < |x - 2| < \delta$, we have $|(x + 2) - 4| < \varepsilon$, or $|x - 2| < \varepsilon$. So we can just take $\delta = \varepsilon$, and this says that whenever $0 < |x - 2| < \varepsilon$, we have $|x - 2| < \varepsilon$. I think we can all agree that that has to be true! So $\lim_{x \to 2} f(x) = 4$, as claimed.

Let's try a slightly trickier one. Soon, we will see how to make this one less tricky, but we'll work from first principles now.

Example. We will show that $\lim_{x \to 1} x^2 = 1$. This means that, for every $\varepsilon > 0$, we can find a $\delta > 0$ such that whenever $0 < |x - 1| < \delta$, then $|x^2 - 1| < \varepsilon$. We get a clue for how to find δ from the factorization of $x^2 - 1$. Note that $x^2 - 1 = (x - 1)(x + 1)$, and the $x - 1$ is related to δ. So, we can solve for $|x - 1|$ in the inequality for ε: we have $|x - 1| < \frac{\varepsilon}{|x+1|}$. So it would be nice to set $\delta = \frac{\varepsilon}{|x+1|}$. However, this doesn't make sense, because δ isn't allowed to depend on x: it's a condition *on* x. One solution to this conundrum is to cut off δ at some point: if we make δ smaller, that makes it easier for $|x^2 - 1| < \varepsilon$ to be true. So, let us require that $\delta \leq 1$. (There's nothing special about 1, except that it's a positive number. Any other positive number will do just as well.) This means that we're only interested in x with $|x - 1| < 1$. By the triangle inequality, we have

$$|x + 1| = |(x - 1) + 2| \leq |x - 1| + 2 < 1 + 2 = 3.$$

Thus instead of letting $\delta = \frac{\varepsilon}{|x+1|}$, which doesn't make sense, we can say that δ is $\leq$ the *smallest possible* value of $\frac{\varepsilon}{|x+1|}$, subject to our restriction that $|x - 1| < 1$. That is, we can take $\delta = \min\left(\frac{\varepsilon}{3}, 1\right)$.

So far, all of that was just reverse-engineering a good choice for δ. We still have to go back and check that it actually works. So, let's assume that $0 < |x - 1| < \min\left(\frac{\varepsilon}{3}, 1\right)$, and we'll check if it's true that $|x^2 - 1| < \varepsilon$. We have

$$\begin{aligned}
|x^2 - 1| &= |x - 1| \cdot |x + 1| \\
&\leq 3|x - 1| \\
&< 3 \cdot \frac{\varepsilon}{3} \\
&= \varepsilon,
\end{aligned}$$

as desired. So we have successfully shown that $\lim_{x \to 1} f(x) = 1$.

Remark 30.2. Sometimes we will encounter the following situation: we will have an interval (a, b) and a number $y \in (a, b)$ and a function f defined on all of (a, b) *except* y. In this case, we can still determine whether $\lim_{x \to y} f(x)$ exists, because the limit does not depend on the value at y, only on the values near y. This is useful because it allows us to determine the most logical value to assign a function at some number where it is not *a priori* defined.

30.2 Limit laws

Clearly, we don't want to have to go through this process of evaluating limits straight from the definition every time. Fortunately, we have a bunch of limit laws that can save us considerable effort.

Proposition 30.3.

(1) If $c \in \mathbb{R}$, then $\lim_{x \to a} c = c$ for all $a \in \mathbb{R}$.
(2) If $\lim_{x \to a} f(x)$ and $\lim_{x \to a} g(x)$ both exist, then $\lim_{x \to a}(f(x) + g(x))$ also exists, and

$$\lim_{x \to a} (f(x) + g(x)) = \lim_{x \to a} f(x) + \lim_{x \to a} g(x).$$

(3) *If $c \in \mathbb{R}$ and $\lim_{x \to a} f(x)$ exists, then $\lim_{x \to a} cf(x)$ also exists, and*

$$\lim_{x \to a} cf(x) = c \lim_{x \to a} f(x).$$

(4) *If $\lim_{x \to a} f(x)$ and $\lim_{x \to a} g(x)$ both exist, then $\lim_{x \to a}(f(x)g(x))$ also exists, and*

$$\lim_{x \to a} (f(x)g(x)) = \lim_{x \to a} f(x) \cdot \lim_{x \to a} g(x).$$

(5) *If $\lim_{x \to a} f(x)$ and $\lim_{x \to a} g(x)$ both exist and $\lim_{x \to a} g(x) \neq 0$, then $\lim_{x \to a} \frac{f(x)}{g(x)}$ also exists, and*

$$\lim_{x \to a} \frac{f(x)}{g(x)} = \frac{\lim_{x \to a} f(x)}{\lim_{x \to a} g(x)}.$$

(6) *If $\lim_{x \to a} g(x)$ exists and is equal to b, and $\lim_{x \to b} f(x)$ also exists and is equal to $f(b)$, then*

$$\lim_{x \to a} f(g(x)) = f(b).$$

There are others as well, but those will be enough for many purposes.

We won't prove all of these, but let's prove (4) as an example of how these things work. To do this, we need to start with a lemma.

Lemma 30.4. *Let $f : \mathbb{R} \to \mathbb{R}$ be a function, and let $a \in \mathbb{R}$. If $\lim_{x \to a} f(x)$ exists, then there exists some $B > 0$ and some $\rho > 0$ such that $|f(x)| \leq B$ for all x with $|x - a| < \rho$.*

Proof. Suppose that $\lim_{x \to a} f(x) = L$, but there do not exist such a B and ρ. Then for every positive integer n, we can find some x_n with $|f(x_n)| > n$ and $0 < |x_n - a| < \frac{1}{n}$. Choose an N such that $N > L + 1$, so that whenever $n \geq N$, then $|f(x_n)| > L + 1$. Then for every $\delta > 0$, we can find an n such that $|x_n - a| < \delta$ and $|f(x_n) - L| > 1$. This contradicts the assumption that $\lim_{x \to a} f(x) = L$. ∎

We can now prove Proposition 30.3(4).

Proof of Proposition 30.3(4). Suppose that $\lim_{x \to a} f(x) = L$ and $\lim_{x \to a} g(x) = M$, and let $\varepsilon > 0$. By Lemma 30.4, there exists a $B > 0$ and a $\rho > 0$ such that $|f(x)| \leq B$ whenever $0 < |x - a| < \rho$. Furthermore, there exist $\delta_1, \delta_2 > 0$ such that $|f(x) - L| < \frac{\varepsilon}{2|M|}$ whenever $0 < |x - a| < \delta_1$, and

$|g(x) - M| < \frac{\varepsilon}{2B}$ whenever $0 < |x - a| < \delta_2$. Let $\delta = \min(\rho, \delta_1, \delta_2)$. Then whenever $0 < |x - a| < \delta$, we have

$$|f(x)g(x) - LM| \leq |f(x)g(x) - f(x)M| + |f(x)M - LM|$$
$$= |f(x)| \cdot |g(x) - M| + |M| \cdot |f(x) - L|$$
$$< B \cdot \frac{\varepsilon}{2B} + |M| \cdot \frac{\varepsilon}{2|M|}$$
$$= \varepsilon.$$

Thus we have

$$\lim_{x \to a} f(x)g(x) = LM,$$

as claimed. ∎

For example, if we believe that $\lim_{x \to a} x = a$ (which is very easy to check, along the lines of the first example in §30.1), then we can easily see that $\lim_{x \to a} x^2 = a^2$, because

$$\lim_{x \to a} x^2 = \lim_{x \to a} x \cdot \lim_{x \to a} x = a \cdot a = a^2.$$

More generally, a very similar argument shows that if $f(x)$ is any polynomial, then $\lim_{x \to a} f(x) = f(a)$.

30.3 Continuity

A closely related notion to that of a limit is that of *continuity*. It is often said that a function being continuous means that you can draw its graph without lifting your pen off the paper. This is true for functions on an interval, but it isn't true much more generally. So, while you can use that for insight now, it may go wrong in the future, when you learn about continuity more generally.[1]

Here is the definition of continuity, easily stated in terms of limits:

Definition 30.5. Let $f : (a, b) \to \mathbb{R}$ be a function, and let $y \in (a, b)$. We say that f is *continuous* at y if

$$\lim_{x \to y} f(x) = f(y).$$

We say that f is *continuous* on (a, b) if it is continuous at all numbers $y \in (a, b)$.

---

[1]The notion of being able to draw something without lifting your pen is called *path-connectedness*.

Another way of expressing the same idea, by restating the limit definition, is as follows:

Definition 30.6. Let $f : (a, b) \to \mathbb{R}$ be a function, and let $y \in (a, b)$. We say that f is *continuous* at y if for every $\varepsilon > 0$, there is a $\delta > 0$ such that whenever $|y - x| < \delta$, we have $|f(y) - f(x)| < \varepsilon$. We say that f is *continuous* on (a, b) if it is continuous at all numbers $y \in (a, b)$.

As we already saw, even before we actually defined the term "continuity," all polynomials are continuous. So are many other functions that you are friends with: rational functions (where the denominators do not vanish), trigonometric functions, exponential and logarithmic functions, and compositions of all of these. So, for instance, $e^{\cos(\sqrt{e^{e^x}}) + x^3}$ is a continuous function on all of $\mathbb{R}$. It wouldn't be much fun to check that directly with the definition, but limit laws spare us from having to do that.

However, from time to time, we do have to use limit laws directly to prove (or disprove) continuity. A famous and surprising example of continuity is the following function, sometimes known as the *Riemann function* or the *Thomae function*. It is defined as

$$f(x) = \begin{cases} 0 & x \notin \mathbb{Q}, \\ \frac{1}{q} & x = \frac{p}{q} \text{ in lowest terms, with } q > 0. \end{cases}$$

So, for instance, $f(\sqrt{2}) = 0$, but $f(\frac{8}{5}) = \frac{1}{5}$. Since $0 = \frac{0}{1}$, we have $f(0) = 1$, and similarly $f(n) = 1$ for all integers n.

Theorem 30.7. *The Riemann function is continuous at all irrational x and discontinuous at all rational x.*

We won't work out all the details of this, since some of them are a bit tiresome, but let's go over the key points. Suppose $a \in \mathbb{Q}$. Then $f(a) > 0$, say with $f(a) = \frac{1}{q}$. Take $\varepsilon = \frac{1}{2q}$, say. (Even $\varepsilon = \frac{1}{q}$ works, but we must remember that there is no prize for bravery and refrain from trying to find the largest possible ε.) Then for any $\delta > 0$, there are irrational numbers x such that $0 < |x - a| < \delta$; and for these values of x, we have $f(x) = 0$, and in particular $|f(x) - f(a)| \not< \varepsilon$. Thus f is discontinuous at all rational numbers.

On the other hand, f *is* continuous at all the irrational numbers. This seems more surprising. The reason is that, for any irrational number a, we can pick a δ small enough such that whenever x is rational and $0 < |x - a| < \delta$, then the denominator of x is large. So, pick some $\varepsilon > 0$, and suppose that q is a positive integer such that $\frac{1}{q} < \varepsilon$. Then pick a δ small enough that whenever x is rational and $0 < |x - a| < \delta$, the denominator of x is at least q. Thus for all x with $0 < |x - a| < \delta$, we have $|f(x)| \leq \frac{1}{q} < \varepsilon$. It is a bit tedious to explain exactly how to select δ such that whenever x is rational and $|x - a| < \delta$, then the denominator of x is at least q. But it is pretty clear intuitively that this can be done. If we mark all the rational numbers whose denominators are less than q, these numbers form a discrete set, so we can choose a δ sufficiently small as to avoid them all.

Remark 30.8. On the other hand, there is no function $f : \mathbb{R} \to \mathbb{R}$ that is continuous at all the rationals and discontinuous at all the irrationals. This is a more difficult theorem that relies on a more advanced celebrated theorem in analysis called the *Baire Category Theorem*. See for instance [Pug15, §4.7]. See also [GO03, I.22 and I.23].

30.4 Problems

(1) Prove directly from the definition that $f(x) = x^2$ is continuous.
(2) Prove directly from the definition that

$$f(x) = \begin{cases} x + 1 & \text{if } x < 0, \\ 2x + 1 & \text{if } x \geq 0 \end{cases}$$

is continuous.
(3) More generally, let $f(x)$ and $g(x)$ be two continuous functions, and suppose that $f(0) = g(0)$. Prove that

$$h(x) = \begin{cases} f(x) & \text{if } x < 0, \\ g(x) & \text{if } x \geq 0 \end{cases}$$

is a continuous function.
(4) Does the limit

$$\lim_{x \to 3} \frac{|x - 3|}{x - 3}$$

exist? If so, find the limit, and prove that your answer is correct. If not, prove it using the definition of a limit.
(5) Suppose that $f : \mathbb{R} \to \mathbb{R}$ is a continuous function, and $a_1, a_2, a_3, \ldots$ is a Cauchy sequence in $\mathbb{R}$. Prove that $f(a_1), f(a_2), f(a_3), \ldots$ is also a Cauchy sequence. What does it converge to?
(6) Suppose that $f : \mathbb{R} \to \mathbb{R}$ is a bounded (not necessarily continuous) function, i.e. there is some real number B such that $|f(x)| \leq B$ for all $x \in \mathbb{R}$. Prove that

$$\lim_{x \to 0} x f(x) = 0.$$

Conclude that

$$\lim_{x \to 0} x \sin\left(\frac{1}{x}\right) = 0.$$

(7) Consider the function

$$f(x) = \begin{cases} x & \text{if } x \in \mathbb{Q}, \\ 0 & \text{if } x \notin \mathbb{Q}. \end{cases}$$

At which numbers is f continuous? Prove that your answer is correct.

(8) Let $f : \mathbb{R} \to \mathbb{R}$ be a function. Prove that f is continuous at some number $a \in \mathbb{R}$ if and only if there exists a real number b such that for every sequence $x_1, x_2, x_3, \ldots$ in $\mathbb{R}$ with $\lim_{n \to \infty} x_n = a$, we have $\lim_{n \to \infty} f(x_n) = b$.

Chapter 31

The Intermediate and Extreme Value Theorems

31.1 The Intermediate Value Theorem

If you think of a continuous function on $\mathbb{R}$ as one whose graph you draw without lifting your pen off the paper (even though, as we mentioned, this is not quite right), then you should expect the following: suppose $f : \mathbb{R} \to \mathbb{R}$ is a continuous function such that $f(-1) < 0$ and $f(1) > 0$. Then there should be some $x \in (-1, 1)$ such that $f(x) = 0$: the graph can't jump from negative to positive without passing through 0. This is something that mathematicians have believed since ancient Greece.

A famous example of its implicit use in Euclid's *Elements* [Euc02, Book 1, Proposition 1] occurs in the explanation of how to construct an equilateral triangle with compass and straightedge, as follows. First, draw two points A and B (which can be completely arbitrary, except that they must be distinct). Then draw a circle centered at A passing through B, and a circle centered at B passing through A. These circles have two intersection points, say C and D. Then both $\triangle ABC$ and $\triangle ABD$ are equilateral triangles. See Figure 31.1.

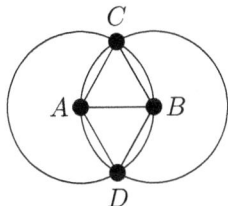

Figure 31.1. Constructing an equilateral triangle.

At a first glance, this seems like a good construction. However, there's a problem with it: how do you know that the two circles have any points of intersection? From the picture, this is clear: after you've drawn the first circle, the second circle contains some points inside the first circle, as well as some points outside of it. So it must contain some points exactly on the first circle, and these are the intersection points. But this is exactly the sort of thing we need the Intermediate Value Theorem for! Here's the precise statement:

Theorem 31.1 (Intermediate Value Theorem). *Let $f : (a, b) \to \mathbb{R}$ be a continuous function, and suppose that $a < a_1 < b_1 < b$. Suppose further that y is some number such that $f(a_1) < y < f(b_1)$. Then there is some $x \in (a_1, b_1)$ such that $f(x) = y$.*

Of course, there's also a version where $f(a_1) > y > f(b_1)$.

The proof of the Intermediate Value Theorem relies on the least upper bound property of $\mathbb{R}$ that we discussed back in Chapter 26. Let's recall what that says:

Theorem 31.2 (Least Upper Bound Property). *Let $S \subseteq \mathbb{R}$ be bounded above and nonempty. Then S has a least upper bound.*

Proof of the Intermediate Value Theorem. Let $S = \{x \in [a_1, b_1] : f(x) \le y\}$. Since $a_1 \in S$, S is nonempty. Furthermore, S is bounded above, since $x \le b_1$ for all $x \in S$. It follows that S has a least upper bound, say u. Note that $u \in (a_1, b_1)$, because continuity of f ensures that there exists some $x > a_1$ such that $f(x) < y$, and similarly that for all x sufficiently close to b_1, we have $f(x) > y$. We claim that $f(u) = y$.

Let $\varepsilon > 0$. Since f is continuous, there is some $\delta > 0$ such that $|f(x) - f(u)| < \varepsilon$ whenever $|x - u| < \delta$. (Since we can freely make δ smaller if desired, let us ensure that $\delta \le \min(u - a_1, b_1 - u)$.) It follows that whenever $u - \delta < x < u + \delta$, we have

$$f(u) - \varepsilon < f(x) < f(u) + \varepsilon.$$

Because $u = \sup(S)$, it follows that there is some $x_1 \in S$ such that $u - \delta < x_1 \le u$ and $f(x_1) \le y$. Since $x_1 \in (u - \delta, u + \delta)$, we have

$$f(u) - \varepsilon < f(x_1) \le y.$$

Now, for any x_2 with $u < x_2 < u + \delta$, we have

$$f(x_2) < f(u) + \varepsilon.$$

Furthermore, $f(x_2) > y$, since $x_2 \notin S$. Thus we have $y < f(x_2) < f(u) + \varepsilon$. Combining these inequalities, we see that

$$f(u) - \varepsilon < f(x_1) \leq y < f(x_2) < f(u) + \varepsilon,$$

or

$$f(u) - \varepsilon < y < f(u) + \varepsilon.$$

This holds for all $\varepsilon > 0$, so $y = f(u)$, as desired. ∎

Corollary 31.3. *Let $f(x)$ be a polynomial of odd degree. Then f has a real root.*

Proof. Let $f(x) = a_n x^n + a_{n-1} x^{n-1} + \cdots + a_0$, where $a_n \neq 0$. If $a_n > 0$, then $f(x) < 0$ for x sufficiently negative and $f(x) > 0$ for x sufficiently positive. Thus the Intermediate Value Theorem guarantees that f has a root. If $a_n < 0$, then replace $f(x)$ with $-f(x)$, which has the same roots. ∎

Another interesting classic application of the Intermediate Value Theorem is to a case of a theorem known as the *Avocado Sandwich Theorem*. The most famous version of the Avocado Sandwich Theorem is as follows:

Theorem 31.4 (Avocado Sandwich Theorem). *Let A, B, and C be three bounded subsets of $\mathbb{R}^3$.[1] Then there is a plane in $\mathbb{R}^3$ that simultaneously cuts A, B, and C into two pieces with equal volumes.*

You should think of A as being the top slice of bread, B as being the avocado, and C as being the bottom slice of bread in the sandwich. Then we can make just one straight cut that slices the sandwich into two equal half-sandwiches. There's also a higher-dimensional version, involving n bounded subsets of $\mathbb{R}^n$.

We won't be able to prove either the 3-dimensional version or the higher-dimensional version of the Avocado Sandwich Theorem, because their proofs rely on a much more complicated higher-dimensional version of the Intermediate Value Theorem called the *Borsuk–Ulam Theorem*. (See for instance my algebraic topology book [BBR21] for a proof.) However, the Intermediate Value Theorem can be used to prove either of two *lower*-dimensional versions of the Avocado Sandwich Theorem.

---

[1] There is a small caveat needed in this and all other versions of the Avocado Sandwich Theorem. See Remark 31.7.

Theorem 31.5 (2-dimensional Avocado Sandwich Theorem). *Let A and B be two bounded subsets of $\mathbb{R}^2$. Then there is a line in $\mathbb{R}^2$ that slices both A and B into two pieces of equal area.*

See Figure 31.2 for an example.

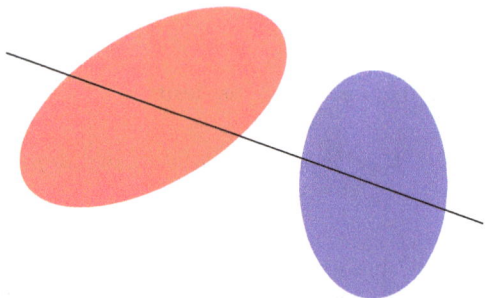

Figure 31.2. A line cutting two bounded regions into pieces of equal areas: the red region above the line has the same area as the red region below the line, and similarly with the blue region.

Theorem 31.6. *Let A be a bounded subset of $\mathbb{R}^2$, and let ℓ be a line in $\mathbb{R}^2$. Then there is a line parallel to ℓ that slices A into two pieces of equal area.*

We'll prove only Theorem 31.6. The proof of Theorem 31.5 relies on Theorem 31.6 together with another application of the Intermediate Value Theorem, plus some mild technical details that require more careful treatment than we want to get into here.

Proof of Theorem 31.6. Let the area of A be $[A]$. Define one of the half-planes obtained by deleting ℓ from $\mathbb{R}^2$ to be the *left half-plane* and one to be the *right half-plane*. If $t > 0$, define ℓ_t to be the line parallel to ℓ inside the right half-plane whose distance from ℓ is t. Similarly, if $t < 0$, define ℓ_t to be the line parallel to ℓ inside the left half-plane whose distance from ℓ is $-t$; let $\ell_0 = \ell$.

Also, for each t, let w_t be the length of $A \cap \ell_t$. Since A is bounded, the numbers w_t are bounded above, say by some $w > 0$. Now, define a function $f : \mathbb{R} \to \mathbb{R}$ by setting $f(t)$ to be the area of A to the left of ℓ_t. (See Figure 31.3.) We claim that f is a continuous function. To see this, pick some $t \in \mathbb{R}$ and some $\varepsilon > 0$. We must show that there is some $\delta > 0$ such that whenever $|t - u| < \delta$, we have $|f(t) - f(u)| < \varepsilon$. Since the length of $A \cap \ell_u$ is at most w for any u, we have $|f(t) - f(u)| \leq w|t - u|$. Thus if $|t - u| < \delta$, then $|f(t) - f(u)| \leq w|t - u| < w\delta$. Thus if we choose $\delta = \frac{\varepsilon}{w}$, then whenever $|t - u| < \delta$, we have $|f(t) - f(u)| < w\delta = \varepsilon$, as desired. Thus f is continuous.

Now, note that for t sufficiently negative, all of A lies to the right of ℓ_t, so $f(t) = 0$. Similarly, for t sufficiently positive, all of A lies to the left of ℓ_t, so $f(t) = [A]$. Thus by the Intermediate Value Theorem, there is some $t \in \mathbb{R}$ such that $f(t) = \frac{[A]}{2}$. For this value of t, the line ℓ_t slices A into two pieces of equal area. ∎

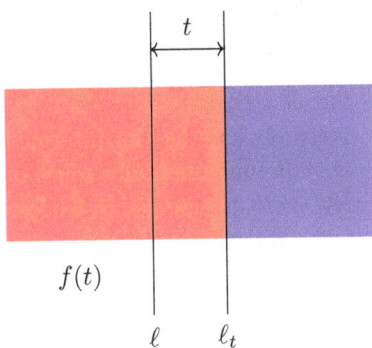

Figure 31.3. The red area is $f(t)$.

Remark 31.7. There is actually one small but important technical detail that we have been completely sweeping under the rug here. There are some extremely strange subsets of $\mathbb{R}^2$ for which it is not possible to assign a well-defined area, at least with the sorts of properties you expect of areas. These are called *non-measurable sets*. They cannot be explicitly described in a reasonable way, yet they exist. None of the theorems in this chapter apply to such sets; generally these theorems are stated with extra conditions so that A, B, C, and so forth are forced to be measurable, and thus notions like area and volume make sense. This isn't something you should worry about at all right now, but at some point later on in your mathematical career, you will want to know that non-measurable sets exist and what some sensible conditions are to avoid them. See for instance [Pug15, Chapter 6] for an introduction to measure theory.

31.2   The Extreme Value Theorem

Another classic theorem, of much the same flavor as the Intermediate Value Theorem, is the Extreme Value Theorem. While the Intermediate Value Theorem says that there is always some c such that $f(c)$ is some specific intermediate value between $f(a)$ and $f(b)$, the Extreme Value Theorem instead says that there is some c such that $f(c)$ is at least as big as any $f(x)$ with $x \in [a, b]$.

Theorem 31.8 (Extreme Value Theorem). *Let $f(x)$ be a continuous function on some interval containing $[a, b]$, where a and b are real numbers with $a \leq b$. (Note that we do not allow $a = -\infty$ or $b = +\infty$ for this theorem.) Then there is some $c \in [a, b]$ such that $f(c) \geq f(x)$ for all $x \in [a, b]$.*

Again, there is a similar version about minima: there is some $d \in [a, b]$ such that $f(d) \leq f(x)$ for all $x \in [a, b]$.

This isn't as intuitively obvious as the Intermediate Value Theorem. One reason for this is that it fundamentally relies on $[a, b]$ being a *closed* interval. Before we prove the Extreme Value Theorem, let's see some examples to show that it doesn't work unless the interval is a finite closed interval.

Example. Let $f(x) = \frac{1}{x}$ on the interval $(0, \infty)$. Then f does not achieve a maximum value, because it goes off to infinity as x gets close to 0.

Example. Let $f(x) = x$ on the interval $(0, 1)$. There is no $c \in (0, 1)$ such that $f(c) \geq f(x)$ for all $x \in (0, 1)$: for any $c \in (0, 1)$, $f(c + \varepsilon) > f(c)$ for any $\varepsilon > 0$ such that $c + \varepsilon < 1$.

In order to prove the Extreme Value Theorem, we first state and prove a weaker version of it, which merely says that a continuous function on a finite closed interval doesn't go off to ∞. This version will be used to prove the full Extreme Value Theorem. So, our strategy will be first to show that f is bounded above, then use the least upper bound property to show that f has a *least* upper bound, and then finally to show that the upper bound is actually achieved.

Lemma 31.9 (Boundedness Theorem). *Let $f(x)$ be a continuous function on some interval containing $[a, b]$, where a and b are real numbers with $a \leq b$. (Note that we do not allow $a = -\infty$ or $b = +\infty$ for this theorem.) Then there exist constants m and M such that $m \leq f(x) \leq M$ for all $x \in [a, b]$.*

Proof. We'll just prove the upper bound; the proof of the lower bound is symmetric. Suppose that f has no upper bound. This means that for all positive integers n, we can find some $x_n \in [a, b]$ such that $f(x_n) > n$. Thus $x_1, x_2, x_3, \ldots$ is a sequence on the bounded interval $[a, b]$, so the Bolzano–Weierstraß Theorem says that there is a convergent subsequence $x_{n_1}, x_{n_2}, x_{n_3}, \ldots$ of $x_1, x_2, x_3, \ldots$ Let $x = \lim_{k \to \infty} x_{n_k}$; by Theorem 29.4, $x \in [a, b]$. Since f is continuous at x and x_{n_k} is a Cauchy sequence converging to x, $f(x_{n_k})$ is also a Cauchy sequence converging to $f(x)$, as you proved in problem 5 in Chapter 30. However, $f(x_{n_k})$ is not a Cauchy sequence, since the terms go off to ∞. It follows that we cannot always find an $x_n \in [a, b]$ such that $f(x_n) > n$. This completes the proof of the Boundedness Theorem. ∎

Now we'll use the Boundedness Theorem to prove the Extreme Value Theorem.

Proof of the Extreme Value Theorem. By the Boundedness Theorem, there is some M such that $f(x) \le M$ for all $x \in [a, b]$. Thus $S = \{f(x) : x \in [a, b]\}$ is bounded above. It is also nonempty, because it contains $f(a)$. Thus, by the least upper bound property, S has a least upper bound, say M_1. Since M_1 is the least upper bound, for any positive integer n, we know that $M_1 - \frac{1}{n}$ is *not* an upper bound, so there exists some $x_n \in [a, b]$ such that $f(x_n) > M_1 - \frac{1}{n}$. For each n, then, we have

$$M_1 - \frac{1}{n} < f(x_n) \le M_1,$$

so

$$\lim_{n \to \infty} f(x_n) = M_1.$$

By the Bolzano–Weierstraß Theorem, the bounded sequence x_n has a convergent subsequence $x_{n_1}, x_{n_2}, x_{n_3}, \ldots$, converging to some $c \in [a, b]$. Thus by the continuity of f, we have

$$\lim_{k \to \infty} f(x_{n_k}) = f(c) = M_1.$$

Since $M_1 \ge f(x)$ for all $x \in [a, b]$ by construction, we have $f(c) \ge f(x)$ for all $x \in [a, b]$, as claimed in the Extreme Value Theorem. ∎

31.3   Problems

(1) Prove that there is some real number x such that $\cos(x) = x$.

(2) Suppose that $f, g : \mathbb{R} \to \mathbb{R}$ are continuous functions. Suppose that $f(0) < g(0)$ and $f(1) > g(1)$. Prove that there is some $x \in (0, 1)$ such that $f(x) = g(x)$.

(3) Let $f : \mathbb{R} \to \mathbb{R}$ be a continuous function such that $f(0) = f(1)$. Prove that there is some $x \in [0, \frac{1}{2}]$ such that $f(x) = f(x + \frac{1}{2})$.

(4) Let $f : \mathbb{R} \to \mathbb{R}$ be a continuous function, and suppose that $f(x) \in [a, b]$ for all $x \in [a, b]$. Prove that there is some $c \in [a, b]$ such that $f(c) = c$.

(5) Let $f : \mathbb{R} \to \mathbb{R}$ be a continuous function, and suppose that there is some closed interval $[a, b]$ such that for all $x \in [a, b]$, there is some $y \in [a, b]$ with $|f(y)| \leq \frac{1}{2}|f(x)|$. Prove that there is some $c \in [a, b]$ such that $f(c) = 0$.

(6) A set $S \subseteq \mathbb{R}^2$ is said to be *convex* if, for every two points p and q in S, the line segment connecting p and q is entirely contained in S. Now, let S be a closed, bounded, convex set in $\mathbb{R}^2$. Prove that there is a line in $\mathbb{R}^2$ that divides S into two pieces that have equal perimeter and area.

(7) Let $f : \mathbb{R} \to \mathbb{R}$ be a continuous function, and let $[a, b]$ be a finite closed interval in $\mathbb{R}$. Show that the image of f on $[a, b]$, i.e. $\{f(x) : x \in [a, b]\}$, is either a single number or a finite closed interval.

(8) Let $f : \mathbb{R} \to \mathbb{R}$ be a continuous function, and let (a, b) be a (finite or infinite) *open* interval in $\mathbb{R}$. What are the possible shapes for the image of f on (a, b), i.e. $\{f(x) : x \in (a, b)\}$? In the case of closed intervals, problem 7 says that the image must be a single number or a finite closed interval, so give an analogous answer here.

Chapter 32

Infinite series

32.1 From sequences to series

Suppose we have a sequence $a_1, a_2, a_3, \ldots$ One tempting thing to do with it is to add up the terms. For instance, instead of considering the sequence $\frac{1}{2}, \frac{1}{4}, \frac{1}{8}, \frac{1}{16}, \ldots$, where $a_n = \frac{1}{2^n}$, we might wish to sum them up, by considering the infinite sum

$$\frac{1}{2} + \frac{1}{4} + \frac{1}{8} + \frac{1}{16} + \cdots = \sum_{n=1}^{\infty} \frac{1}{2^n}.$$

Note that at the moment, it's not necessarily clear what this means. We know how to add together two real numbers, and we can extend from there to add up any *finite* number of real numbers. For instance, we define $a+b+c$ to be $(a + b) + c$. That is, we first add a and b to get a real number, and then we add the resulting number to c. Continuing on in this way, we can add together n real numbers for any positive integer n. However, this does not tell us how to add up *infinitely* many real numbers.

In order to do that, we need a new definition. Instead of adding up *infinitely* many terms at once, we'll just add together a larger and larger *finite* number of them. In this way, we'll construct a sequence: the sequence of partial sums. Given a sequence $a_1, a_2, a_3, \ldots$, construct a new sequence $s_1, s_2, s_3, \ldots$ by setting

$$s_n = \sum_{k=1}^{n} a_k = a_1 + a_2 + a_3 + \cdots + a_n.$$

The sum of all infinitely many of the a_n's is then defined to be the limit of the sequence $s_1, s_2, s_3, \ldots$, if it exists.

Definition 32.1. Given a sequence $a_1, a_2, a_3, \ldots$ of real numbers, the corresponding *series* $\sum_{n=1}^{\infty} a_n$ is defined to be $\lim_{n \to \infty} s_n$ if it exists, where

$$s_n = \sum_{k=1}^{n} a_k.$$

If the limit exists, then the series is said to *converge*, and otherwise it is said to *diverge*. We call s_n the nth *partial sum* of the series.

It will often be useful to use the following easy formulae relating the a_n's and the s_n's:

$$a_n = s_n - s_{n-1}, \qquad s_n = s_{n-1} + a_n.$$

Using the first of these formulae, we can recover the original sequence a_n from the sequence s_n of partial sums.

Example. Consider the sequence $a_n = \frac{1}{2^n}$ discussed above. We claim that the corresponding series $\sum_{n=1}^{\infty} \frac{1}{2^n}$ converges, with a sum of 1. One way to do this is to derive a closed-form expression for s_n. We claim that $s_n = 1 - \frac{1}{2^n}$. We can prove this by induction. The base case is $n = 1$, where we have $s_1 = \frac{1}{2} = 1 - \frac{1}{2^1}$, as claimed. For the inductive step, suppose that $s_n = 1 - \frac{1}{2^n}$. We will show that $s_{n+1} = 1 - \frac{1}{2^{n+1}}$. We have

$$s_{n+1} = s_n + a_{n+1}$$
$$= \left(1 - \frac{1}{2^n}\right) + \frac{1}{2^{n+1}}$$
$$= 1 - \left(\frac{1}{2^n} - \frac{1}{2^{n+1}}\right)$$
$$= 1 - \frac{1}{2^{n+1}},$$

as claimed.

Thus to determine whether the sum converges (and to what value), we need only investigate the limit $\lim_{n \to \infty} \left(1 - \frac{1}{2^n}\right)$. This limit is clearly equal to 1. (As a challenge to yourself, make sure you can actually prove this using the formal definition of the limit of a sequence.)

One thing we quickly notice about series is that if $\sum_{n=1}^{\infty} a_n$ is going to converge, then the terms a_n must go to zero. This is indeed true:

Proposition 32.2. *Suppose that $\sum_{n=1}^{\infty} a_n$ converges. Then $\lim_{n \to \infty} a_n = 0$.*

Proof. If $\sum_{n=1}^{\infty} a_n$ converges, then $s_1, s_2, s_3, \ldots$ must be a Cauchy sequence. Thus for any $\varepsilon > 0$, there must be some N such that whenever $m, n > N$, we have $|s_m - s_n| < \varepsilon$. In particular, taking $m = n + 1$, we must have $|s_{n+1} - s_n| < \varepsilon$ for all $n > N$. But $s_{n+1} - s_n = a_{n+1}$, so this means that $|a_{n+1}| < \varepsilon$ for all $n > N$. Thus $\lim_{n\to\infty} a_n = 0$, as desired. ∎

32.2   The harmonic series

It is natural to wonder about whether the converse to Proposition 32.2 is true: If the terms go to 0, does the series converge? Alas, the answer is no, as was shown by Nicole Oresme all the way back in the 14th century. He constructed a series whose terms go to zero, but that nonetheless diverges. This series is the *harmonic series*.

Definition 32.3. The *harmonic series* is the series $\sum_{n=1}^{\infty} \frac{1}{n}$.

Theorem 32.4. *The harmonic series diverges.*

There are many proofs of Theorem 32.4, and you can find many of them in [KS06]. We'll first look at Oresme's justifiably famous original proof.

Proof. We show that the limit does not exist by showing that we can make the partial sums greater than any positive number we like. To do this, observe that

$$1 + \frac{1}{2} + \frac{1}{3} + \frac{1}{4} + \frac{1}{5} + \frac{1}{6} + \frac{1}{7} + \frac{1}{8} + \frac{1}{9} + \cdots + \frac{1}{16} + \cdots$$

$$> \frac{1}{2} + \frac{1}{2} + \underbrace{\frac{1}{4} + \frac{1}{4}}_{\frac{1}{2}} + \underbrace{\frac{1}{8} + \frac{1}{8} + \frac{1}{8} + \frac{1}{8}}_{\frac{1}{2}} + \underbrace{\frac{1}{16} + \cdots + \frac{1}{16}}_{\frac{1}{2}} + \cdots$$

$$= \frac{1}{2} + \frac{1}{2} + \frac{1}{2} + \frac{1}{2} + \frac{1}{2} + \cdots.$$

Grouping the terms in the second line by their denominators, we find that the sum of all the $\frac{1}{4}$ terms is $\frac{1}{2}$, as is the sum of all the $\frac{1}{8}$ terms, and all the $\frac{1}{16}$ terms, and so forth. Thus for any n, we have

$$\sum_{k=1}^{2^n} \frac{1}{k} > \frac{n+1}{2}.$$

Now, let N be any positive integer. We can find a partial sum s_k of the harmonic series such that $s_k > N$: we have

$$s_{2^{2N-1}} = \sum_{k=1}^{2^{2N-1}} \frac{1}{k} > \frac{(2N-1)+1}{2} = N.$$

Since the partial sums are unbounded, the harmonic series diverges.   ∎

Now let's look at another proof. This one relies on the possibility of manipulating terms of a series. In particular, if a series whose terms are all positive converges to some sum s, and a subsequence converges to some other sum t, then we may remove the terms of the subsequence, and the remaining terms converge to $s - t$. Another ingredient is that if we have a series converging to s, and we divide all the terms by 2, then the resulting series converges to $\frac{s}{2}$. Both of these things are "obvious," of course, but we really ought to prove them rather than relying on our intuition. We won't do that though, as they are straightforward and fairly boring.

Proof 2. Suppose, on the contrary, that the harmonic series were to converge, say to s. Then we would have

$$s = 1 + \frac{1}{2} + \frac{1}{3} + \frac{1}{4} + \frac{1}{5} + \frac{1}{6} + \frac{1}{7} + \frac{1}{8} + \frac{1}{9} + \frac{1}{10} + \cdots. \qquad (32.1)$$

Now, divide (32.1) by 2, to get

$$\frac{s}{2} = \frac{1}{2} + \frac{1}{4} + \frac{1}{6} + \frac{1}{8} + \frac{1}{10} + \frac{1}{12} + \frac{1}{14} + \frac{1}{16} + \frac{1}{18} + \frac{1}{20} + \cdots. \qquad (32.2)$$

Note that in this way we end up with all the terms of the harmonic series with even denominators. If we subtract (32.2) from (32.1), we get

$$\frac{s}{2} = s - \frac{s}{2} = 1 + \frac{1}{3} + \frac{1}{5} + \frac{1}{7} + \frac{1}{9} + \frac{1}{11} + \frac{1}{13} + \frac{1}{15} + \cdots,$$

where the even-denominator terms get canceled. Since the sum of the even-denominator and odd-denominator terms are both equal to $\frac{s}{2}$, they are equal to each other, so we have

$$1 + \frac{1}{3} + \frac{1}{5} + \frac{1}{7} + \frac{1}{9} + \cdots = \frac{1}{2} + \frac{1}{4} + \frac{1}{6} + \frac{1}{8} + \frac{1}{10} + \cdots.$$

However, this is absurd, since every term on the left side is greater than the corresponding term on the right side. (Or, if you prefer, move everything to one side, to get

$$1 - \frac{1}{2} + \frac{1}{3} - \frac{1}{4} + \frac{1}{5} - \frac{1}{6} + \frac{1}{7} - \frac{1}{8} + \cdots = 0.$$

But then if we group each pair of terms, to get

$$\left(1 - \frac{1}{2}\right) + \left(\frac{1}{3} - \frac{1}{4}\right) + \left(\frac{1}{5} - \frac{1}{6}\right) + \left(\frac{1}{7} - \frac{1}{8}\right) + \cdots = 0,$$

then we end up with a sum of positive numbers equaling zero, which is impossible.) Thus we have contradicted our initial assertion that the harmonic series converges. ∎

Remark 32.5. Remember that, when we first discussed Cauchy sequences, we remarked that it was important to make sure that *every* pair of sufficiently far out terms in the sequence are close together, rather than just every *consecutive* pair. In terms of series, the condition that every consecutive pair is close together just says that $\lim_{n \to \infty} a_n = 0$, i.e. Proposition 32.2. As we can now see, that is not enough to conclude that every pair of sufficiently far out terms are close together.

32.3   Geometric series

The problem of determining whether a general series converges or diverges can be very challenging. For instance, it is an open problem to determine whether the series

$$\sum_{n=1}^{\infty} \frac{\csc^2(n)}{n^3}$$

converges, and this question is related to some beautiful theorems in the theory of Diophantine approximation and transcendental number theory. (See [Ale11].) However, we can classify certain series that converge, and we can determine their sums.

Definition 32.6. A sequence $a_0, a_1, a_2, a_3, \ldots$ is said to be a *geometric sequence* if there is some real number r such that $\frac{a_{n+1}}{a_n} = r$ for all $n \geq 0$. If $a_0, a_1, a_2, a_3, \ldots$ is a geometric sequence, the corresponding series $\sum_{n=0}^{\infty} a_n$ is said to be a *geometric series*.

It follows from the definition that if $a_0, a_1, a_2, \ldots$ is a geometric sequence, then there exist real numbers a, r such that $a_n = ar^n$. The number a is just a_0, and $r = \frac{a_{n+1}}{a_n}$ (which, by the definition, is independent of n).

Theorem 32.7. *The geometric series $\sum_{n=0}^{\infty} ar^n$ converges if and only if $|r| < 1$. When it converges, then it converges to $\frac{a}{1-r}$.*

Proof. We begin by finding the values of the partial sums. By a straight-forward induction, we can show that $s_n = \frac{a(1-r^{n+1})}{1-r}$. By the limit laws for sequences, $\lim_{n \to \infty} s_n$ exists if and only if $|r| < 1$; when this happens, $\lim_{n \to \infty} r^{n+1} = 0$, so $\lim_{n \to \infty} s_n = \frac{a}{1-r}$, as claimed. ∎

There is a nice intuitive way to come up with the answer $\frac{a}{1-r}$. It requires a proof to make it rigorous, but it's always useful to know what the correct answer is before we start proving something. Let us suppose that $\sum_{n=0}^{\infty} ar^n = s$. Then $\sum_{n=0}^{\infty} ar^{n+1} = sr$. Subtracting these two equations, we have

$$
\begin{array}{ccccccccccccc}
a & + & ar & + & ar^2 & + & ar^3 & + & ar^4 & + & ar^5 & + & \cdots & = & s \\
 & & ar & + & ar^2 & + & ar^3 & + & ar^4 & + & ar^5 & + & \cdots & = & sr \\
\hline
a & & & & & & & & & & & & & = & s(1-r).
\end{array}
$$

Thus we have $a = s(1-r)$, or $s = \frac{a}{1-r}$.

The problem here is that it requires some care to get the correct conditions on r to make this work. Why doesn't this argument give the right answer when $r = 2$, for instance, so that we would have $1 + 2 + 4 + 8 + 16 + 32 + \cdots = \frac{1}{1-2} = -1$? That's because this method doesn't see any of these analyses: there are no ε's and no limits, so how can it possibly detect the conditions on r? Indeed, there are other number systems, such as the 2-adic integers $\mathbb{Z}_2$ that we looked at in Chapter 11, in which it really is true that $1 + 2 + 4 + 8 + 16 + 32 + \cdots = -1$. But this is certainly not the case in $\mathbb{R}$!

32.4   Comparison tests

Sometimes we can determine whether a series converges or diverges by comparing it with another series whose convergence or divergence is already understood, such as a geometric series or the harmonic series. This was actually already the idea in Oresme's proof of the divergence of the harmonic series. He compared the harmonic series to the series $\frac{1}{2} + \frac{1}{2} + \frac{1}{2} + \frac{1}{2} + \cdots$, which clearly diverges. Then he showed that, with a suitable grouping of terms, the harmonic series is termwise greater than that one. In what follows, we will assume that $a_n \geq 0$ for all n, since series with both positive and negative terms present some additional complications.

Theorem 32.8. *Let $a_1, a_2, a_3, \ldots$ and $b_1, b_2, b_3, \ldots$ be two sequences.*

(1) If $\sum_{n=1}^{\infty} b_n$ converges and $0 \leq a_n \leq b_n$ for all n, then $\sum_{n=1}^{\infty} a_n$ converges.

(2) If $\sum_{n=1}^{\infty} b_n$ *diverges and* $0 \le b_n \le a_n$ *for all* n, *then* $\sum_{n=1}^{\infty} a_n$ *diverges.*

Proof.

(1) Note that a series converges if and only if its sequence of partial sums forms a Cauchy sequence. Let $s_n = \sum_{k=1}^{n} a_k$ and $t_n = \sum_{k=1}^{n} b_k$. Pick an $\varepsilon > 0$. Since $\sum_{n=1}^{\infty} b_n$ converges and so the sequence of t_n's is Cauchy, there is some N such that whenever $m, n > N$, we have $|t_m - t_n| < \varepsilon$. Suppose, without loss of generality, that $m > n$. Then

$$|t_m - t_n| = |b_{n+1} + b_{n+2} + \cdots + b_m| < \varepsilon.$$

Since $0 \le a_k \le b_k$ for all k, we have

$$|s_m - s_n| = |a_{n+1} + a_{n+2} + \cdots + a_m| \le |b_{n+1} + b_{n+2} + \cdots + b_m| < \varepsilon$$

as well. Thus $\sum_{n=1}^{\infty} a_n$ converges.

(2) This proof is very similar: since $\sum_{n=1}^{\infty} b_n$ diverges, the sequence $t_1, t_2, t_3, \ldots$ is not Cauchy, so there is some $\varepsilon > 0$ such that for any N, there exist $m, n > N$ such that $|t_m - t_n| > \varepsilon$. For the same values of m, n, we have $|s_m - s_n| \ge |t_m - t_n|$, so the sequence $s_1, s_2, s_3, \ldots$ is not Cauchy either. Therefore $\sum_{n=1}^{\infty} a_n$ diverges. ∎

Here are some examples of how to use this comparison test:

Example. The series $\sum_{n=1}^{\infty} \frac{1}{\sqrt{n}}$ diverges, because it is termwise $\ge$ the harmonic series.

Example. The series $\sum_{n=1}^{\infty} \frac{1}{n^2}$ converges. To see this, we first investigate the series $\sum_{n=1}^{\infty} \frac{1}{n(n+1)}$. Observe that $\frac{1}{n(n+1)} = \frac{1}{n} - \frac{1}{n+1}$, so we have

$$\sum_{n=1}^{\infty} \frac{1}{n(n+1)} = \sum_{n=1}^{\infty} \left(\frac{1}{n} - \frac{1}{n+1} \right)$$

$$= \left(1 - \frac{1}{2} \right) + \left(\frac{1}{2} - \frac{1}{3} \right) + \left(\frac{1}{3} - \frac{1}{4} \right) + \cdots$$

$$= 1,$$

since everything else cancels. We can't use the comparison test yet, because $\frac{1}{n^2} > \frac{1}{n(n+1)}$. However, we can multiply by 2, to get

$$\sum_{n=1}^{\infty} \frac{2}{n(n+1)} = 2.$$

Now we can use it: $\frac{1}{n^2} \le \frac{2}{n(n+1)}$, so we have a series which is termwise $\le$ a convergent series, so it converges. Determining the actual sum is a more difficult problem, one that was solved by Euler. (The sum, rather surprisingly, is $\frac{\pi^2}{6}$. How did the π get in there, and even more surprisingly, since when does π^2 show up anywhere?[1] See [Cha03] for several proofs that this series sums to $\frac{\pi^2}{6}$.)

32.5   Problems

(1) Does the series $\sum_{n=1}^{\infty}(-1)^{n-1}$ converge or diverge? Prove that your answer is correct.

(2) Does the series $\sum_{n=1}^{\infty}\frac{\log(n)}{n}$ converge or diverge? Prove that your answer is correct.

(3) Suppose that $a_1, a_2, a_3, \ldots$ is a sequence of positive numbers, and

$$\lim_{n\to\infty}\frac{a_{n+1}}{a_n} = r$$

for some r.

   (a) Prove that if $r < 1$, then there is some r' with $r \le r' < 1$ and an integer N such that $a_n < (r')^n$ for all $n > N$. Prove that if $r > 1$, then there is some r' with $r > r' > 1$ and an integer N such that $a_n > (r')^n$ for all $n > N$.

   (b) If $r < 1$, then prove that $\sum_{n=1}^{\infty} a_n$ converges. If $r > 1$, then prove that $\sum_{n=1}^{\infty} a_n$ diverges. This result is known as the *ratio test*.

(4) Does the series $\sum_{n=1}^{\infty}\frac{\sqrt{n}}{2^n}$ converge or diverge? Prove that your answer is correct.

(5) Suppose that $a_1 \ge a_2 \ge a_3 \ge \cdots \ge 0$ is a sequence of nonnegative numbers such that $\lim_{n\to\infty} a_n = 0$.

   (a) Prove that $\sum_{n=1}^{\infty}(-1)^{n-1}a_n = a_1 - a_2 + a_3 - a_4 + \cdots$ converges.

   (b) Conclude that $\sum_{n=1}^{\infty}\frac{(-1)^{n-1}}{\sqrt{n}}$ converges.

   (c) Using a computer or calculator, compute the sum to 3 digits of accuracy by taking a sufficiently high partial sum. How can you guarantee that you have gone far enough?

(6) Suppose that $a_1, a_2, a_3, \ldots$ is a sequence of positive numbers, and

$$\lim_{n\to\infty}\sqrt[n]{a_n} = r$$

for some r.

---

[1] Actually, it shows up in plenty of places. For instance, the volume of a torus, or a 4-dimensional ball.

(a) Prove that if $r < 1$, then there is some r' with $r \le r' < 1$ and an integer N such that $a_n < (r')^n$ for all $n > N$. Prove that if $r > 1$, then there is some r' with $r > r' > 1$ and an integer N such that $a_n > (r')^n$ for all $n > N$.

(b) If $r < 1$, then prove that $\sum_{n=1}^{\infty} a_n$ converges. If $r > 1$, then prove that $\sum_{n=1}^{\infty} a_n$ diverges. This result is known as the *root test*.

(7) Determine whether the following series converge or diverge. In each case, prove that your answer is correct.

(a) $\displaystyle\sum_{n=1}^{\infty} \frac{n^n}{5^n + 3}$.

(b) $\displaystyle\sum_{n=1}^{\infty} \frac{n^n}{5^{n^2}}$.

(c) $\displaystyle\sum_{n=1}^{\infty} \left(\frac{3n^2 + 6n + 1}{5n^2 + 1} \right)^n$.

(8) Find a series $\sum_{n=1}^{\infty} a_n$ that converges, but such that $\sum_{n=1}^{\infty} a_n^2$ diverges. Prove that if $a_n \ge 0$ for all n and $\sum_{n=1}^{\infty} a_n$ converges, then $\sum_{n=1}^{\infty} a_n^2$ converges.

Bibliography

[AGP94]   William R. Alford, Andrew Granville, and Carl Pomerance. There are infinitely many Carmichael numbers. *Annals of Mathematics*, 139(3):703–722, 1994.

[Ale11]   Max A. Alekseyev. On convergence of the Flint Hills series. *arXiv preprints*, 2011. `arXiv:1104.5100`.

[Apo76]   Tom M. Apostol. *Introduction to Analytic Number Theory*. Undergraduate Texts in Mathematics. Springer-Verlag, 1976.

[Art04]   Rob D. Arthan. The Eudoxus real numbers. *arXiv preprints*, 2004. `arXiv:0405454`.

[AZ18]   Martin Aigner and Günter M. Ziegler. *Proofs from The Book*. Springer, sixth edition, 2018.

[Bak19]   Matt Baker. The Stern–Brocot tree, Hurwitz's theorem, and the Markoff uniqueness conjecture, 2019. URL: `https://mattbaker.bl og/2019/01/28/the-stern-brocot-tree-hurwitzs-theorem-and-t he-markoff-uniqueness-conjecture/`.

[BBR21]   Clark Bray, Adrian Butscher, and Simon Rubinstein-Salzedo. *Algebraic Topology*. Springer, 2021.

[BE02]   Manuel Benito and J. Javier Escribano. An easy proof of Hurwitz's theorem. *American Mathematical Monthly*, 109(10):916–918, 2002.

[BT04]   Edward B. Burger and Robert Tubbs. *Making Transcendence Transparent: An Intuitive Approach to Classical Transcendental Number Theory*. Springer Science & Business Media, 2004.

[CF97]   John Horton Conway and Francis Y. C. Fung. *The sensual (quadratic) form*. Number 26 in The Carus Mathematical Monographs. The Mathematical Association of America, 1997.

[Cha03]   Robin Chapman. Evaluating $\zeta(2)$, 2003. URL: `https://empslocal. ex.ac.uk/people/staff/rjchapma/etc/zeta2.pdf`.

[Cox13]   David A. Cox. *Primes of the Form $x^2 + ny^2$: Fermat, Class Field Theory, and Complex Multiplication*, volume 119. John Wiley & Sons, second edition, 2013.

[CW00]   Neil Calkin and Herbert S. Wilf. Recounting the rationals. *American Mathematical Monthly*, 107(4):360–363, 2000.

[Els10]   Christian Elsholtz. A combinatorial approach to sums of two squares and related problems. In *Additive number theory. Festschrift in honor of the sixtieth birthday of Melvyn B. Nathanson*, pages 115–140. Springer, 2010.

[Euc02]   Euclid. *Euclid's Elements*. Green Lion Press, Santa Fe, NM, 2002. All thirteen books complete in one volume, The Thomas L. Heath translation, Edited by Dana Densmore.

[Fef05]   Solomon Feferman. *The Number Systems: Foundations of algebra and analysis*. American Mathematical Society, second edition, 2005.

[FS09]   Philippe Flajolet and Robert Sedgewick. *Analytic Combinatorics*. Cambridge University Press, 2009.

[GKP94]   Ronald L. Graham, Donald E. Knuth, and Oren Patashnik. *Concrete Mathematics*. Addison–Wesley Publishing Company, second edition, 1994.

[GO03]   Bernard R. Gelbaum and John M. H. Olmsted. *Counterexamples in Analysis*. Dover Publications, Inc., 2003. Corrected reprint of the second (1965) edition.

[Gou20]   Fernando Q. Gouvêa. *p-adic Numbers: An Introduction*. Universitext. Springer Nature, third edition, 2020.

[Isa02]   Daniel C. Isaksen. A cohomological viewpoint on elementary school arithmetic. *American Mathematical Monthly*, 109(9):796–805, 2002.

[Kob12]   Neal Koblitz. *p-adic Numbers, p-adic Analysis, and Zeta-Functions*, volume 58 of *Graduate Texts in Mathematics*. Springer Science & Business Media, second edition, 2012.

[KS06]   Steven J. Kifowit and Terra A. Stamps. The harmonic series diverges again and again. *The AMATYC Review*, 27:31–43, 2006.

[Len08]   Hendrik W. Lenstra, Jr. Solving the Pell equation. In *Algorithmic number theory: lattices, number fields, curves and cryptography*, volume 44 of *Mathematical Sciences Research Institute Publications*, pages 1–23. Cambridge University Press, 2008.

[Mac37]   Holbrook Mann MacNeille. Partially ordered sets. *Transactions of the American Mathematical Society*, 42(3):416–460, 1937.

[Mat70]   Yuri V. Matiyasevič. The Diophantineness of enumerable sets. *Doklady Akademii Nauk SSSR*, 191:279–282, 1970.

[Mat19]   Benjamin Matschke. Proofs by example. *arXiv preprints*, 2019. `arXiv: 1909.00480`.

[May15]   James Maynard. Small gaps between primes. *Annals of Mathematics*, 181(1):383–413, 2015.

[Meš12]   Romeo Meštrović. Euclid's theorem on the infinitude of primes: a historical survey of its proofs (300 BC–2017) and another new proof. *arXiv preprints*, 2012. `arXiv:1202.3670`.

[Mil17]   Steven J. Miller. *The Probability Lifesaver: All the tools you need to understand chance*. Princeton Lifesaver Study Guide. Princeton University Press, 2017.

[Niv47]   Ivan Niven. A simple proof that π is irrational. *Bulletin of the American Mathematical Society*, 53:509, 1947.

[NZM91] Ivan Niven, Herbert S. Zuckerman, and Hugh L. Montgomery. *An Introduction to the Theory of Numbers*. John Wiley & Sons, Inc., fifth edition, 1991.

[Pol14] D. H. J. Polymath. New equidistribution estimates of Zhang type. *Algebra & Number Theory*, 8(9):2067–2199, 2014.

[Pug15] Charles C. Pugh. *Real Mathematical Analysis*. Undergraduate Texts in Mathematics. Springer, Cham, second edition, 2015.

[PWZ96] Marko Petkovšek, Herbert Wilf, and Doron Zeilberger. $A = B$. A K Peters, 1996. Foreword by Donald Knuth.

[Roy90] Ranjan Roy. The discovery of the series formula for π by Leibniz, Gregory and Nilakantha. *Mathematics Magazine*, 63(5):291–306, 1990.

[Rub18] Simon Rubinstein-Salzedo. *Cryptography*. Springer Undergraduate Mathematics Series. Springer, 2018.

[Sho09] Victor Shoup. *A Computational Introduction to Number Theory and Algebra*. Cambridge University Press, second edition, 2009.

[Sil14] Joseph H. Silverman. *A Friendly Introduction to Number Theory*. Pearson Education, 2014.

[Spi07] Alexander Spivak. Крылатые квадраты (Winged squares), 2007. URL: `mmmf.msu.ru/lect/spivak/summa_sq.pdf`.

[SS07] Judith D. Sally and Paul J. Sally, Jr. *Roots to Research: A Vertical Development of Mathematical Problems*. American Mathematical Society, 2007.

[Sta99] Richard P. Stanley. *Enumerative Combinatorics. Volume 2*, volume 62 of *Cambridge Studies in Advanced Mathematics*. Cambridge University Press, Cambridge, 1999. With a foreword by Gian-Carlo Rota and appendix 1 by Sergey Fomin.

[Sta12] Richard P. Stanley. *Enumerative Combinatorics. Volume 1*, volume 49 of *Cambridge Studies in Advanced Mathematics*. Cambridge University Press, Cambridge, second edition, 2012.

[Sta15] Richard P. Stanley. *Catalan Numbers*. Cambridge University Press, 2015.

[Sta20] Richard P. Stanley. A survey of parking functions, 2020. URL: `http://www-math.mit.edu/~rstan/transparencies/pf.pdf`.

[Su99] Francis Edward Su. Rental harmony: Sperner's lemma in fair division. *American Mathematical Monthly*, 106(10):930–942, 1999.

[Wei15] Ittay Weiss. Survey article: The real numbers—a survey of constructions. *Rocky Mountain Journal of Mathematics*, 45(3):737–762, 2015.

[Wil05] Herbert S. Wilf. *Generatingfunctionology*. CRC Press, third edition, 2005.

[Zag90] Don Zagier. A one-sentence proof that every prime $p \equiv 1 \pmod 4$ is a sum of two squares. *American Mathematical Monthly*, 97(2):144, 1990.

[Zha14] Yitang Zhang. Bounded gaps between primes. *Annals of Mathematics*, 179(3):1121–1174, 2014.

Notation Index

Subject Index

www.ingramcontent.com/pod-product-compliance
Lightning Source LLC
Chambersburg PA
CBHW061619220326
41598CB00026BA/3808